SMART SCIENCE, DESIGN & TECHNOLOGY

Smart Science, Design and Technology

ISSN: 2640-5504
eISSN: 2640-5512

Book series editors

Stephen D. Prior
Faculty of Engineering and Physical Sciences, University of Southampton, Southampton, UK

Siu-Tsen Shen
Department of Multimedia Design, National Formosa University, Taiwan, R.O.C.

Volume 2

Smart Science, Design & Technology

Editors

Artde D.K.T. Lam
Fujian University of Technology, P.R. China

Stephen D. Prior
University of Southampton, UK

Siu-Tsen Shen, Sheng-Joue Young & Liang-Wen Ji
National Formosa University, Yunlin, Taiwan

CRC Press
Taylor & Francis Group
Boca Raton   London   New York

CRC Press is an imprint of the
Taylor & Francis Group, an **informa** business

A BALKEMA BOOK

CRC Press/Balkema is an imprint of the Taylor & Francis Group, an informa business

© 2020 Taylor & Francis Group, London, UK

Typeset by Integra Software Services Pvt. Ltd., Pondicherry, India

Library of Congress Cataloging-in-Publication Data

Applied for

Published by: CRC Press/Balkema
            Schipholweg 107C, 2316XC Leiden, The Netherlands

First issued in paperback 2023

ISBN: 978-1-03-257069-3 (pbk)
ISBN: 978-0-367-17867-3 (hbk)
ISBN: 978-0-429-05812-7 (ebk)

DOI: https://doi.org/10.1201/9780429058127

Publisher's Note
The publisher has gone to great lengths to ensure the quality of this reprint but points out that some imperfections in the original copies may be apparent.

Smart Science, Design & Technology — Lam et al. (Eds)
© 2020 Taylor & Francis Group, London, ISBN 978-0-367-17867-3

Table of contents

Computer science & information technology

Cultural & creative research

Preface

We have great pleasure in presenting this conference proceeding for technology applications in engineering science and mechanics from the selected articles of the International Conference on Applied System Innovation (ICASI 2019), organized by Taiwanese Institute of Knowledge Innovation, from 10th to 18th April, 2019 in Fukuoka, Japan.

The ICASI 2019 conference was a forum that brought together users, manufacturers, designers, and researchers involved in the structures or structural components manufactured using smart science. The forum provided an opportunity for exchange of the research and insights from scientists and scholars thereby promoting research, development and use of computational science and materials. The conference theme for ICASI 2019 was "Computational Science & Engineering" which tried to explore the important role of innovation in the development of technology applications. Contributions dealing with design, research, and development studies, experimental investigations, theoretical analysis and fabrication techniques relevant to the application of technology in various assemblies, ranging from individual to components to complete structures, were presented at the conference. The major themes on technology included Material Science & Engineering, Communication Science & Engineering, Computer Science & Engineering, Electrical & Electronic Engineering, Mechanical & Automation Engineering, Architecture Engineering, IOT Technology, and Innovation Design. About 300 participants representing 10 countries came together for the 2019 conference and made it a highly successful event. We would like to thank all those who directly or indirectly contributed to the organization of the conference.

The articles presented at the ICASI 2019 conference have been published as a special issue in various journals. In this conference proceeding we present some selected articles from various themes. A committee consisting of experts from leading academic institutions, laboratories, and industrial research centres was formed to shortlist and review the articles. The articles in this conference proceeding have been peer reviewed according to the predefined standards. We are extremely happy to publish this conference proceeding and dedicate it to all those who have made their best efforts to contribute to this publication.

Artde D.K.T. Lam, Stephen D. Prior,
Siu-Tsen Shen, Sheng-Joue Young & Liang-Wen Ji

Smart Science, Design & Technology — Lam et al. (Eds)
© 2020 Taylor & Francis Group, London, ISBN 978-0-367-17867-3

Editorial board

Advanced optical systems design

Smart Science, Design & Technology — Lam et al. (eds)
© *2020 Taylor & Francis Group, London, ISBN 978-0-367-17867-3*

A study of optical design for glasses with 3D infrared microscopic ophthalmoscope

Yi-Chin Fang & Yih-Fong Tzeng
Department of Mechanical and Automation Engineering, National Kaohsiung University of Science and Technology, Kaohsiung, Taiwan

Kuo-Ying Wu*
Ph.D. program in Engineering Science and Technology, National Kaohsiung University of Science and Technology, Kaohsiung, Taiwan

Tzu-Chyang King
Department of Applied Physics, National Pingtung University, Pingtung, Taiwan

Che-Wei Lin
Department of Mechanical and Automation Engineering, National Kaohsiung University of Science and Technology, Kaohsiung, Taiwan

ABSTRACT: In this work, we design a three-dimensional miniature ophthalmoscope that can be worn on glasses. The purpose of this device is to monitor eye conditions in real time and record fundus images for direct transmission via Internet to medical providers. After accumulating sufficient data, ophthalmology departments can apply big-data analysis techniques to improve medical treatment. In this work, we design a set of images ranging in size from large to small to test the ability of the human eye to recognize images shown on two- and three-dimensional displays. A statistical method based on the mean is used to determine eye recognition, and the results in turn determine the design of the wearable miniature ophthalmoscope. The design of optics is based mainly on the requirements and specifications of the ophthalmoscope. A large aperture (F/# = 1.4) is used to increase the light input into the device, without increasing the power of the light source. We also design an image-transfer system based on 2.5× fixed-focus telecentric right-angled prisms. The optical design was based on visible light from a simulated light source, with the addition of near-infrared radiation. According to the literature, the infrared spectrum (965 nm) penetrates human eye tissue better and can thus be used by physicians to support and solidify their diagnoses.

Keywords: lens design, ophthalmoscope, human eye recognition, telecentric optical system

1 INTRODUCTION

Today's computer, communication, and consumer products are light, thin, and easy to carry. Image information has been transferred from TV sets to smart phones. Unfortunately, the shorter viewing distance and longer time spent watching causes alarming rates of eye diseases such as macular degeneration and cataracts, which are irreversible. To protect public eye health, traditional ophthalmoscopes need to use a mydriatic agent when inspecting the fundus, which obliges the patient to endure intensive illumination during the procedure. Such equipment is bulky and must be used by doctors, which complicates the detection of such ophthalmic diseases for the average family. To combat this problem, ophthalmoscopes should be made smaller and more economical, which requires the design of a new generation of these devices. Thus, better detection and prevention of such eye diseases require the popularization of these devices on the mass market.

Currently, instruments for detecting fundus are found only in ophthalmology clinics. Furthermore,

*Corresponding author: 0115907@nkust.edu.tw

when a patient is diagnosed with an eye condition, they likely have to be transferred to a hospital. However, if the fundus can be inspected by a portable device, and if results may be transmitted over the Internet, a patient's eye condition could be continuously monitored at all times. The fundus images would be transmitted to the ophthalmology authority through the cloud, and big-data analysis techniques could be exploited to establish a self-contained family medical monitoring system to enhance the diagnosis of the ophthalmologist.

With this motivation, we present herein a design for three-dimensional (3D) miniature ophthalmoscope glasses. Based on eye recognition experiments with human subjects, we first determine whether two-dimensional (2D) or 3D images are preferable for human eye recognition. The results of these experiments determine the optical design of the miniature ophthalmoscope. [1, 2]

2 EYE-RECOGNITION EXPERIMENT WITH HUMAN SUBJECTS

An experiment was conducted to compare 2D and 3D positive square image recognition and define the optotype based on the graph area [3, 4]. Table 1 shows the optotypes defined from 1.0 to 1.5. The visual standard defined at 1.0 is the largest, with a projected area on the screen of 9 cm^2. Furthermore, we define the subsequent visual targets according to the proportional reduction law (i.e., the projected area of each visual graphic screen is reduced to that of the previous screen). During the experiment, a projector was used to project the naked-view 2D image and the shutter-type 3D stereo image. The projector provided a brightness of 2500 lux, and the ambient illumination was limited to 10 lux or less to reduce the contrast error caused by ambient lighting. The distance between the participant and the projection screen was 5 m.

During the experiment, the graphs viewed by a participant were played at random by the program, and each screen presented only a single graph. Each participant initially viewed the 2D image with naked eyes; after completing the cycle test, the 3D image was used for the next cycle test. The 3D image experiments required the participants to wear shutter-type 3D stereo glasses for viewing 3D images. During each test, each participant was monitored for each optotype. In addition, each participant rested for 10 s after observing the graphics to avoid eye fatigue and errors.

The 2D and 3D graphic-recognition experiments involved 20 participants, each of whom were subjected to five sets of tests, with each group acquiring 60 data (6 per view mark). Table 2 shows the average test results of all the participants. The experimental results (mean values) for 2D and 3D recognition are plotted in Figure 1 and show that human eye recognition is better for 3D graphics than for 2D graphics.

We use the t-test to verify these results. The t-test involves an experiment that tests the interaction between the mean values. Each mean value is calculated by applying the t-test,

Table 2. Experimental results.

Optotype		Mean	SD	t	p value
1.0	2D	0.89	0.11	−2.45	0.024
	3D	0.92	0.08		
1.1	2D	0.83	0.1	−3.66	0.002
	3D	0.87	0.1		
1.2	2D	0.63	0.1	−7.37	0
	3D	0.73	0.1		
1.3	2D	0.46	0.13	−3.8	0.001
	3D	0.54	0.15		
1.4	2D	0.27	0.17	−5.09	0
	3D	0.34	0.17		
1.5	2D	0.12	0.14	−9.53	0.53
	3D	0.16	0.16		
Avenge	2D	0.53	0.1	−9.53	0
	3D	0.60	0.1		

Table 1. List of optotype images.

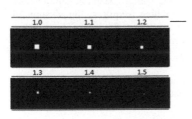

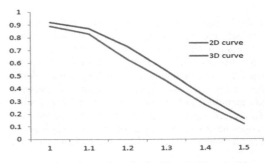

Figure 1. Experimental results for 2D and 3D recognition.

following which the p value may be obtained. A smaller p value indicates a lower mutual influence between mean values. The comparison between mean values is credible. The mean of each optotype in this experiment has a relatively low p value, which indicates a low degree of interaction between the 2D and 3D mean values in this experiment. Therefore, the 3D display mode is preferable over the 2D display mode, and this conclusion is statistically significant.

3 DESIGN OF THREE-DIMENSIONAL MINIATURE OPHTHALMOSCOPE

Thus, the results of 2D and 3D pattern-recognition experiments indicate that visual recognition by human eyes is superior for 3D images than for 2D images. We therefore design a miniature ophthalmoscope with a 3D display mode.

3.1 Design of three-dimensional system

The design goals for the miniature ophthalmoscope are that it be light, thin, short, and small. In addition, the ophthalmoscope must be simple, and the way in which the components attach to the frame of the glasses must also satisfy the design requirements (see Figure 2). The ophthalmoscope contains two sets of telecentric optical cameras, so the 3D image can be synthesized from a picture taken in dual-lens mode. In addition, the fundus mirror contains an infrared light source, which allows the ophthalmoscope to capture infrared images. This is a great help in diagnosing diseases of the retina.

We follow the stereo-image design for flat screens proposed by Liu [5] in 2001 to determine the best angles for acquiring 3D images. The two image sensors were angularly separated by 12° between their optical axes, so the resulting 3D image has a maximum depth-of-field ratio (~99.8%). The sensor was calculated geometrically to be 20 mm from the object, which was measured at an angle of 12°. The two sensors were separated by 4 mm, as shown in Figure 3, to obtain the best 3D images. This is the optimal positional configuration for the proposed 3D image sensor.

3.2 Design of optical system

The telecentric optical system uses triangular prisms to transfer images. These elements can rotate the optical axis 90°, which improves the degrees of freedom of the design. Figure 4 shows the architecture of the telecentric optical system and a ray-tracing simulation for this design.

This design uses a large aperture to increase the incident light (F/# 1.4). In addition to visible light, the optical system was equipped with a 965 nm near-infrared light source, allowing the micro-eye lens to receive near-infrared images, which adds another diagnosis mode. The transmission of near-infrared light through a crystal and through the

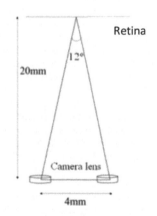

Figure 3. Position of dual lenses.

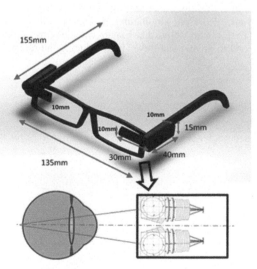

Figure 2. Photograph and draft plot of 3D miniature ophthalmoscope.

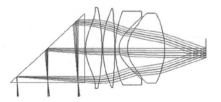

Figure 4. Telecentric optical system and ray-tracing simulation.

5

Table 3. Specifications of optical system.

Initial Conditions of Design	
Image Height	2.4 mm
Source wavelength	Photonic 5 with 481, 546, 656, 965 nm
Focal Length	10 mm
Magnification	2.5x
F/#	2.0
Overall Length	20 mm
Optical Distortion	< 3%
MTF	> 20–40% (100 lp/mm)

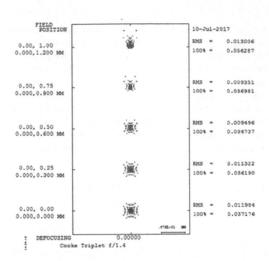

Figure 6. Spot diagrams from simulation of optical system.

human eye to the retina increases the macular image of the retina and allows the symptoms to be monitored in the near-infrared, thereby exploiting its penetration capabilities [6]. Table 3 lists the specifications of the optical system.

3.3 Optical system simulation and analysis

We used Code V optical simulation analysis software to design the lens and implement the simulation. CODE V is well known for the accuracy of its pupil grid function, which allows it to predict the pupil curve more accurately than other software. The important point is that the pupil curve directly reflects the aberrations of the entire system; except for chromatic aberrations. This optical model was repeatedly analyzed, so that the optical components were repeatedly corrected and optimized to obtain the expected results, as explained below.

The interference of electrical noise from the image sensor made it difficult for the subjects to clearly see the images with a modulation transfer function (MTF) < 20%. Figure 5 shows a MTF simulation of the optical system. The MTF performs well in the range 20% – 40%, which satisfies the requirements of the optical system.

Because the telecentric optical system is used to miniaturize the device structure, the problem of aberration inherent in the design is difficult to overcome. Fortunately, the lens group at the rear end can reduce

the aberration of the optical system. Figure 6 shows spot diagrams from a simulation of the optical system. Although not ideal, the spots fall within an acceptable range. The optical system also has an optimized space.

The optical system introduces a distortion of less than 2% and is maximized for human vision (see Figure 7). If the distortion exceeds 2%, the human eye can distinguish the deformation of the image. Thus, the distortion introduced by the proposed optical system falls within acceptable limits. To reduce the degree of distortion, one may increase the volume of the mechanism, reduce the wavelength of the light source, and/or incorporate more-advanced optics.

Figure 8 shows diagrams of the chromatic aberration from simulations of the optical system, and Figure 9 shows diagrams of various other aberrations, also from simulations of the optical system. These results show that longer wavelengths correspond to greater stability. Thus, reducing the range of

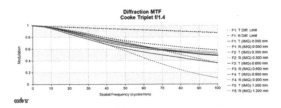

Figure 5. Results of MTF simulation of optical system.

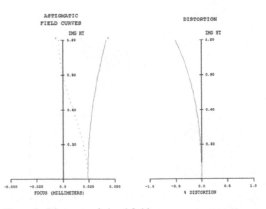

Figure 7. Diagrams of visual field.

6

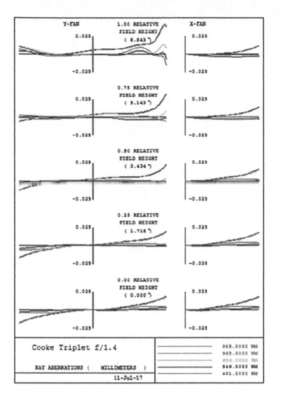

Figure 8. Diagrams of chromatic aberrations.

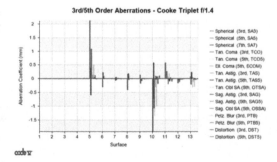

Figure 9. Diagrams of various aberrations.

wavelengths used should make the optical system more stable and reliable.

Thus, simulations of the optical system for the miniature ophthalmoscope produce results that satisfy standard specifications and are in line with design expectations.

4 CONCLUSIONS AND RECOMMENDATIONS

This study bases the design of a wearable miniature ophthalmoscope on human eye recognition.

The goal of this device is to provide data on individual retinas that can be transferred via Internet to medical providers for treatment by big-data technology, which would be of great help for combating eye diseases.

Human vision produces different results for 2D and 3D images. A large number of experimental data and statistical methods indicate that 3D displays are superior to 2D displays for obtaining 3D retinal images. Therefore, we designed a 3D ophthalmoscope to obtain 3D retinal images to facilitate the diagnosis of various eye diseases.

The experiment with human eye recognition uses statistical data to show that 3D displays lead to a better recognition rate than 2D displays. In this study, we plot the mean and standard deviation to determine the quality, and verify the mean by using the *t*-test. Follow-up experiments with other identification methods should allow further development of the miniature ophthalmoscope.

The miniature ophthalmoscope exploits a telecentric optical system to miniaturize the device, making it light, thin, short, and small. The device attaches to normal glasses, thereby freeing the user from unnecessary encumbrances. This device thus responds to the growing demand for home medical equipment.

The human eye recognition experiment provides an identification method that determines which display mode is most effective. It may be applied over a broad range of applications, from military to medical and other fields. Summarizing the data from such experiments into an indicator would determine the quality of the image display method in real time, which is highly desirable for automated identification technology. Moreover, the miniature 3D ophthalmoscope uses a telecentric optical system model, whose feasibility is demonstrated in this study. The optical design contributes to the miniaturized lens design.

REFERENCES

[1] B.W. Wu, Y.C. Fang, L.S. Chang. "Studies of human vision recognition: Some improvements." Journal of Modern Optics, 57. 107–114, 2010.
[2] B.W. Wu, Y.C. Fang, L.S. Chang, "Study on human vision model of the multi-parameter correction factor." Proceedings of SPIE - The International Society for Optical Engineering. 7496. 10.1117/12.829954, 2009
[3] Fairhurst, A.M., and Lettington, A.H., "Method of predicting the probability of human observers recognizing targets in simulated thermal images." Optical Engineering, vol. 37, no. 3, pp. 744–7751, 1998.
[4] Fairhurst, A.M., and Lettington, A.H., "The effect of visual perception on the required performance of imaging systems." Journal of Modern Optics, vol. 47, no. 8, pp. 1435–1446, 2000.
[5] Liu, R.Z. "Stereo-Imaging Design for Flot-Screen." Master's Thesis, Instiule of Optical Sciences, National Central University, Taoyuan, Taiwan, 2001.
[6] Cohen, J. "Statistical Power Analysis for the Behavioral Science." Academic Press: New York, NY, USA, 1977.

Applications of concurrent engineering in design

Smart Science, Design & Technology — Lam et al. (eds)
© 2020 Taylor & Francis Group, London, ISBN 978-0-367-17867-3

Design thinking and practical application of kitchen waste machines

Hsuhong Lee*
Graduate School of Design, Master & Doctoral Program, National Yunlin University of Science and Technology, Taiwan

Lihsun Peng
Department of Creative Design, National Yunlin University of Science and Technology, Taiwan

ABSTRACT: This study discusses the problem of food waste, clarifying the long-standing waste of food resources and improper treatment of agriculture; exploring cross-disciplinary and design thinking to promote a friendly natural environment, environmentally friendly life, sustainable resources, and organic waste recycling. Juxtaposition of various product design solutions can convert food waste into sustainable energy. Finally, from the case and strategy, this study evaluates the product vision that balances the circular economy and sustainable development of the economy and the environment.

Keywords: green design, sustainable design, organic waste, circular economy

1 INTRODUCTION

This study aims to solve environmental problems and sustainable development, and meanwhile, help consumers understand the expression "recycling and classification of food waste first." This study also hopes to start with "family life" as the source, that is, to solve the problem of sustainable recycling and reuse of food waste, so that the problems of environmental pollution and garbage can be decreased and reduced. Therefore, from the product design of many kitchen waste machines, the most distinctive one, "Taiwan Kitchen Waste Machine - Electronic Pig," is used as a key case study. The study will evoke discussions through a real case and solve the entire problem in the end. In this way, it will promote "design for the real world" and promote "cradle-to-cradle design" in an important part of the "circular economy."

2 ENVIRONMENTAL ISSUES AND DISCUSSION

According to estimates of the United Nations Food and Agriculture Organization (FAO) in its 2011 report, global kitchen waste was as high as 1.3 billion tons per year. Moreover, food or agricultural products for human consumption were wasted or damaged, accounting for one-third of total food production. The report also said that consumers in rich countries waste about 222 million metric tons of food per year, almost equal to the total production of all of sub-Saharan Africa. Vegetables, fruits, and ground provisions were the major waste (Gustavsson et al., 2011). After combining the FAO's and multi-party reports, the author found that the best way to deal with the waste of food resources is Reduction, Reuse, Recycling, and Buried. "Reduction" refers to savings from production and consumption. "Reuse" refers to the sharing of over-produced food for humanity as a food bank or for use by animals. "Recycling" refers to the conversion of food waste into reusable forms of energy, such as electricity, soil, and enzyme liquids. "Buried" refers to the method of composting. However, the way in which countries deal with food waste is related to the importance of environmental sustainability in each country; or, there are differences in procedures because of the problem of hoof-and-mouth disease.

About 800,000 tons of kitchen waste has been produced in Taiwan each year. Although 70% of the kitchen waste was cooked at high temperature, it would be reused in the pork industry to feed the pigs or given to the private composting operators (Chen & Zhang, 2014). However, according to a study commissioned by the Environmental Protection Administration, between 2003 and 2012, an average of about 33% of kitchen waste was mixed into

---

* Corresponding author. E-mail: d10330009@yuntech.org.tw

general waste and incinerated or buried (Environmental Protection Administration, ROC. 2013). More than 80% of these food wastes were raw kitchen waste, such as uncooked food including leaves and peels (Shen, 2014). In addition to garbage generated by households, fruit and vegetable residues in the fruit and vegetable market were also sent to incinerators and landfills in large quantities. Between 2007 and 2011, the wholesale market produced an average of about 130,000 metric tons of fruit and vegetable residues per year (Jiang et al., 2015). However, these raw kitchen wastes tend to have too high a water content, which lowers the furnace temperature of the incinerator. In order to make the incinerator burn completely, the industry may add plastic and paper to mix and burn together. When the plastic is burned with salty kitchen waste, it is easy to produce dioxin. If people consume dioxin from the food chain, they will develop various malignant tumors and lesions, and pregnant women will produce deformed children (Qiu, 2016). Moreover, the current average of 90% of waste in Taiwan will still be disposed of through incinerators. One million tons of slag will be produced each year, and dioxin is hidden in it. According to reports, there have been four illegal dumpings in Taiwan, with a cumulative dumping rate of 100,000 tons. The dumping site is usually the water source, and toxic substances may easily enter animals and plants through the soil or the water source, and finally return to the human body (101 News Media, 2016). In addition, some landfills that were "specialized for slag" were also filled with saturated and volatile toxic substances, and air pollution also affected local agriculture, which made residents resentful (Xu, 2018). Some landfills that were "collecting garbage" were facing a full-fledged crisis in the garbage mountain. Because kitchen waste was mixed into plastic garbage, the resulting stench and sour taste made the residents complain (Taiwan Environmental Information Center, 2017). In the composting field, although the compost is decomposed, decomposition can be done gradually by microorganisms to reduce the kitchen waste. This process, though, can produce methane and carbon dioxide, which contribute to the warming room effect (Chen, 2007). It also caused local residents to dislike or protest against the generation of stench, sewage, and mosquitos (Taiwan Environmental Information Center, n.d.).

In summary, the following is briefly described. If the "leftovers" in Taiwan couldn't be reused by people or eaten for animals, then the organic matter would be mixed and became "food waste." If these food wastes enter the incinerator and the landfill, the slag will be plagued by dioxin; or, the incinerator may not burn hot enough, causing a large amount of stench in the garbage mountain of the landfill. The composting field was indeed able to digest and reduce the amount of food waste that was concentrated into organic soil, but the conversion process also produced methane and carbon dioxide that contributed to the greenhouse effect. The food waste that was mixed into the general garbage inevitably had to be incinerated by the incinerator. These conditions existed only because people initially lacked the habit of "recycling food waste as a resource." Even in terms of kitchen waste and garbage diversion, Taiwan's performance has been prominent in the world. Nevertheless, Dr. Chen Wenqing from the Environment and Development Foundation suggested that the recycling of food waste should grasp the multiple applications of "Feedization, Fertilization, Energyization" and try to adapt to local conditions in various counties and cities (Chen & Zhang, 2014). Therefore, it is more effective to reduce the freight rate and carbon footprint and agent removal fee for entrusting the incinerator manufacturer (PTS, 2015).

3 RESEARCH METHODS

Therefore, after exploring the product design solutions of various kitchen waste machines in various countries, this study uses the household "electronic pig" kitchen machine by Chen Fong Environmental Engineering CO.LTD as a "single case research" with the three principles of uniqueness, inspiration, and criticalness (Pan, 2004). The research source integrates the literature analysis method and the interview method (Guan et al., 2010). After clarifying the problems and status of the whole social phenomenon, it proposes innovative methods and strategies to solve the problem. And, from this case, the thinking background of design theory and the development vision of practical application are arranged. The kitchen waste machine is the beginning of self-discipline recycling, reuse, and classification from a small family. It not only provides a source for organic soil, but also educates consumers to promote planting to create a green culture. It can help solve, reduce, and curb the level of waste disposal problems that will be accumulated by each unit in the future. Therefore, through this research, the author can gain insight from the perspective of green sustainable design. It is also hoped that this research can promote the recovery of environmentally sustainable education, and also solve the dilemma of environmental pollution and garbage flooding, while achieving a balanced solution of resources and resources in addition to achieving a balance between the economy and the environment.

4 THEORETICAL LITERATURE REVIEW

4.1 *The difference between circular economy, reuse economy, and linear economy*

The chairman of the Taiwan Circular Economy Network consolidated the economic development plan of the Dutch government and pointed out the differences in these three economic developments

(Huang, 2017). He mentioned that "circular economy" is a resource-recoverable economic and industrial system that uses renewable energy and refuses to use toxic chemicals that cannot be reused. Recycling materials, products, processes, and business models are used to achieve waste-free results, thereby ensuring that the Earth's limited resources can be recycled in a recyclable and sustainable manner. Among them, the two systems —the "biological cycle" of biochemical raw materials and the "industrial cycle" of chemical metal materials—each make resources a closed and sustainable cycle process (Michael & William, 2002). If Professor Michael Braungart's C2C concept is used to incorporate developmental differences (Masugata et al., 2016), the author believes that "circular economy" means "cradle to cradle design." The idea is to turn resources into resources again. "Linear economy" means "cradle to grave." It is a one-time production model with planned removal of old products. The point is that resources can only become garbage. The "reuse economy" is a transitional state between the two. The idea is to make garbage as useful as possible.

4.2 Decomposition method for returning organic waste and kitchen waste to organic soil

"Food waste" refers to food residue or vegetable and fruit residues removed during the cooking process, including expired food, leftovers, and vegetable waste (Young, 2018). Yang Qiuzhong, an academician of the Academia Sinica of Taiwan, developed the "free-handling treatment," stating that the treatment can quickly convert kitchen waste and various organic wastes into organic fertilizers with enzymes within three hours. He outlined "organic waste," the bodies of plants and animals and, more broadly, crop waste, livestock manure, dregs, food processing waste, fruits and vegetables, food waste, and horticultural plant waste. He has long studied composting and biotechnology, and he has proposed that the rate of "enzyme decomposition" will be much better than "microbial decomposition" for decomposing food waste. Actually, both methods of decomposition are better than traditional composting. The traditional composting method would take two or three months, and there would be problems of smelling, mosquitoes, and dirty sewage (Industrial Development Bureau-Ministry of Economic Affairs, 2005). The "microbial decomposition" method provides two ways: "aerobic microorganisms" and "anaerobic microorganisms." The rate of decomposition of aerobic microorganisms is more than 10 times that of anaerobic microorganisms (Tu et al., 2017). However, anaerobic microorganisms are used for closed biogas power generation because the water content is excessive and the kitchen waste is sticky and will become stinky. Aerobic microorganisms use fermentation to provide oxygen for several steps during the accumulation process, which has no odor and is more efficient for decomposition (Michael & William, 2002). According to research results, when the moisture content of the deposit is 40–70%, it is more suitable for the activities of aerobic microorganisms, and the population of microorganisms can be stable at 45–65°C (Tu et al., 2017). Therefore, if the design of the kitchen machine is maintained under such optimal growth conditions, aerobic microorganisms and good bacteria will survive and promote the most efficient decomposition, and produce higher-quality organic soil.

4.3 Green design thinking and sustainable design of energy cycle

Taiwan green design scholar Ke Yaozong (2017), believes that "the process of thinking about green design" should focus on three aspects: 1. The perception of green energy; 2. "Energy storage" of green energy; 3. Green "renewable energy." He believes that with the three directions as the goal, combined with the innovative thinking of forward-looking technology and green life, it can achieve the goal of humanized energy (Ryoichi, 2001). If we look at the kitchen waste machine from this perspective, "perception" is to spontaneously reduce waste and use ethical actions to complete resource sustainability. "Energy storage" is the endless use of gardening and leftover food after collecting food waste and planting it. "Renewable energy" is the production of organic soil that can contain, store, and share energy. Therefore, in the daily activities of planting gardens, building farms, and gardening, the need to deal with kitchen waste does not require humans to chase after the garbage truck, but rather to decompose and convert kitchen waste into organic soil. "Perpetual Design" is a design that simultaneously measures environmental protection and economic development while considering the sustainable use of energy and resources, and can meet the needs of current generations while meeting the real needs of future generations (Ryoichi, 2001). In "Green Design," Ryoichi (2001) proposes that it is also called "Eco Design" and enumerates 15 elements; for example, purification of the environment, prevention of pollution, reuse, waste reduction, recycling, soil fertility, decomposable renewable cycle, and energy conservation (Tu, 2014). If we use these two theoretical perspectives to examine the kitchen machine, then the design of the product is in line with the design thinking. Yet in the different cases of the subsequent kitchen waste machines, the "concurrent engineering" approach is not used in the process of product development; thus, the resources used in the process are saved as much as possible to achieve resource reduction, light weight, or modularization. Instead, the theory of "sustainable design" emphasizes that the product itself can convert the functions and services of renewable energy to achieve subsequent resource reduction. This is needed to solve some long-term environmental problems in society.

4.4 *Environmentally sustainable lifestyle design*

In recent years, as society and the environment continue to change, design has also changed. Each era has different visions of environmental protection, social issues, and economic models. People who share the same values and ideas will also hope to incorporate these values into their lifestyles and situations (Funatsu, 1996). When designers are designing, they have an eye toward the environment, society, and culture, so that product design can reflect the needs of the times. Like the "lifestyle design process" proposed by the Japanese scholar Funatsu Kunio (1996), the designer fully understands social needs and studies the life experience of the people, thereby proposing a product design suited to the life situation (Kang, 2015). More importantly, designers should face up to the long-term environmental pollution problems in society and take the circular economy as their responsibility. By designing, people can also fully live the life of environmental sustainability.

5 CASE DISCUSSION: TAIWAN KITCHEN WASTE MACHINE - ELECTRONIC PIG

5.1 *The introduction of the company and the concept of the background*

Chen Fong Environmental Engineering CO.LTD was founded in 1997 by the company's chairman, Ms. Kang Ru-Zhen. At the beginning of her business, her husband, Mr. Cai Cai-Zhou, actively invested in wastewater treatment engineering design and industrial sludge treatment and other related equipment and construction techniques to help the plant discharge sludge for reduction. But then he unfortunately suffered from a myocardial infarction. On the one hand, Ms. Kang observed that the problem of food waste in the world is serious. On the other hand, Ms. Kang measured that if she helped industrialists prevent waste sludge from the end, it would be better to trace the problem from the source of the kitchen waste, that is, "every family." In the third aspect, she focused on her husband's health and hoped to promote organic and vegetable diets. This practice could benefit more people with healthy and nontoxic diets while at the same time providing environmental protection (Zhu, 2019).

5.2 *Introduction of the product*

Therefore, Ms. Kang initiated the establishment of a product design that can solve family kitchen waste, so she began to consult various professionals, including Pierre H. Loisel, a pioneer in organic composting in Taiwan and a pesticide-free farmer; Professor Zhong Renci, who specializes in fertilizer and soil science; and Professor Yang Qiuzhong, who invented the world's first three-hour enzyme composting method (Zhu, 2019). Then, in conjunction with the company's own industrial equipment research and development technology, in 2011 Ms. Kang launched the "Electronic Pig" Earth System, the most friendly home appliance product in the global environment. Thus, family food waste in daily life could be easily recycled through the "electronic pig" and easily converted into organic fertilizer to create a clean living environment and a green organic garden.

5.3 *The characteristics and use of technology*

The key decomposition technology of product design is achieved by using the pure natural forces of "microbial decomposition." The four materials, "wood chips, enzymes, powdered sugar, and water," are the core organic substances that are good for cultivating and breeding "aerobic microorganisms and good bacteria," and these decompose the food waste by generating enzymes (Zhu, 2019). Moreover, the internal design of the product is also based on "mechanical tumbling into oxygen, sealing and maintaining the temperature at 36 to 50 degrees" (Zhu, 2019). All of the above conditions are the growth environment in which aerobic microorganisms can best maintain stable populations. In use, the above materials can be mixed and stirred in the kitchen machine one day, and the food waste can be put in, up to 2 kg at any time, the next day. It takes three to four days for the user to use the soil for the first time (depending on the kitchen allowance, it takes seven days if waste is too much). After taking the soil, it takes one or two days to decompose 2 kg of food waste, which can be decomposed into fertile organic soil. The food waste that can be put in may include fruits and vegetables, starch, raw and cooked meat, and soft shells. In particular, pineapple skin and papaya skin help to form enzymes that break waste down faster (Zhu, 2019).

5.4 *Comparison and analysis of household kitchen waste machines in various countries*

After visiting the director Zhu (2019), with the aid of the data, the differences between household kitchen machines in various countries can be obtained. The kitchen waste machine produced in Japan and South Korea uses a mold or a high-temperature drying water absorption method and belongs to the "kitchen waste drying and crushing machine" type. This method produces a kitchen waste powder that is not very fertile organic soil. Moreover, while the device is in operation, users cannot put in food waste at any time. And because of the drying function, it has high power consumption (up to 2,000 watts, universal 800 W, and minimum 335 W). However, the electronic pig can be put into the kitchen waste at any time, and the maximum power consumption is only 275

W. It is a "machine that turns the kitchen waste into organic soil." Nonetheless, it is necessary to keep the "fasting period" of the electronic pig at two days before taking any soil. This way, it is more convenient for obtaining 100% organic soil without mixing the kitchen waste (Zhu, 2019). Based on the above analysis, it is also the reason why the author picks this as a key case study.

6 DISCUSSION, INFLUENCE, AND VISION

If food waste is used in the "linear economy" model, it is used in agriculture for rapid and prolific production, so it uses a lot of chemical fertilizers, herbicides, and pesticides. As a result, pesticides remain in fruits and vegetables, endangering peoples' health. This can also cause serious chemical pollution in the soil and also affect the serious pollution of surface water and groundwater (Masugata et al., 2016; Kang, 2015). This mode is: self-planting → harvesting fruits and vegetables → people → food → waste of kitchen waste and leftover food → pig raising → incinerator and landfill → garbage mountain + land to human poison residue + dioxin + methane and CO_2 (Huang, 2017). However, if kitchen waste is under the "circular economy" of sustainable design, the biological resources brought into it are all enclosed in the renewable energy of the biological cycle: planting → harvesting fruits and vegetables → people → food → kitchen waste and leftover food → food bank → Electronic Pig → organic fertilizer → planting (Huang, 2017). Resources are also recycled from the cycle, not only allowing the people to grow their own pesticide-free food, but also expanding "local agriculture" from small families to community cooperation, or the vision of the sky farm and garden on the top floor of the apartment. The situation and environment will be totally in line with the green design elements advocated by Yamamoto Ryoichi; in particular, the use of kitchen waste machines to create nutritious organic soils that can be used to cultivate "oxygenated plants," which can green and beautify the home environment, and also help the indoor air-conditioning system that absorbs gas and produces oxygen. If the people cultivate "local agriculture," they can grow their own herbal tea materials, fruits, and vegetables, which can save money. Especially on typhoon days, fruits and vegetables are still available, which reduces the carbon footprint that was originally produced by transportation. If this concept is used in "horticultural planting treatment," "parental farm education," and "campus green farm," it can not only help the elderly to have a healthy and sustainable culture, but they can also teach children, students, and urban people that, by personally planting, they are closer to knowing the source of their food. Due to personal planting, people may understand the hardships of the peasants, and they will not waste or discard food anymore. It can also increase people's awareness of plants, who will not just think that the foods are on the supermarket shelves. Moreover, because of the popularity of kitchen waste machines in the home or community, housewives, especially those elderly with injured feet, won't need to chase garbage trucks every day. If every household has a kitchen waste machine, in the daily life of each family, it will implement recycling of waste resources and sustainable environmental protection, and expand into a social circle of life to jointly establish a community farm. Then, kitchen waste in towns and villages in this area will be greatly reduced. And under the multi-practice of organic fruits and vegetables, more people will have healthy and green bodies as well as minds, and they will also teach each other to share fruits and vegetables and organic soil. Furthermore, we can pay more attention to avoidance of pollution and garbage, and achieve the goal of educating on nature and sustaining resources on earth with actual planting.

REFERENCES

J. Gustavsson et al., 2011. *Global food losses and food waste – Extent, causes and prevention*. Rome: FAO, 6. Available at: http://goo.gl/NFA2v5

J. Chen & G. Zhang, 2014. *Food waste find out the way*, Taiwan: Foundation of the Public Television Culture Foundation (PTS). Available at: https://goo.gl/UNMLFX

Environmental Protection Administration, ROC. 2013. *Environmental Protection Statistics Monthly*, Taiwan, No. 298 64.

B. Shen, 2014. *We send food to the incinerator (upper and lower)*. Taiwan: Housewives United Foundation. Available at: https://goo.gl/nVPxWh & https://goo.gl/Nt5cdM.

K. Jiang et al., 2015. *102-104 General Waste Disposal Before the final disposal of general waste, the sampling and analysis work commissioned project plan*, Taiwan: Executive Yuan Environmental Protection Agency. Available at: https://goo.gl/MR1nt9

H. Qiu, 2016. Taiwan also has a hidden Dioxin crisis, Taipei: World Mag. No.218. Available at: https://goo.gl/ZXLm6i

101 News Media, 2016. *If the kitchen waste is not burned, the PCDD/Fs will be born*. Available at: https://goo.gl/h2wS9d

R. Xu, 2018. The landfill will be saturated, Taiwan: United Daily News. Available at: https://goo.gl/X8RZaT

Taiwan Environmental Information Center, 2017. 8000 metric tons of garbage acid wind near Yunlin. Available at: https://goo.gl/zMUCWz

I. Chen, 2007. *Measurement of carbon dioxide and methane emissions from composting processes*, Ph.D. Thesis, Institute of Microbiology and Biochemistry, NTU. Available at: https://goo.gl/9Z2gbG

Taiwan Environmental Information Center, n.d., Composting field news. Available at: https://goo.gl/nczAWN

PTS, 2015, *Incinerators can't be eaten*, PTS-Our Island, March 16, 2015, available at: https://youtu.be/gUFxP7rVZqg

S. Pan, 2004. *Quality Research: Theory and Application*. Taipei: Psychology Press 255.

X. Guan et al., 2010, *Design Research Method*. Taipei: Quanhua Book Co., Ltd.

C.Y. Huang, 2017. *Circular Economy*. Taiwan: World Magazine 41–44, 172–175.

B. Michael & M. William, 2002. *Cradle to Cradle: Remaking the Way We Make Things*, NYC: North Point Press.

T. Masugata et al., 2016. *Food Sociology: From the perspective of life and place*, Taipei: Opening Learning Publishing Co Ltd. 184–201.

C. Young, 2018. *Kechuang Lecture Hall: New Agriculture-Magic Rapid Production of Organic Fertilizers*, Science American Magazine. Available at: https://youtu.be/lCs2BULJoXU

Industrial Development Bureau-Ministry of Economic Affairs (Taiwan Green Productivity Foundation), 2005. *Composting Technology and Equipment Manual and Case Compilation, Taiwan*: IDB-MOEA 9-11. Available at: https://goo.gl/6zr2fk

J. Tu et al., 2017. *Green Design (innovation and Design Praactice*. Taichung, 10, 53, 200~201.

Y. Ryoichi, 2001. *Environmental materials*. Beijing.

J. Tu, 2014. *Life Style Design: culture, life, consumption and product design*. Taipei: Asia-Pacific Book 53–55.

K. Funatsu, 1996. *Electrical Design Department*, Director of Life Software Research Office, Japan: Matsushita Electric Industrial Co. Ltd.

R. Kang, Interview with J. Lai, 2015. *[Treasure Island is interesting] Electronic pig-housewife's counterattack, family transformation magic!* Taiwan: Bao-Dao Network (Greatest Idea Strategy Co. Ltd). Available at: https://youtu.be/9pUqP-ONOzk.

Z. Zhu, 24 Feb. 2019. *Personal interview.*

Applications of information & management

Smart Science, Design & Technology — Lam et al. (eds)
© 2020 Taylor & Francis Group, London, ISBN 978-0-367-17867-3

A study of the knowledged employee losing analysis – by using analytic hierarchy process

Chechang Chang*
School of Internet Economics and Business, Fujian University of Technology, PR China

Hsiuya Chang
Department of Business Administration, Soochow University, Taipei, Taiwan

ABSTRACT: People are the most critical resources for enterprises to maintain their competitiveness. Talents, namely knowledge employees, are the most urgent human resources in enterprises. Chinese enterprises have faced to change themselves to cope with rapid economic development. The present study was to investigate the factors affecting the loss of knowledge workers in terms of enterprise attributes, organizational management, internal relations, personal needs and job attributes. Through the measurement of relative weights of all factors, the results indicated the key factors of leaking knowledge workers. The suggestions are provided to enterprises to reduce the flow away in order to maintain sustainable development of employees.

Keywords: knowledged employee, employee leaking, Aanalytic Hierarchy Process

1 INTRODUCTION

Peter Drueker said that there is only one real resource in an enterprise, and that is human resource. The competitive advantage of human resource and its potential becomes increasingly important. In recent years, the loss of knowledge personnel in many companies is drawing much attention of researchers. The loss of talents and technicians appears to be a chain reaction, which not only lowers the morale of workers, but also seriously affects the competitiveness of enterprises. Therefore, it is necessary to investigate the factors influencing the loss of knowledgeable talents.

The loss of knowledge employees is complex involving many factors. Multiple factors considered in the present study was by comparing the relative importance of all factors and concluding the most important factors that impact the loss of talents. The results of data analysis will be proposed to business managers to improve management policies.

2 FACTORS AFFECTING THE LEAKING OF KNOWLEDGE-BASED WORKERS

In terms of the definition of knowledgeable workers, different scholars provide different definitions. Zhang Qian thinks that whether a talent is valuable depends on his/her contribution. The value the employee creates with his wisdom is assumed to be higher than the value he creates with his physical strength. The present study adopted Peter Drueker's definition of knowledgeable employees. According to Drueker, Knowledgeable workers refer to those who can master and take advantage of symbols and concepts and utilize information and knowledge to work. The authors in the present study consider that knowledgeable employees have the characteristics of independence, innovation, motivation and self-realization.

Turnover means voluntary resignation of employees. The loss of employees is inconsistent with the company's expectation, in which enterprises are helpless to stop the resignation of employees. The loss of important employees will miss their positive role if their resignation is not what the company expected. Mobley said that the resignation of knowledgeable employees means the termination of their relationship with the company.

March and Simon put forward a model to explain employee's turnover. It means that psychological or personality factors resulted in the loss of employees is related to economic, business and environmental factors. The model consists of two modules: employee's satisfaction with their company and possibility of facilitating mobility among different companies induced by the resignation of employees. The realization of employees' self-worth in their work, their relationships with others in the company, and their

*Corresponding author: E-mail: 1794298328@qq.com

competence to deal with their jobs are all closely related to their satisfaction with jobs. The style of a leader, the remuneration, and the degree of involvement in the job, employee's educational level, promotion and salary all influence the realization of employees' self-worth.

At the earliest stage of Industrial Revolution, economists put forward the impact of the factors of employee training, wage, labor market, industrial environment, and unemployment rate on employee turnover. Due to the relevant studies about management, from the perspective of organizational commitment, job satisfaction and psychological contract, some scholars concluded the factors affecting the loss of knowledgeable employees. With the development of globalization, scholars have put forward some new significant factors, such as company's geographical position, leaders' competence, and cost of personal turnover. In terms of geography, the company having superior location with good traffic conditions and high economic vitality will inherently attract talents. Leader's competence refers to a leader's ability in coordination, control, decision-making and planning. Strong leadership can make the enterprise cohesive enough to motivate employees to contribute themselves to enhance the competitiveness of companies. Cost of personal turnover refers to material and non-material losses that employees have to undertake when quitting the job. If the lossing and damage of resignation is serious enough to damage the future of employees, employees will not resign.

3 RESEARCH FRAMEWORK

In order to understand the factors affecting the loss of employees, the present study considered 30 factors in the questionnaire with each item having one Likert scale with 1 representing strongly disagree and 5 representing strongly agree. A total of 100 questionnaires were distributed to those who had experiences of resignation. 91 questionnaires were collected, of which 87 were valid. KMO = 0.733 > 0.7. Bartlett's sphericity is 724.692, with a significance = 0.000 < 0.01. It showed that the data were suitable for further factor analysis. 23 factors were selected through factorial analysis, and the framework of this study is presented.

4 THE RESULTS OF ALL ASPECTS ANALYSIS

From the results, it is obvious that enterprise attributes are the most important factor affecting the loss of knowledgeable workers, its weight is 36.0%, accounting for more than 1/3 of all factors. Organizational management ranks the second important factor with a weight of 23.8%.

Enterprise attributes and organizational management are directly related to enterprises. It

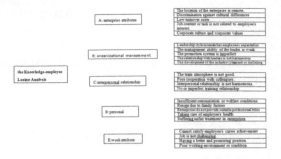

Figure 1. The research aspect and object.

indicated that most employees care most about whether the enterprises can help them realize self-worth in jobs.

The last two factors are job attributes and personal needs, both affecting employees' needs and job achievement. Once employees' psychological needs are satisfied, they will pursue self-realization in the job, which positively enhance the overall operational quality of enterprises.

The results of the present study show that among the factors influencing the resignation of employees. Before quitting, many factors would cross over the employee' mind. If the cost of resignation is high, they will refrain from quitting jobs. The turnover costs are affected by many factors such as work experiences, organizational factors, environment, and interpersonal network. Besides the factors of birth place and graduation place, the economic strength and traffic convenience of the city are also the factors affecting workers' willingness in retaining a job.

Because of the weakness of local economy in China, the lack of first-hand information, the lag of data acquisition, the inconvenience of communication or the lack of inter-industry exchanges, the enterprises in cities with remote location will have crisis of losing employees, while the enterprises in cities with superior geographical position with convenient transportation and attractive local economy will potentially attract employees. Work content or task is valued important by knowledgeable workers, which is related to personal competence and job requirement. Knowledgeable workers were primarily comprised of those who were born after 1980s and 1990s, because they are more competitive to fulfill challenging job requirements.

Traditionally, the benefits such as salaries, welfare, job titles, promotion opportunities, and better positions that employees have attached great importance but not of great concern on it. However, knowledgeable workers are very concerned about the location, training system, and leadership style of enterprises, which are not considered to be important by general employees.

Knowledge-workers focus on information exchange and innovation. This means that while people's living standards are improving, they will pay more attention to self-realization. This positive position is conducive to the achievement of knowledgeable workers' achievement goals. The factor of working environment belongs to the category of working factors, which will affect employees' recognition of the enterprise. If enterprises cannot provide employees with favorable conditions to ensure the employees' development, employees will resign easily. From the perspective of career development, employees' career has different stages of development with each stage having its own characteristics, tasks and goals.

From the perspective of career development, employees' career has different stages of development with each stage having its own characteristics, tasks and goals. This was divided career development planning into four stages: exploration, establishment, maintenance and recession. This could be shown as Table 2.

The result of the present study found that the ranking of salary and welfare is lower than other factors, such as job challenge and job interest. Among all the participants, aged 31-40 was the major group, accounting for 45.4%. They are at the top of their career development; however, salary, welfare, and promotion opportunities are not what they concern. On the contrary, knowledgeable workers care more about self-achievement and non-material factors than ordinary workers, under the condition of enjoying favorable income. Salary and welfare can certainly affect the willingness of knowledgeable workers to stay in enterprises. Enterprises should improve the salary and welfare in order to attract knowledgeable workers to keep contributing themselves to enterprises.

Among all factors, the last two factors were unfair treatment and health, ranking at 22 and 23 respectively. It shows that these two factors are the least important reasons to make knowledgeable workers quit their jobs. Unfair treatment has a low impact on the weight, because knowledgeable workers generally stay at an important position in the enterprise, and are less likely to encounter unfair treatment. Health is the most important factor in the present study. Knowledgeable workers surveyed in the present study are young, so their healths have not played a significant role yet.

Table 1. Results of weight analysis of the criteria.

Aspect	Criteria	Weight	Ranking	Hierarchical series	Overall ranking
Enterprise attribute (0.360;1)	Secluded position	0.250	2	0.090	2
	Cultural discrimination	0.118	4	0.043	7
	Turnover cost	0.334	1	0.120	1
	Interest in work	0.196	3	0.070	4
	corporate culture	0.101	5	0.036	13
Organizational management (0.238;2)	Leadership style	0.311	1	0.074	3
	Management ability	0.238	2	0.057	6
	Promotion system	0.174	3	0.042	9
	Relationship with leaders	0.152	4	0.036	13
	Industrial environment	0.125	5	0.030	16
Interpersonal relation ship (0.146;3)	Team atmosphere	0.127	4	0.018	21
	Work cooperation	0.255	2	0.037	12
	Interpersonal relationship	0.164	3	0.024	18
	Training system	0.454	1	0.066	5
Personal needs (0.120;5)	Remuneration and welfare	0.184	3	0.022	19
	family factors	0.358	1	0.043	7
	Post	0.218	2	0.026	17
	Health	0.136	4	0.016	22
	Unfair treatment	0.106	5	0.013	23
Work attributes (0.137;4)	Career achievement	0.258	3	0.035	15
	Job challenges	0.306	1	0.042	9
	Better position	0.145	4	0.020	20
	Work guarantee	0.290	2	0.040	11

Source: the authors in the present study.

Table 2. Career development stages for knowledgeable workers.

Age	Career	Major needs and tasks
22-29	Exploratory	Establishing working objectives, understanding self-value and ability, defining one's own needs, and choosing an appropriate lifestyle.
29-35	Developmental	To constantly improve one's own ability to make career reach the best status, some people try to change the environment or job to achieve their desired life and establish their own value.
36-44	Dangerous	Often encountering difficulties when adapting to social changes, peer pressure and social values.
Over 45	Mature	Have fully exhibit personal abilities, satisfied with their life and readjust their living interest and lifestyle.

Source: the authors in the present study.

5 CONCLUSION AND SUGGESTIONS

By comparing the relative weights of all factors, it was found that enterprises can reduce the loss of knowledgeable employees through effective management strategies.

First, according to relevant studies in the past, salary is the primary factor affecting the loss of employees. The psychological support and encouragement provided by enterprises can help knowledgeable workers form their self-esteem and sense of belonging. To strengthen employees' sense of belonging, enterprises can work on the following: (1) The enterprise has to build a fair, respectful, meaningful, and friendly environment. The enterprise have to find talents to construct a team in order to achieve sustainable development. The key for achieving it is to select appropriate talents based on strict standards. (2) The enterprise should care about the employees' psychological development by caring about their feelings. If employees have a sense of belonging and security, they will devote themselves to the job.

Secondly, working on staff training and career development, once employees' personal development matches organizational development, they will have positive attitudes towards the operation of organization; otherwise, they will have negative attitudes and even have intention to resign. Enterprises have to understand employees' needs at different stages, satisfy their needs by improving working conditions in order to reduce their attempt to resign. Enterprises should manage to meet their needs by providing challenging tasks and trainings to help employees develop new skills or professional competencies. This can help improve employees' loyalty.

Lastly, building up a perfect promotion system, perfect promotion system can motivate employee's motivation. It is the enterprise's commitment to employees and the employees' expectation to the enterprise. Promoting people who have leadership and management ability can optimize the quality of management and timely motivate other knowledgeable workers, under which the manager's leadership and management ability will be improved.

ACKNOWLEDGMENT

This paper has been supported by the following project. Fund No.: 2018B02 (the Fujian Administrative Management Institution Project, Fujian Province, PR China).

REFERENCES

H. Hong, 2017. *A study of organizational commitment, job satisfaction and turnover intention using network analysis - A case study of a company.* Yuanzhi University.

Q. Xin, 2015. *Research on the reasons and countermeasures on the loss of knowledgeable workers.* Jinan: Shandong University.

S. Huang, 2015. *Scientific and Economic Market 15*(4) 88–89.

X. Lu, 2014. *Study on the relationship among job stress, self-efficacy and turnover intention of knowledgeable workers.* Chengdu: Southwest University of Finance and Economics.

Y. Hi, 2016. *The study on the loss of knowledgeable workers and Countermeasures.* Yangzhou: Yangzhou University.

Y. Hu, 2016. *Research on the reasons of the loss of knowledgeable workers.* Yangzhou: Yangzhou University.

The research on quality of employment and congruous extent of college graduates' employment

Xingyan Du*, Suping Chen & Chechang Chang
Fujian University of Technology, Fuzhou, Fujian, PR China

ABSTRACT: The present study recruited college graduates of 2016 by considering the factors of their capability, knowledge, professionalism and characteristics to analyze the correlation between each factor in order to find out the factors influencing the employment. The results of regression analysis showed that there was a positive correlation between their income and capability, knowledge, professionalism and characteristics. The results of data analysis verified the hypotheses of quality of employment and precise employment congruity.

1 INTRODUCTION

There have been numerous studies on the quality of employment and college graduates' precise employment. However, the relevant studies concerning the quality of employment and college graduates' precise employment have yielded inconsistent results. The results of previous studies can be summarized as follows.

In terms of quality of employment, influencing factors, evaluation systems, and countermeasures, Zhang (2013) elaborated on the theories concerning quality of college graduates' employment, and constructed an evaluation system composed of the opinions from the government, graduates, and employers. Based on the objective situations and individual's subjective attitudes towards their jobs, Wang (2015) analyzed the factors influencing the quality of graduates' employment. The studies with respect to the evaluation system and countermeasures with regard to the quality of graduates' employment have been widely explored (Qin, 2007; Ke, 2007; Yang & Zhang, 2013; Song, 2016; Wu, 2017; Li & Cheng, 2017; Zhou & Huang, 2018). A few scholars also attempted to investigate the evaluation method with respect to the quality of graduates' employment by means of innovative research methods. Wei (2013), Yang (2015), and Du (2017) used AHP to analyze the quality of jobs that college graduates engaged in. Some scholars used second hand data to explore this issue (Zhu & Zhen, 2013; Wu & Wang, 2015). Some scholars attempted to explain the quality of employment with a quantitative analysis paradigm. Ke (2010) used a linear regression to analyze the relationship between graduates' professionalism and the quality of employment. Zhong & Liu (2015) used a linear regression and binary logistic regression to

explore the relationship between graduates' working attitudes and quality of employment. Huang (2012) and Shi & Ding (2017) investigated the impacts of human capital and social capital on the quality of employment. Yue & Tian (2016) also explored the impact of human capital on the quality of college graduates' employment, and examined the moderating factor of professional identity.

Secondly, in terms of the relation of precise employment and the quality of employment, the research focus was primarily on how to realize precise employment through precise employment services. Most scholars adopted the perspectives of ideological guidance of precise employment services, team building, platform constructions, and security mechanisms to facilitate college graduates' job seeking. Wang (2016), based on the big data, put forward the idea of precise employment service in colleges and universities, in which precise control of job opportunities, precise employment, precise control of information, precise platform of employment, and precise information transmission were involved. Some scholars took some schools as examples to discuss and explain this issue.

All the above studies have provided valuable suggestions for college graduates and also for scholars who are interested in the study of precise employment of college graduates. The studies concerning the quality of college graduates' employment are many; however, the relevant studies with respect to college graduates' precise employment are scant. In terms of the research issues, most studies primarily investigated the practical operation of precise employment from the perspective of students. Generally speaking, previous studies failed to apply theories to explain the connotations, essences, influencing factors and precise employment patterns. In terms of research methods, previous

---

*Corresponding author: E-mail: dxy@fjut.edu.cn

studies mostly adopted qualitative methods but ignored quantitative method. The present study intended to fill in the gap by recruiting college graduates who graduated one year as participants to investigate whether there was a relationship between students' knowledge, professionalism, and characteristics and precise employment by considering the factors of knowledge congruity, professionalism congruity, and expectation congruity.

2   RESEARCH DESIGN AND HYPOTHESES

There are many evaluation criteria and factors influencing the quality of employment. However, according to the Theory of Person-Vocation Fit and the Theory of Resource Restriction, precise job congruity is the key factor to measure the quality of employment. This type of congruity includes the congruity between personal capability and job position, and the congruity between personal characteristics and job types. That is, the job position is assumed to be congruous with personal capability and personality. Therefore, the present study adopted the perspective of precise employment congruity by analyzing the factors in order to explore their impact on the quality of employment mechanism. To address the unresolved issues, the theoretical hypotheses in the present study are as follows:

Hypothesis 1: the more positive correlation between personal knowledge proficiency and job requirements, the better the quality of employment.

Hypothesis 2: the more positive correlation between personal expectation and job types, the better the quality of employment.

Hypothesis 3: the more positive correlation between personal professionalism and job positions, the better the quality of employment.

3   DATA COLLECTION AND RESEARCH METHODS

3.1   Data collection

The data of the present study had gained approval from the third-party, MyCOS, which had been winning a good reputation from the government, academic fields, business organizations and public. The data used in the present study, collected from the students in Fujian University of Technology, was further examined by MyCOS. The survey was conducted in June, 2017. The participants were all the graduates of Fujian University of Technology in 2016. That is, the author in the preseny study intended to understand their employment situation one year after graduation. The total number of graduates' mailboxes in 2016 was 4036, and 4000 qualified mailboxes were preserved after preliminary evaluation. Among them, 0 was found to be problematic. The number of valid mailboxes = (the number of qualified mailboxes - the

number of wrong mailboxes) = 4000. Among them, the number of unsubscribed mailboxes was 107. The number of mailboxes = (the number of valid mailboxes - the number of unsubscribed mailboxes) = 3839. The effective response rate was 52.7% (effective response rate = number of returned questionnaires/number of valid mailboxes), involving 47 majors in 12 colleges. According to the purposes of the present study, 852 valid questionnaires were collected by excluding invalid questionnaires, involving 44 majors in 12 colleges, in which there were 521 males and 331 females.

3.2   Variables

The present study included independent variables, dependent variables and control variables. Quality of employment was an independent variable. Monthly income (Y) represented the sum of average monthly income of the participants. Only those who graduated from school one year were the target participants in the present study. The extent of precise employment was an independent variable, which represented the discrepancy between personal capability and job requirements, namely "capability congruity extent" (X1). Secondly, the discrepancy between personal knowledge proficiency and the knowledge required by job position is "knowledge congruity extent" (X2). Thirdly, whether the job was in line with personal expectation was referred to "expectation congruity extent" (X3), in which 1 representing conformity and 0 representing nonconformity. Fourthly, whether the job corresponded with personal professionalism was "professional congruity extent" (X4), in which 1 representing conformity and 0 representing non-conformity. Gender was regarded as a control variable. In the current social context, gender still played an important role in employment, so the present study used "gender" as a control variable (M1), in which 1 representing girls and 0 representing boys. The formula used in the analysis was as follows:

$$Y_i = \alpha + \beta X_i + \gamma M_i + \varepsilon \; Y_i = \alpha + \beta X_i + \gamma M_i + \varepsilon$$
(1)

Among the variables, Y_i represented the quality of the ith graduate's employment; X_i represented the congruous extent of the ith graduate; M_i represented the gender of the ith graduate, α represented the constant, β represented the coefficient of the independent variable X and dependent variable Y; γ represented the coefficient of the control variable M and dependent variable Y; and ε represented a random error.

3.3   Research methods

The present study adopted linear regression to analyze all data. The relationship between independent variables and dependent variables was analyzed by using the least square function in linear regression. The results of linear regression were aimed to answer the research hypotheses.

4 DATA ANALYSIS AND RESULTS

4.1 Data analysis

The present study used SPSS 24 software to analyze the collected data. All test items were examined in order to test their validity and reliability. Generally speaking, if Cronbach's alpha is greater than 0.7, it means the test item is highly reliable. The overall coefficient of the test items in the present study was 0.85. A factorial analysis was conducted to analyze the validity of the questionnaires. The Kaiser-Meyer-Olkin Measure of Sampling Adequacy (KMO) should be greater than 0.7, and the Bartlett's tests of sphericity should be less than 0.0001. The KMO of the questionnaire in the present study was 0.795, the Bartlett's tests of sphericity was 0.00. Therefore, through the above tests, the reliability and validity of the test items were highly reliable and valid. In addition, the author also carried out a descriptive analysis, and the results exhibited normal distribution; therefore, linear regression was used to carry out further analysis.

4.2 Linear regression analysis

In order to investigate the relationship between monthly income and four independent variables (e.g., capability congruity extent, knowledge congruity extent, characteristics congruity extent, and professionalism congruity extent), an independent sample T-test was used. The significant value was set at 0.05. According to the results of data analysis, the four independent variables were significantly correlated with monthly income, the significant value was respectively 0.004, 0.002, 0.003 and 0.002 (see Table 1).

$$Y = 3498.929 + 1157.652X1 + 1154.237X2 \\ + 365.857X3 + 331.242X4 - 548.227M1 \quad (2)$$

4.3 Result of data analysis

According to the above analysis, the independent variables representing the factors affecting precise employment were significantly positively correlated with the dependent variable. The control variable is significantly negatively correlated with the dependent variable. That is to say, "capability congruity extent", "knowledge congruity extent", "expectation congruity extent", and "professional congruity extent" were significantly positively correlated with "monthly income". To be more specific, when "capability congruity extent" increased one point, personal "monthly income" increased 1157.625 RMB. When "knowledge congruity extent" increased one point, personal "monthly income" increased 1154.237 RMB regularly. When "expectation congruity extent" increased one point, the "monthly income" increased 365.857 RMB correspondingly. When "professionalism congruity extent" increased one point, the "monthly income" increased 365.857 RMB accordingly. At the same time, "gender" is significantly negatively correlated with "monthly income". Specifically, when compared with boys, girls' monthly income was about 548.227 less than that of boys' under the same employment situation. To sum up, the results of data analysis all supported the hypotheses.

5 CONCLUSION

The theoretical assumption of precise employment matching can completely explain the quality of employment. The present study intended to explore how to improve the quality of graduates' employment by explaining the factors influencing the congruity of employment through an empirical analysis. According to the results of data

Table 1. The regression analysis between independent and dependent variables.

Coefficients[a]

Model	Unstandardized coefficient		Standardized coefficient		
	B	SD	Beta	t	Sig.
1 (Constant)	3498.929	332.847		10.512	.000
X1	1157.652	405.669	.100	2.854	.004
X2	1154.237	362.759	.111	3.182	.002
X3	365.857	124.803	.097	2.931	.003
X4	331.242	107.251	.104	3.088	.002
M1	-548.227	108.318	-.167	-5.061	.000

a dependent variable: y

analysis, the results could be concluded as follows. First, the "capability congruity" was one of the most important factors influencing the quality of employment. When one's personal capability was outstanding, he/she would secure a better job yielding a high quality of employment, which matched the theoretical assumption of precise employment matching. Secondly, in terms of "knowledge congruity", when one's personal knowledge proficiency was favorable, he/she would find an appropriate job consistent with his/her knowledge and resulted in a desirable quality of employment, in which the theoretical assumption of precise employment matching was supported. Thirdly, in terms of "expectation congruity", when the job position corresponded with personal expectations and characteristics, he/she would be satisfied with his/her jobs and led to a better quality of employment; thereby the theoretical assumption of precise employment matching was supported as well. Fourthly, in terms of "professionalism congruity", when personal professional level was consistent with the job position, he/she would be satisfied with the job and yielded a high quality of employment, which supported the theoretical assumption of precise employment matching. The results of data analysis all supported the hypotheses.

ACKNOWLEDGEMENTS

The author acknowledge the support by the Junior Teachers Education and Research Project of Foundation of the Education Department of Fujian Province (JAS170301).

REFERENCES

D. Yue and Y. Tian, 2016. *Jiangsu Higher Education 16*(1) 101–104.

H. Shi and Y. Ding, 2017. *Population and Economy 17*(5) 90–97.

J. Huang, 2012. *Beijing Social Science 12*(6) 52–58.

J. Li and L. Cheng, 2017. *China Youth Research 17*(10) 92–99.

J. Qin, 2007. *Chinese Youth Research 07*(3) 71–74.

J. Zhou and Y. Huang, 2018. *School Party Construction and Ideological Education 18*(1) 88–90.

L. Song, 2016. *Heilongjiang Higher Education Research 16*(5) 118–121.

M. Wang, 2016. *Ideological and Theoretical Education 16*(6) 84–88.

Q. Yang and J. Zhang, 2013. *Journal of Southwest Normal University (Natural Science Edition) 13*(1) 151–155.

Q. Zhong and K. Liu, 2015. *Exploration of Higher Education 15*(3) 107–113.

R. Zhu and Y. Zhen, 2013. *China Adult Education 13*(3) 117–120.

T. Wang, 2015. *Higher Education Explorations 15*(11) 104–109.

T. Wei, 2013. *Education and Economics 13*(2) 43–47.

X. Du, 2017. *Educational Review 17*(2) 66–69.

X. Wu and S. Dong, 2017. *Scientific Progress and Countermeasures 17*(2) 140–144.

Y. Ke, 2007. *China Higher Education Research 07*(7) 82–84.

Y. Ke, 2010. *China Youth Research 10*(7) 98–100.

Y. Wu and Z. Wang, 2015. *Jiangsu Higher Education 15*(1) 100–104.

Y. Yang and J. Li, 2015. *Chinese Journal of Education 15*(6) 148–149.

Y. Zhang, 2013. *China Higher Education Research 13*(5) 82–86.

Z. Chen, 2017. *School Party Building and Ideological Education 17*(4) 81–83.

Smart Science, Design & Technology — Lam et al. (eds)
© 2020 Taylor & Francis Group, London, ISBN 978-0-367-17867-3

The service quality study of catering industry's-based on SERVQUAL model

Xie He
School of Management, Fujian University of Technology, PR China

Chechang Chang*
School of Internet Economics and Business, Fujian University of Technology, PR China

ABSTRACT: Development of the catering industry promotes the development of the local economy. Competition between restaurants has changed from product competition and price competition into service competition. More and more customers gain additional value through consumption and personalized service so as to enjoy the greatest satisfaction. In this paper, the SERVQUAL model was used to investigate the quality of service perceived by customers in the catering industry. With the five major dimensions and twenty-one criteria as the factors analyzed through the Analytic Hierarchy Process (AHP), the factors of customer service quality were observed. Based on the results of data analysis, some suggestions were provided to the catering industry to encourage them to provide even better services instead of marketing promotion in order to survive in a competitive market.

1 INTRODUCTION

China's catering industry is developing like wildfire. According to analysis of the relevant food and beverage industries, the total number of restaurants nationwide has exceeded 6 million, of which Chinese restaurants account for 75%. The number of employees in China's catering industry continued to grow, reaching 16.08 million in the third quarter of 2016. Catering business turnover in the past 20 years has grown dramatically and is expected to grow up to 18% in the future. With the rapid development of the catering industry, whether the service quality is improving along with it is still questioned by customers. The present study is based on the SERVQUAL (Service Quality) by adopting important criteria to investigate customers' perception of service quality and then using the analytic hierarchy process (AHP) to analyze the relative weight of all factors. Finally, appropriate suggestions were put forward to the catering industry by encouraging them to adopt integrated marketing strategy.

2 SERVQUAL SCALE OF SERVICE QUALITY FOR CATERING INDUSTRY

SERVQUAL is a customer-oriented quality assessment tool based on the theory of customer's perceived service quality (Wang & Yu, 2016; Wei, 2014; Yang,
2015; Qu et al. 2008). The basic idea is to judge the quality of service by considering customers' expectation and perception. It is a multivariable customer perceived service quality measure. Regarding the catering industry, five dimensions of perceived service quality are proposed: visible, reliability, responsibility, assurance, and empathized. Visible is the appearance of facilities, equipment, and personal service. This could be shown as Table 1.

First of all, the SERVQUAL (Service Quality) was used on this research in the present study. Each item of the questionnaire is in Likert scale with 5 representing strongly agree and 1 representing strongly disagree. A total of 100 questionnaires were distributed, of which 85 were valid. KMO sampling appropriateness adequacy was 0.733 > 0.7. The Bartlett test sphericity was 724.692 and significance = 0.000 < alpha = 0.01, which shows that the data were valid after factorial analysis. Twenty-one items were left in the formal experiment as shown in Figure 1.

3 RESEARCH METHODS: ANALYTIC HIERA-RCHY PROCESS

The purpose of this study was to analyze the factors that affect the quality of service in the catering industry and to calculate the relative weight of all factors in the two aspects. Twenty questionnaires were distributed with 14 valid questionnaires left.

---

*Corresponding author: E-mail: 1794298328@qq.com

Table 1. The aspect and criteria of SERVQUAL

Aspect	Criteria	connotation
Empathized Empathized refers to the idea that enterprise can take care their customers, provide personal services, reflect humanity of the company, and make the service personalized and humane.	C11 Considerate to customers	When it comes to problems in the process of providing services, customers' interests and rights are the priority to be considered.
	C12 Coordinate with each other	Refers to the effective communication and cooperation between colleagues in the workplace.
	C13 Providing consultation	When a customer is confronted with a problems, he/she can get an immediate reply.
	C14 Courtesy	Refers to strictly abide by the etiquette code in the process of providing services, providing accurate and unmistakable friendly service.
	C15 Customer rights and interests	When providing welfare, we will do our best to win the rights and interests of our clients.
	C16 Exclusive service	To provide customers with one to one services, and meet customer service needs according to individual requirements of customers.
Responsibility Responsibility is the ability of an enterprise to quickly respond to customer needs and provide services. This also reflects to the effectiveness of problem solving.	C21 Patience	In the process of service, regardless of the customer's purchase intention, maintain a service attitude, patiently answer the doubts of customers.
	C22 Proper arrangement	When faced with a large number of customers, even if the individual is in a busy state, customers can still be arranged, so that customers will not be overlooked.
	C23 Professional competence	It is reflected that the service personnel have sufficient knowledge and professional skills to provide services.

(Continued)

Table 1. (Continued)

Aspect	Criteria	connotation
	C24 Rapid reaction	It reflects to the prompt response of the customer when the customer requests the service demand, and makes corresponding solutions.
Reliability Reliability refers to the ability of an enterprise to provide a correct and timely service in accordance with commitments.	C31 Complete information	When providing customer related project auxiliary material information, the information is confirmed and is not misleading to customers.
	C32 Timely service	It reflects the ability to solve customer problems in committed service hours and contents.
	C33 Accurate service	It is to provide appropriate service when consumers first propose requests
	C34 Service flow	It is clear that there is a clear process guide in providing services to customers and completing transactions.
Assurance Assurance refers to the professional skill knowledge and confidence which all the employees present.	C41 moderate service	It means that every service customer can keep courtesy and integrity.
	C42 Service stronghold	It has enough service space and comfortable space to park conveniently.
	C43 Customer commitment	Something that promises to be done will have to be done
	C44 Customer assistance	When customers need services, they can timely help and promote the smooth solution of the problem according to the needs of customers.
Visible Visible means the enterprise has perfect hardware, facilities, equipment and appearance of the	C51 Reception facilities	It refers to the reception and reception facilities that are comfortable, perfect and humanized.

(Continued)

28

Table 1. (Continued)

Aspect	Criteria	connotation
staff. Thus, customers can have perceive comfortable towards services.	C52 Recognition and publicity	A clear and attractive corporate identification and publicity campaign allows customers to identify clearly.
	C53 Environmental design	It refers to the rationality of the overall environment design of customers from the beginning of the enterprise, from geographical location to mobile line.

Data sources: organized by authors.

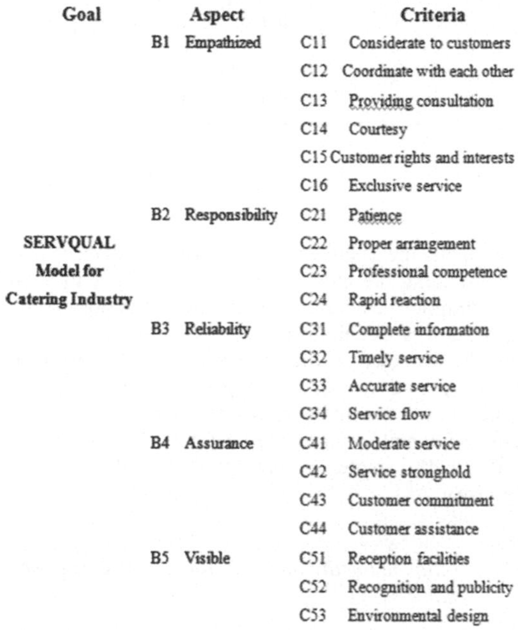

Goal	Aspect		Criteria
	B1 Empathized	C11	Considerate to customers
		C12	Coordinate with each other
		C13	Providing consultation
		C14	Courtesy
		C15	Customer rights and interests
		C16	Exclusive service
	B2 Responsibility	C21	Patience
SERVQUAL		C22	Proper arrangement
Model for		C23	Professional competence
Catering Industry		C24	Rapid reaction
	B3 Reliability	C31	Complete information
		C32	Timely service
		C33	Accurate service
		C34	Service flow
	B4 Assurance	C41	Moderate service
		C42	Service stronghold
		C43	Customer commitment
		C44	Customer assistance
	B5 Visible	C51	Reception facilities
		C52	Recognition and publicity
		C53	Environmental design

Figure 1. Research framework of catering service.

Analytic Hierarchy Process (AHP) was developed by Thomas L. Saaty in 1971, which is used to deal with uncertainty or multiple evaluation criteria. It is characterized by a combination of qualitative and quantitative research. Several factors were categorized into different categories and analyzed. The results of data analysis would be proposed to the decision makers to help them make correct strategies.

The AHP stratified the related factors in a hierarchy. All factors were analyzed to examine the internal consistency. When C.I. ≤ 0.1, it shows all items reached internal consistency. When C.I. > 0.1, it shows the internal consistency was weak.

Geometric mean was used to calculate the valid items to obtain the relative weight of all items. Super Decisions software was used to analyze the relative weight of the selected items in each criterion so as to calculate the priority and relative weight of all factors.

4 RESULTS AND ANALYSIS

The results of the above analysis illustrated that returns to scale under family long-term contract made great contribution to the overall efficiency of rubber industry. The improvement of efficiency mainly depended on the expanding of scale and its effect was small, but pure technical efficiency has not fully played its role in overall efficiency improvement. So in order to improve the overall efficiency, the importance of technology should be fully paid attention to. Scientific and technological innovation should be promoted to achieve the application and transformation of scientific technological achievements.

4.1 Relative weight of the aspect

From the results shown in Table 2, the values of the first three aspects are more than 20%, which represents they are equally important to customers. Consumers can easily perceive how hardworking the catering industry is from the dining environment and services provided in the restaurant. In fact, this may also fully reflect the status of China's catering industry. Some catering industries really care about price and service quality, but only a few are concerned about dining atmosphere or the professional ability of the service staff, and of course few industries pursue customer loyalty. Probably no one will care if customers come to dinner every day.

Reliability ranks fourth and empathized ranks fifth, with value less than 20%. Consumers' purchase desire can be activated by advertisement, marketing activities, or sales activities. Similarly, it is difficult for the catering industry to have SOP, because excessive SOP in cooking tends to lose its individuality. No consumer pays attention to whether the service information is complete or accurate. When consumers are dissatisfied, they will not complain positively, and will choose not to consume again or tell their relatives and friends to make a flexible boycott. On the contrary, no one will pursue long-term relationships with customers and provide exclusive or customized services for customers. This can explain why the weights of these two factors are relatively low.

4.2 Relative weight of different criteria

From Table 2, publicity activities, rapid response, proper arrangement, and customer promise were the top factors. The market in the catering industry is in full swing, and the domestic catering industry can be said to have been outcompeted by outgoing services.

Table 2. Relative weight of different criteria.

Aspect	Criteria	Weight	Ranking	level	Ranking
Empathy	Providing advices	0.140	6	0.021	21
0.152	Considerate to customers	0.185	2	0.028	15
	Coordination	0.173	3	0.026	17
	Courtesy	0.187	1	0.028	15
	Customer rights and interests	0.168	4	0.025	18
	Exclusive service	0.148	5	0.022	20
Responsiveness	Patience	0.171	4	0.042	12
0.243	Proper arrangement	0.262	2	0.064	4
	Professional competence	0.234	3	0.057	7
	Rapid reaction	0.333	1	0.081	2
Reliability	Complete information	0.292	2	0.054	8
0.183	Timely service	0.251	3	0.0460	11
	Accurate service	0.330	1	0.060	6
	Service flow	0.127	4	0.023	19
Assurance	Appropriateness	0.299	1	0.065	3
0.217	Service stronghold	0.251	3	0.054	8
	Customer promise	0.292	2	0.063	5
	Assisting customers	0.158	4	0.034	13
Tangibility	Reception facilities	0.264	2	0.054	8
0.205	Publicity activities	0.572	1	0.117	1
	Environmental design	0.165	3	0.034	13

Source of data: organized by author.

If restaurants cannot send food to customers out of restaurant, their services would be regarded as poor. When restaurants get involved in takeaway service, attention will be paid to the issues of transportation, storage, and display of foods in order to keep foods from deteriorating. Nowadays, most people pursue convenience and service quality but ignore table manners and catering culture.

Mutual coordination, customer rights and interests, service processes, exclusive services, and providing consultations are the last five factors. This shows that only a few catering industries pursue custom-oriented and individualized services. No wonder these factors are not getting too much attention, and no one has thought to establish a friendly relationship with their customers. This is a truly customized and personalized service. Creating a comfortable, relaxed, and enjoyable dining environment for customers will make them feel satisfied and strengthen their loyalty. It is clear that the cost of keeping a customer will be much lower than developing a new customer.

5 CONCLUSIONS AND SUGGESTIONS

5.1 *Using integrated marketing to enhance brand reputation*

The catering industry should integrate the hardware service and software service, form the synergy effect, and change with the customers' needs. Food and beverage industry is a consumer commodity, emphasizing the independence and integrity of each consumption. Therefore, when the reputation of the restaurant is established, more and more consumers will be attracted to the restaurant, and finally form an integrated marketing propaganda. In the mean time, the dishes and services can adapt to seasonal, environmental changes and customers' needs, which will finally create greater profits for the catering industry.

5.2 *Enhance loyalty and maintain a good customer relationship*

The loyalty of customers here refers to dependence and reconsumption expectation formed by customers' loyalty, trust, and experience of commodity or service. Treating customers as family members while providing foods and services is the best way to increase the customers' loyalty. This enhances customer loyalty, not only winning reputation but also reducing the cost of customer maintenance. Restaurants should build customer loyalty by providing personalized services, such as providing special tableware and delicacies for children and the elderly or send greetings and blessings regularly through the company system to bless customers to make them feel respected. They should maintain customers' loyalty through more communication.

5.3 Strengthening training of service staff with integrated marketing

Currently, the requirement from talent is no longer just to fit a single position. As for the food and beverage industry, the marketing needs not just a salesperson, but a talent with the ability to integrate a variety of resources. Regarding the catering industry, they are looking for people who may able to understand food etiquette, the dishes, cooking, the atmosphere, the layout of the restaurant, and marketing activities. Therefore, it is necessary to train employees through job rotation or cross-departmental training. They should not be applied only in one department; they should be trained in many key positions, such as sales, customer service, and catering. By doing this, not only can they broaden their knowledge, but they also can motive sustainable development of the catering business.

ACKNOWLEDGMENT

This paper has been supported by the following project. Fund No.: 2017FZA01 (13-5 the Fujian high education innovation Project, Fujian Province, PR China).

REFERENCES

R. Wang and T. Yu, 2016. *Development and application of community education service quality evaluation tool based on SERVQUAL.* Beijing: Tsinghua University press.

D. Wei, 2014. *Research on integrated marketing communication strategy based on customer relationship.* Nanjing University of Science and Technology.

W. Yang, 2015. *Shanghai Management Science* 37(3), 51–54.

X. Qu, G. Xiang, T. Song, and Z. Wang, 2008. *Management Observation* 8(10), 169–170.

Smart Science, Design & Technology — Lam et al. (eds)
© 2020 Taylor & Francis Group, London, ISBN 978-0-367-17867-3

The study of evaluation of service quality of home delivery by the nonadditive model

I-Hsiang Lin*
School of Economics and Management, Xiamen University of Technology, Xiamen, PR China

Chien-Hua Wang
School of Management, Fujian University of Technology, Fuzhou, PR China

ABSTRACT: This paper uses literature collection and analysis. The initial factors are of service quality and are defined, and online questionnaires are filled by consumers. In the empirical research, first, the evaluative dimensions are obtained by related literatures and relative weights between the independent dimensions are obtained by the analytic hierarchy process (AHP). Finally, six home delivery companies in Taiwan are selected in the paper. The fuzzy measure and fuzzy integral are utilized to find the research results of measurement of service quality and to rank them from best to worst. The results can be provided to home delivery companies to quickly respond to market needs and reinforce the existing competitive advantages and improve competitiveness.

1 INTRODUCTION

The home delivery industry is part of the logistics industry. Logistics provides rapid circulation of goods and shortens the shopping time and shopping distance of consumers. Home delivery enables consumers to enjoy immediate, fast, and convenient home service after consumption activities. The emergence of home delivery also reflects the evolution of the market and consumption habits.

The service range of the home delivery industry includes functions such as delivery to home, specified delivery time, fresh preserving service, and cash on delivery. Therefore, the measurement of the service quality of home delivery is not a single criterion or a single construct; it is instead a problem between multiple criteria decisions. And the service itself has intangible features that are difficult to measure so that the evaluator's perception of service quality involves subjective cognition and the concept of fuzziness (Bellman et al. 1970; Carman, 1990). So, if we use accurate value, we may not be able to express the evaluator's perception of the service quality of home delivery. Therefore, this study first sorted out relevant literature on the service quality of home delivery and drafted initial service quality factors, and conducted research by means of questionnaires. Second, depending on the research needs, we discussed separately the service quality of home delivery companies, related literatures, analytic hierarchy process (AHP), fuzzy measure,

and fuzzy integral. Finally, the feasibility of the mode was verified by case analysis and the final results were analyzed and discussed.

2 LITERATURE REVIEW

Mentzer et al. (1999) survey the employees of logistics service companies, which were provided by the military, then combines the results with the research of Parasuraman et al. to jointly develop the scale of logistics service quality. Chen (2002) applies a multicriteria evaluation method to evaluate the service quality of the home delivery industry in Taiwan. He intended to explore the perspectives of business management, customer service, and government supervision in order to establish a service quality assessment model for the domestic home delivery industry. Zhang et al. (2002) analyze a set of index systems of quality evaluation relevant to logistics service and compare the traditional logistics service target with the service dimension defined in the PZB model. Zhao et al. (2012) adopt the service quality theory of PZB model to establish the logistics service quality index system and the AHP to design the comprehensive evaluation model of logistics service quality. Liao et al. (2013) investigate the related factors that influence the adoption of home delivery service, and apply the analysis network process and importance performance analysis to

*Corresponding author. E-mail: 2016000008@xmut.edu.cn

analyze the service quality and customer satisfaction of home delivery service providers in the Taiwan market. Although many researchers discuss service quality, the dimensions or criteria of measurement are existing interaction, so this paper adopts fuzzy integral to calculate performances and fuzzy measure to analyze critical dimensions or criteria.

3 FUZZY INTEGRAL

Let g be a fuzzy measure defined on a power set, $P(x)$, which satisfies the following characteristics:

$$\forall A, B \in P(X), \ A \cap B = \emptyset, \text{ then,}$$

$$g_\lambda(A \cup B) = g_\lambda(A) + g_\lambda(B) + \lambda g_\lambda(A)g_\lambda(B), \quad (1)$$
$$for -1 \leq \lambda \leq \infty$$

Setting $X = \{x1, x2, ..., x_n\}$, fuzzy density $gi = g\lambda(\{x_i\})$ can be formulated as follows:

$$g_\lambda\{(x_1, x_2, ..., x_n)\} = \frac{1}{\lambda} | \prod_{i=1}^{n} (1 + \lambda g_i) - 1| \quad (2)$$

Next, let h be a measureable set function defined on the fuzzy measureable space. Assuming that $h(x1) \geq h(x_2) \geq \cdots \geq h(x_n)$, then the fuzzy integral of fuzzy measure $g(\cdot)$ with respect to $h(\cdot)$ can be defined as follows (K. Ishii et al. 1985):

$$\int h \cdot dg = h(x_n)[g(H_n) - g(H_{n-1})] + h(x_{n-1}) \cdot$$
$$[g(H_{n-1}) - g(H_{n-2})] + \cdots + h(x_1)g(H_1) \quad (3)$$

where $H_1 = \{x_1\}$, $H_2 = \{x_1, x_2\}$, ..., $H_n = \{x_1, x_2, ..., x_n\} = X$.

4 RESEARCH RESULTS

The research object of this paper was based mainly on six home delivery companies in Taiwan. The investigation of the performance value of each home delivery company was mainly carried out on the Internet, and the questionnaire survey lasted for one month. A total of 178 questionnaires were collected, and 43 were deducted. Therefore, there were 135 valid questionnaires. Respondents were required to use services from at least one of these home delivery companies. The home service company service quality satisfaction scale content is scored from 0 to 100 points, where 100 is the highest satisfaction, decreasing in order. In addition, in order to obtain the relative importance of each evaluation criterion, the authors also conducted an expert survey. In addition to the rich knowledge of the home delivery companies investigated, we

expect that the experts also have a certain degree of understanding of the general consumers. The subjects of this expert survey were mainly scholars and managers of home delivery companies, with a total of 10 people. However, after a consistency check by the AHP questionnaire, only eight valid questionnaires were left for subsequent analysis and discussions.

Next, this study collects related literatures to select appropriate criteria. There is a total of 20 criteria. These criteria can establish five dimensions by factor analysis, which achieves the purpose of criteria reduction and consolidation. The first dimension is "Service convenience" (D_1) including popularization (C_1), timeliness (C_2), convenience (C_3), open all year round (C_7), and small delicate items (C_8). The second dimension is "Freights and payment methods" (D_2) including clear price (C_5) and multiple payment methods (C_6). The third dimension, "Service content" (D_3), includes guaranteed delivery (C_9), home delivery (C_{10}), night delivery (C_{11}), specified delivery time (C_{12}), cash on delivery (C_{13}), and fresh preserving service (C_{14}). The fourth dimension is "Pre-service and follow-up service" (D_4) including traceability (C_4), reasonable and quick compensation standards (C_{18}), provide consumer complaints channel (C_{19}), and prior notice before delivery (C_{20}). The fifth dimension is "Home delivery staff" (D_5) including profession of home delivery staff (C_{15}), service of home delivery staff (C_{16}), and instant notification of problems (C_{17}). By factor analysis, the questions of this study had reduced 20 variables to five principal component constructs in order to identify several common factors which are conceptually meaningful and independent of each other that can affect raw materials among a group of relevant and difficult-to-interpret materials. The interpretations of the five principal component constructs are 18.18%, 13.89%, 6.42%, 5.88%, and 5.59%, respectively. The cumulative interpretation variable also reached 49.98%.

Then this study used fuzzy measure and fuzzy integral to obtain the integrated performance value from each criterion within the integrated construct for different scenarios. Finally, we calculated the comprehensive performance value of each alternative. The details were shown in Table 1 and Table 2.

In addition, the weight of each common factor was obtained by the AHP method. Then, by means of fuzzy integrals, the integrated performance value of each common factor was calculated. Finally, we calculated the comprehensive performance value of each alternative, as shown in Table 2. According to Table 2, the best comprehensive performance value is H5 home delivery company, followed by H1 home delivery company, H3 home delivery company, H4 home delivery company, and H2 home delivery company. The worst is H6 home delivery company.

Table 1. The λ value of each dimension and fuzzy measure of each subset.

Fuzzy measure $g(\cdot)$			
Service convenience ($\lambda = -0.75$)			
$C_{11} = 0.21$	$C_{12} = 0.20$	...	$\{C_{11}, C_{12}, C_{13}, C_{14}, C_{15}\} = 1$
Freights and payment methods ($\lambda = -0.86$)			
$C_{21} = 0.85$	$C_{22} = 0.54$	...	$\{C_{21}, C_{22}\} = 1$
Service content ($\lambda = -0.88$)			
$C_{31} = 0.21$	$C_{32} = 0.15$	...	$\{C_{31}, C_{32}, C_{33}, C_{34}, C_{35}, C_{36}\} = 1$
Preservice and follow-up service ($\lambda = -0.72$)			
$C_{41} = 0.23$	$C_{42} = 0.25$	...	$\{C_{41}, C_{42}, C_{43}, C_{44}\} = 1$
Home delivery staff ($\lambda = -0.94$)			
$C_{51} = 0.35$	$C_{52} = 0.38$	...	$\{C_{51}, C_{52}, C_{53}\} = 1$

Table 2. The integrated performance for common factor's combining fuzzy integral.

	D1	D2	D3	D4	D5	
	$W_1 =$ 0.211	$W_2 =$ 0.194	$W_3 =$ 0.235	$W_4 =$ 0.183	$W_5 =$ 0.177	Performance
H1	81.06	78.70	72.35	76.91	77.49	77.16
H2	76.66	75.40	77.72	74.08	72.75	75.50
H3	79.56	71.98	78.59	73.08	76.37	76.11
H4	74.70	73.89	79.96	72.24	76.81	76.07
H5	80.55	76.88	81.07	77.56	75.62	78.54
H6	77.78	67.75	74.19	75.14	73.3	73.71

5 CONCLUSION

In many previous studies, it was often assumed that the criteria are independent of each other; in fact there is instead a certain degree of relationship. In order to achieve comprehensive performance between the standards, this paper first applied factor analysis and AHP to explore the service quality of home delivery. Then, fuzzy measure and fuzzy integral were utilized in order to calculate the overall performance value of each alternative in each of the independent common factors. Finally, we used the fuzzy integral method along with the weights calculated by AHP to obtain the ranking of the six home delivery companies in Taiwan. Finally, the results of the nonadditive model can assist managers in identifying their competitive advantages relative to other competitors. In the future, we will apply this method to deal with various problem domains.

ACKNOWLEDGEMENT

The author would like to acknowledge financial support from Xiamen University of Technology (grant No. YSK170003R).

REFERENCES

R.E. Bellman, L.A. Zadeh, 1970. Decision making in a fuzzy environment. *Management Science*. 141–146.

J.M. Carman, 1990. Consumer perceptions of service quality: a assessment of the SERVQUAL dimensions. *Journal of Retailing*. 33–55.

J.T. Mentzer, D.J. Flint, and J.L. Kent, 1999. Developing a logistics service quality scale. *Journal of Business Logistics*. 9–32.

S.C. Chen, 2002. A study of the application of the multiple-attribute evaluation method on the service quality of the home delivery industry in Taiwan. Master's thesis. Dept. of Transportation Management, Tamkang University.

C.G. Zhang, J.Z. Zheng, 2002. Index system of quality evaluation in logistics service. *Logistics Technology*. 74–76.

N. Zhao, T. Yu, 2012. Study on construction and evaluation of logistics service quality index systems. *Logistics Technology*. 140–143.

T.Y. Liao, Y.C. Lou, 2013. Service Quality and Customer Satisfaction for Home-delivery Service Providers—Case Study of Five Providers in Taiwan. *J. E-Bus*. 15, 461–490.

K. Ishii, M. Sugeno, 1985. A model of human evaluation process using fuzzy measure. *International Journal of Man-Machine Studies*. 19–38.

Smart Science, Design & Technology — Lam et al. (eds)
© 2020 Taylor & Francis Group, London, ISBN 978-0-367-17867-3

A study of promoting characteristic beautiful villages with resource evaluation of environmental design

Paohua Yang*
School of Design, Fujian University of Technology, Fuzhou, Fujian, PR China

ABSTRACT: With the continuous development trend of beautiful rural construction and the cultural tourism industry, local governments and commercial organizations are actively promoting the construction of beautiful villages with leisure and industrial characteristics, and then promoting local characteristic industrial towns. Their intentions are eager and diverse. However, the resource problem still exists, and the actual effect has not been revealed. At the same time, cultural tourism has attracted the attention of tourism consumers and their ancillary service facilities. There is a need to know how to combine agricultural and cultural, ecological, leisure, and holiday industries with multi-functional characteristics. Environmental planning and design propose the staged requirements of the type. This study defines twenty environmental planning design resource assessments from the influential factors of resource integration, environmental planning, and design resource assessment for sustainable management to create the optimal value orientation of beautiful rural life. Decision-making considerations are presented, followed by the survey method for the evaluation of the design of different professional positions.

Keywords: resource assessment, beautiful countryside, characteristic towns, environmental planning and design

1 INTRODUCTION

There is current interest in the development of beautiful rural construction, the core value and the deep interest in research motivation. For most of the rural areas looking forward to the beautiful rural builders and environmental planning and design, the value-oriented cross-industry combination is constructed with leisure agriculture, holiday tourism, cultural and creative characteristics, health and wellness, homestay development and integrated investment real estate, or entering the new market field in order to rely on poverty alleviation policy, etc., for the sustainable ecological protection of villages and rural areas towards beautiful rural construction and poverty alleviation of villagers. The creative and entrepreneurial model and the creation of cultural and creative products such as local specialties are less obvious. The formation of the people after the construction is lost, and there are bad and ruined scenes. These are due to the lack of resource allocation indicators constructed with the core values of beautiful villages. Therefore, this study intends to explore the internal and external environment of beautiful rural construction from the evaluation system of resource planning and design and use the content analysis method and the focus group method to discuss the in-depth interviews in the qualitative research of the academic and academic circles. Systematic research constructs resource assessments.

2 RESEARCH METHODS

This study collects relevant research literatures such as "Entrepreneurial Resource Assessment," "Characteristics of Beautiful Rural Construction," "Resource Planning and Design," and "Business Environment Investment Factors" and refers to the evaluation criteria of entrepreneurial resources in the design industry by Yang (2014). Bruno and Tyebjee (1984) conducted an empirical study of assessment criteria for investment cases as the basis for model building. Based on theoretical research and practical research, this study develops the object of evaluation into a beautiful rural construction resource project, analyzes the interrela-

---

*Corresponding author. E-mail: lucapaoyang@qq.com

tionship between them, and summarizes the research structure of the overall evaluation model. In the early stages of the study, in-depth interviews were conducted to visit four beautiful rural builders, including designers, village leaders, and investment operators. The total number of staff is fifteen people. During the interview, the respondents' personal opinions were used mainly to understand the considerations of the different professional backgrounds in decision-making for the beautiful rural construction resources, depending on the considerations of the constructor's personnel. The access data uses the content analysis method to draw out the considerations of the planning and design resource evaluation decision, and is listed as the reference for the design of the resource evaluation qualitative questionnaire project of the focus group method in the second stage. It is worth noting that the environmental planning and design resources and investment and management environment included in the beautiful rural construction

The focus group method of the next stage is based on different backgrounds and professions (seven rural leaders, three designers, five investment operators), and the quality questionnaires are aimed at the overall consideration of beautiful rural construction and different are often mutually causal. In the early stage of designing environmental planning resources, the designers themselves are often the cultural and creative design and development resources.

This research structure first analyzes the relevant literatures on the evaluation criteria of beautiful rural construction resources in the past and abroad, and based on the evaluation criteria of various scholars, distinguishes four representative resource assessment criteria (see Figure 1), which remove poverty from rural areas. The characteristics of factors such as policy and farmers' cooperation and agricultural production and marketing are included in the evaluation framework of entrepreneurial resources, and the design and development of Cultural and Creative's characteristics and the design of the image are the characteristics of the resource planning and design, and the relevant aspects are not considered. The evaluation factor is included in the characteristics of

other factors. According to this architecture, all evaluation pointers are divided according to different attributes and divided into four facets. The proposed model is shown in Figure 1.

People include the perspectives of the designer, the rural leader, and the investment operator. The qualitative content analysis results are considered in the overall consideration of beautiful rural construction and the position of designers, rural leaders, and investment operators. Finally, the content analysis results of qualitative questionnaires are described by statistical analysis, correlation analysis, and cluster analysis.

3 THE RESEARCH RESULTS AND DISCUSSION

From the initial interviews (see Table 1), it was found that the designer's point of view did not include personal preference factors and the issue of not taking into account the timeliness of upgrading and development of agricultural products, while the rural leaders did not take into account market trends and style. Identify the problems of design and cultural and creative design and development; the view of investment operators does not consider the inheritance of rural folk culture and the promotion of local specialties. This gives a preliminary understanding of the views of designers, village leaders, and investment operators. In the consideration of rural resource development decision-making, the considerations for leisure tourism and homestay, natural resource attractions planning, infrastructure construction, and management capabilities are taken into account. There are six factors, such as public support and restrictions, and the use of the Internet and technology. The designer's point of view is less concerned with issues such as integrated real estate investment, cost and capital, agricultural product marketing and upgrading, and other operational management issues, but more concerned about investment operators. The characteristics of cultural and creative, which are neglected by viewpoints, are convenient, operational, and ecological environment safety.

According to the results of the preliminary discussion and the evaluation criteria of the related literatures for environmental planning and design resource evaluation, this study also divides the factors of environmental planning and design resource evaluation into four categories: design attributes, management attributes, marketing attributes, and consumption attributes.

The design attributes considered in the category are (1) cultural and creative characteristic design and development, (2) natural resources scenic spot planning, (3) folk culture and customs inheritance, (4) ancient building repair and protection, and (5) scene image recognition design.

In the management attributes consideration category are (1) ecological environmental protection, (2) infrastructure construction, (3) local special product

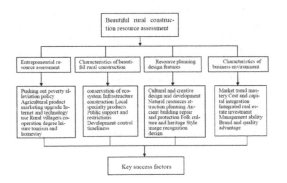

Figure 1. Research architecture data source.

Table 1. Considerations for the assessment of beautiful rural construction resources in different professional backgrounds (initial interview results). Δ table is less important.

Designer perspective	Rural leader perspective	Investment operator perspective	Investment operator perspective	■Listed as a survey item
Management ability	Management ability	Management ability	■Management ability	■
Internet and technology use	Internet and technology use	Internet and technology use	■Internet and technology use	■
X	Agricultural product marketing upgrade	Δ	X	X
Rural villagers cooperation degree	Δ	Rural villagers cooperation degree	■Rural villagers cooperation degree	■
Δ	Pushing out poverty alleviation policy	Pushing out poverty alleviation policy	X	X
Conservation of ecosystem	Conservation of ecosystem	Δ	■Conservation of ecosystem	■
Infrastructure construction	Infrastructure construction	Infrastructure construction	■Infrastructure construction	■
Local specialty products	Local specialty products	X	■Local specialty products	■
Public support and restrictions	Public support and restrictions	Public support and restrictions	■Public support and restrictions	■
X	Δ	Development control timeliness	■Development control timeliness	■
Cultural and Creative design and development	X	Cultural and creative design and development	■Cultural and creative design and development	■
Natural resources attraction planning	Natural resources attraction planning	Natural resources attraction planning	■Natural resources attraction planning	■
Ancient building repair and protection	Ancient building repair and protection	Δ	■Ancient building repair and protection	■
Folk culture and heritage	Folk culture and heritage	X	■Folk culture and heritage	■
Style image recognition design	X	Style image recognition design	X	X
Market trend mastery	X	Market trend mastery	■Market trend mastery	■
leisure tourism and homestay	leisure tourism and homestay	leisure tourism and homestay	■leisure tourism and homestay	■
Δ	Integrated real estate investment	Integrated real estate investment	■Integrated real estate investment	■
Brand and quality advantage	Δ	Brand and quality advantage	■Brand and quality advantage	■
Δ	Cost and capital integration	Cost and capital integration	■Cost and capital integration	■

promotion, (4) public support and restrictions, and (5) timeliness of development management and control.

The investment operation attribute consideration categories are (1) market trend mastery, (2) cost and capital integration, (3) integrated property investment, (4) management and management capabilities, and (5) brand and quality advantages.

User attribute consideration categories are (1) pushing out poverty alleviation policies, (2) upgrading of agricultural products and marketing, (3) internet and technology use, (4) rural villagers' cooperation, (5) leisure tourism and homestays, etc. There are a total of twenty survey items for the next questionnaire survey, in the initial stage. The agricultural product marketing and marketing upgrades, the poverty alleviation policy, and the image design of the styles involved in the interviews were excluded from the survey items

because they did not involve the overall viewpoint. The priority of the rural construction resource assessment considerations (see Table 2) shows that among the top five factors, the designer's perspective emphasizes the ranking: (1) cultural and creative design and development, (2) natural resources and scenic spots planning, (3) folk culture and inheritance of the customs, (4) repair and protection of ancient buildings, and (5) the image recognition design of the style, which is the same as that of the above design attribute consideration category, and the management ability of the user attribute consideration category is valued by the designer, the rural leadership manager and the investment operation attribute, and the rural leadership manager. Paying attention to the management of ecological property protection and infrastructure in the management attribute category, it also attaches

Table 2. Priorities of considerations for resource assessment of resource design for beautiful rural construction resources.

Attributes Sort	User attribute Rural villagers	Management attribute Rural leader perspective	Design attribute Designer perspective	Investment operation attribute Investment operator perspective
1	Pushing out poverty alleviation policy	Conservation of ecosystem	Cultural and creative design and development	Market trend mastery
2	Agricultural product marketing upgrade	Infrastructure construction	Natural resources attraction planning	Cost and capital integration
3	Internet and technology use	Local specialty products	Folk culture and heritage	Integrated real estate investment
4	Rural villagers cooperation degree	Public support and restrictions	Ancient building repair and protection	Management ability
5	leisure tourism and homestay	Development control timeliness	Style image recognition design	Brand and quality advantage
6	Management ability	Management ability	Management ability	Conservation of ecosystem
7	Conservation of ecosystem	Natural resources attraction planning	Internet and technology use	Internet and technology use
8	Natural resources attraction planning	Pushing out poverty alleviation policy	Public support and restrictions	Public support and restrictions
9	Local specialty products	Internet and technology use	Local specialty products	Infrastructure construction
10	Infrastructure construction	Agricultural product marketing upgrade	Conservation of ecosystem	Rural villagers cooperation degree
11	Ancient building repair and protection	Rural villagers cooperation degree	Rural villagers cooperation degree	Development control timeless
12	Folk culture and heritage	leisure tourism and homestay	leisure tourism and homestay	leisure tourism and homestay
13	Integrated real estate investment	Cost and capital integration	Infrastructure construction	Natural resources attraction planning
14	Public support and restrictions	Ancient building repair and protection	Market trend mastery	Cultural and creative design and development
15	Cost and capital integration	Integrated real estate investment	Development control timeliness	Ancient building repair and protection
16	Development control timeliness	Folk culture and heritage	Integrated real estate investment	Folk culture and heritage
17	Style image recognition design	Style image recognition design	Brand and quality advantage	Agricultural product marketing upgrade
18	Brand and quality advantage	Brand and quality advantage	Agricultural product marketing upgrade	Pushing out poverty alleviation policy
19	Cultural and creative design and development	Cultural and creative design and development	Cost and capital integration	Local specialty products
20	Market trend mastery	Marker trend mastery	Pushing out poverty alleviation policy	Style image recognition design

importance to the investment operation attribute consideration category of leisure tourism and housing and cost and capital integration from the perspective of investment operators; the investment operator's perspective is in addition to cost and capital. Integrating and integrating real estate investment also attaches importance to the design and attribute of the design attributes and the design and development of cultural and creative features. Twenty considerations are generally valued by designers, rural leaders, and investment operations. However, rural leaders like to upgrade their agricultural products to the final ranking of designers and investment operators. From the perspective of rural leadership managers, cultural and creative's characteristic design and development is listed as the last item. Among the five considerations of user attribute categories, except for pushing the Internet and technology applications are the three concerns and ranked in the top five, the designers and village leaders are ranked in the bottom ten in the other four. The investment operators listed the folk culture and the local special products in the last ten. This shows that the designer and the rural leadership managers seem to place less emphasis on the development and control timeliness than the investment operator.

In addition, twenty variables for evaluation considerations are clustered separately for designers, rural leaders, and investment operators. To understand the various types of different professional backgrounds in product design evaluation, see the

38

Table 3. Analysis of different professional background clusters.

Designer perspective	Cluster naming	Management perspective	Cluster naming	Investment operator perspective	Cluster naming
15 other considerations	Fully integrated evaluator	15 other considerations	Fully integrated evaluator	18 other considerations	Fully integrated evaluator
Agricultural product marketing upgrade Leisure Travel and Homestay Local specialty products Internet and technology use Cost and capital integration Integrated property investment	Value-oriented user-oriented evaluator	Leisure Travel and Homestay Local specialty products Cultural and creative characteristic design and development Internet and technology use	Target-oriented use-oriented evaluator	Agricultural product marketing upgrade Poverty reduction policy	Individualist evaluator
Natural resources attraction planning Infrastructure construction	Leader's opinion follows the evaluator	Natural resources attraction planning Infrastructure construction	Leader's opinion follows the evaluator	Cultural and creative characteristic design and development Leisure Travel and Homestay Internet and technology use	Innovative follow-up evaluator

analysis results (Table 3). It has a focus on the evaluation perspectives of designers, village leaders, and investment operators. Most clusters of evaluation considerations (more than fifteen) can be called "comprehensive evaluators"; in the case of designers and village leaders, there are decision makers who like natural resource attractions planning, and the infrastructure construction is called "Leader Opinion Followers"; the other focuses on user property upgrades, cost and capital integration, Internet and technology applications, leisure travel and home-stays, and local specialty products. The value of integrated real estate investment is called "value-oriented user-oriented evaluator"; the user attribute factor and target user attribute are "target-oriented use-oriented evaluator"; the emphasis on agricultural product marketing and sales is "individualism." Type evaluator"; focusing on leisure travel and homestay, Internet and technology use, and cultural and creative's characteristic design and development is "innovative follow-up evaluator." The evaluation styles of people with different professional backgrounds seem to be quite similar in terms of designers and rural leadership managers, but there are differences in the perspective of investment operators. Among the six types, most of the respondents are mostly comprehensive evaluation evaluators, but other types of evaluators may have an impact on the evaluation results when they enter the evaluation system, and they cannot be ignored.

4 CONCLUSIONS

The validity of beautiful rural construction depends on whether it matches the original set of restrictive and objective needs (Ansrews, 1984). The degree of coincidence depends on the actual implementation and correct evaluation of environmental planning and design evaluation. The trend of people after rural construction is lost and improper, and the solution is in line with the goals and restrictions required in the resource evaluation. According to interviews and survey results, the main conclusions of this study are as follows:

(1) Most of the twenty evaluation considerations are valued by the respondents. Only the decision makers' preference is ranked as twentieth in the perspective of rural leaders and users, and the other nineteen can be classified as beautiful villages. There are references for the construction of resource and environmental planning and design resource assessment.

(2) Design attribute, management attribute, marketing attribute, and use attribute, the four evaluation categories, can be used as the consideration direction when expanding the design evaluation benchmark, while the designer and the rural leadership manager should strengthen the emphasis on the use of attribute evaluation categories.

(3) The different positions of the respondents and the level of the leaders of the interviewed villages are not much different for the design evaluation factors of the professional backgrounds, only slightly different. This is because the interviewed rural leaders and respondents have gained relative experience and recognition in the practice of beautiful rural construction. Under the circumstance that the people involved in the evaluation have considerable consensus, it is the current beautiful rural construction environment planning and design. This is a good foundation for the use and development of resource assessment.

(4) In the beautiful rural construction environment planning and design experience, many have greater design decision-making intensity, but in their duties, although the decision-making intensity is not much different, but the investment operator has greater design decision-making intensity, the management times And the user is the smallest.

(5) The overall subject's perspective on the three professional backgrounds of designers, rural leadership managers, and investment operators and the overall perspective of beautiful rural construction. There is no difference in the assessment of environmental planning and design resources, designers, and investment. The operator's perspective tends to be the same, but the designer's perspective is different from that of the rural leadership manager.

(6) Design background. The assessment of environmental planning and design resources by the testee from the perspective of the designer, the perspective of the rural leadership manager, and the angle of the investment operator, whether it is their own point of view or different from those of other backgrounds. Designers should further understand the environmental planning design resource assessment perspectives of other job evaluators.

(7) Evaluation of six different types of clusters: "Comprehensive evaluators," "Leader's opinions follow evaluators," "Value-oriented consumer-oriented evaluators," "Target-oriented consumer-oriented evaluators," "Individualism The evaluator and the "innovative follow-up evaluator." Among them, designers, rural leadership managers, and investment operators have a comprehensive and integrated evaluation group.

The assessment of the environmental planning and design resources of different positions and different professional backgrounds has been confirmed by quite a few research institutes. Although the overall evaluation considerations are relevant, in the case of different professional backgrounds, the evaluation factors are prioritized. There are differences. Therefore, in order to improve the evaluation efficiency, it seems that the weighting method of professional people can be increased, and the function of professional skills in evaluation can be exerted.

REFERENCES

A. Topalin, 1980. *The Management of Design Projects*. Associated Business Press, London.

B. Turner, 1990, Design Evaluation, in: Oakley, M. (Ed.) *Design Management: A Handbook of Issues and Methods*. Basil Blackwell Ltd., Cambridge, 345–358.

D. Andrews, 1984. Principles of Project Evaluation, in: Langdon, R. and Gregory, S. (Ed.). *Design Policy: Evaluation*. The Design Council, London, 66–71.

D. Rivett, 1992. Project Management, in: Lydiate, L. (Ed.) *Professional Practicing Design Consultancy*. The Design Council, London, 125–137.

G. Hollins and B. Hollins, 1991. *Total Design: Managing the Design Process in the Service Sector*. Pitman Publishing, London.

J.C. Jones, 1963. A Method of Systematic Design, in Jones and Thornley (Ed.), *Conference on Design Method*.

M. He and Z. Zhang, 1999. *Research on the Basic Framework of Product Design Evaluation*. Cross-Century International Symposium, Ming Chuan University, Taipei.

M. Star, 1988. *Commercial design, creative culture*. Taipei.

P. Yang, 2014. *The evaluation index of entrepreneurial resources in the design industry*. Ph.D. thesis of the Doctoral Program of the School of Design, Yunlin University of Science and Technology.

W. Zhang, 1996. *Product design evaluation research*. Report of the results of the special research project of the National Science Committee of the Executive Yuan, plan No.: NSC-85-2213-E-011-011.

Z. Huang, 2004. *Establishment of the evaluation model for the development potential of leisure agriculture districts*. M.S. thesis of Tourism Management Institute of Nanhua University.

Smart Science, Design & Technology — Lam et al. (eds)
© 2020 Taylor & Francis Group, London, ISBN 978-0-367-17867-3

Analysis on the efficiency of natural rubber production in Hainan State farms

Sufang Zheng* & Chechang Chang
School of Internet Economics and Business, Fujian University of Technology, PR China

ABSTRACT: Since the reform and open-up policy was carried out, Hainan State Farm underwent three main operational modes of Family Long-term Contract, Shareholding Cooperative System and Workers Self-support Economy. The operation efficiency differed in different farms under the three operational modes. This study chose two output variables the total products and the products value of rubber and two input variables the rubber tapping area and the workers number. The technical efficiency of the three modes was calculated and decomposition analyzed. The results showed that, under the first two modes, the improvement of technical efficiency depended on scale efficiency, whereas, under the third mode, it depended on the improvement of pure technical efficiency.

1 INTRODUCTION

Hainan State Farms was founded in 1952 and had 49 farms (branch). After the construction and development for over 50 years, the rubber industry had been greatly developed and its economic power had been strengthened and made a significant contribution to the economic development in Hainan Province. Since 1952, the planting area of rubber increased 12 times from 20,500 hectare to 252,000 hectare. The total output of dry rubber increased 5400 times from 34 tons to 180,000 tons. The product value of rubber products increased 11 times from 0.4 billion RMB in 1978 to 4.6 billion RMB. The total number of workers of rubber industry was 187,900 and overall labor productivity was 45,044 RMB per person (Liu, 2011).

Since 1960s, the purpose of every reform on the operational mode was to make a great breakthrough to the operational management system and guarantee the income of worker families. From its foundation to the Hainan Rubber Group going public in 2010, Hainan State Farm has been at the forefront of scale development, conglomeration, demutualization and industrialization of modern agriculture (Kuang, 2003).

2 MODEL AND METHODS

2.1 Data sources

There were no input and output data of farms in Hainan Agricultural Reclamation Statistics, so in order to objectively and overall reflect the efficiency level of the samples, this study chose two output variables: the total output of dry rubber, its unit was ton and the value of output based on the current year's price and its unit was ten thousand RMB. Two input variables were chosen: the area of tapping rubber, its unit was acre and labor input, the total labor including

rubber planting, fertilization, irrigation and tapping, but it was difficult to collect these data, so the number of workers was chosen. Based on the nature and characteristics of natural rubber industry and combined the existing statistics data and information, the hypothesis conditions were as follows:

(1) Natural rubber industry was a land-constraint and labor intensive industry, so the hypothesis was the output was only related to the labor input and land areas.
(2) The influence of labor on output was the amounts of tapping workers, that is, the technology and labor were homogeneous and the influence of planting, fertilization and management was not considered.
(3) The output could be measured by the total product and the dry rubber of every farm was homogeneous.
(4) The rubber planting in different areas was influenced by the same natural condition, that is, the productivity of rubber production in different farms was not related to the fertility of soil and natural climate.

The choice of input and output met the requirement of DEA method. That is, it should reflect the competitive level of evaluation objects should have management controllability and consider the availability of variable data. Based on the experiences, the total number of input and output could not more than half of the amount of DMU. The less input index, the better and the more output index, the better (Goel, 2015). Based on the above hypothesis, there were no input and output data of farms in Hainan Agricultural Reclamation Statistics, so in order to objectively and overall reflect the efficiency level of the samples, this study chose two output variables: the total output of dry rubber, its unit was ton and the value of output

---

*Corresponding author: E-mail: 401390435@qq.com

based on the current year's prices and its unit was ten thousand RMB. Due to the unified price was the average market price in 2010, the comparability was improved. Two input variables were chosen: the area of tapping rubber, its unit was acre and labor input, the total labor including rubber planting, fertilization, irrigation and tapping, but it was difficult to collect these data, so the number of workers was chosen.

2.2 Model building and method selection

This study chose DEA method and its calculation results as main method.

Model 1 CRS model:

$$\text{Min } \theta_c \tag{1}$$

$$\text{s.t.} \begin{cases} \theta_c x_0 - \sum_{j=1}^{N} \lambda_j x_j \geq 0 \\ -y_0 + \sum_{j=1}^{N} \lambda_j y_j \geq 0 \\ \lambda_j \geq 0; \ j = 1, 2, \cdots, N; \ \theta_c \geq 0 \end{cases}$$

Respectively, x_0 and y_0 stands for the input and output vector of rubber planting. x_j and y_j stands for j^{th} input and output vector of rubber planting. λ_j stands for weight of every DMU, θ_c stands for the overall efficiency under the assumption of fixed returns of scale, the value is between 0 and 1, which reflects the extent of input-output efficiency of rubber planting. If $\theta_c = 1$, it shows that the input-output is completely effective. That is, both technical and scale efficiency is effective. The technical efficiency means that rubber farmers fully make use of resources to achieve maximum output and the best operational conditions. The scale efficiency means that rubber planting is on the stage of fixed returns of scale, i.e., the output expands or reduces as the same ratio with input. If $\theta_c < 1$, it means that the existing technical usage and the configuration of production factor is not on the best condition (Liu, 2006).

Model 2 VRS model:

$$\text{Min } \theta_v \tag{2}$$

$$\text{s.t.} \begin{cases} \theta_v x_0 - \sum_{j=1}^{N} \lambda_j x_j \geq 0 \\ -y_0 + \sum_{j=1}^{N} \lambda_j y_j \geq 0 \\ \lambda_j \geq 0; j = 1, 2, \cdots, N \end{cases}$$

θ_v stands for the pure technical efficiency. It can calculate what extent the pure technical efficiency caused the overall inefficiency under the guiding of input in rubber planting. The value of θ_v is from 0 to 1, the meaning of other variable is the same as CRS model (Wei, 1986).

The relation between the overall efficiency θ_c under the condition of fixed returns of scale, the

scale efficiency θ_y under the condition of variable returns of scale and scale efficiency θ_s is as follows:

$$\theta_c = \theta_v \times \theta_s \tag{3}$$

$$\theta_s = \theta_c / \theta_v \tag{4}$$

By analyzing the conversion relation between overall efficiency, pure technical efficiency and scale efficiency, the scale efficiency of every DMU can be further measured. It can measure whether the scale of rubber planting is optimal under the condition of fixed input (Charnes, 1978).

3 RESULTS AND DISCUSSION

The analysis on the research data showed that there were 44 farms with complete data in 2018, so this study chose them as examples. After Hainan Rubber Group went public in 2018, the related data was confidential documents and not available. Another reason was that many farms were emerged or regrouped in 2008-2018, so the data was in chaos. So the stable year of 2018 was chosen and all data of farms used the unified standard and statistical caliber.

Based on the input-oriented VRS model, the calculation result of production efficiency by using DEAP2.1 software was shown in Table 1.

Table 1. Technical efficiency, pure technical efficiency and scale efficiency of the state-owned rubber in 2010.

Farm	Overall Efficiency	Pure Technological Efficiency	Scale Efficiency	Returns to Scale
Hongming	0.425	0.461	0.921	Irs
Dongchang	0.506	0.562	0.900	Irs
Donglu	0.373	0.559	0.667	Irs
Nanyang	0.491	0.598	0.822	Irs
Dongtai	0.580	0.663	0.875	Drs
Donghong	0.554	0.577	0.960	Irs
Dongsheng	0.690	0.705	0.979	Irs
Dongxing	0.643	0.672	0.957	Drs
Donghe	0.616	0.746	0.826	Irs
Xinzhong	0.516	0.606	0.851	Drs
Zhongrui	0.553	0.574	0.964	Irs
Nanhai	0.433	0.459	0.943	Irs
Jinjiling	0.476	0.555	0.858	Irs
Zhongjian	0.614	0.617	0.995	Drs
Zhongkun	0.719	0.720	0.999	Irs
Hongguang	0.685	0.706	0.970	Irs
Xida	0.521	0.565	0.923	Drs
Ji'nan	0.303	1.000	0.303	Irs
Honghua	0.641	0.646	0.992	Drs
Xipei	0.946	1.000	0.946	Drs

(Continued)

Table 1. *(Continued)*

Farm	Overall Efficiency	Pure Techno-logical Efficiency	Scale Efficiency	Returns to Scale
Xilian	0.777	1.000	0.777	Drs
Lingmen	0.669	0.711	0.941	Irs
Nanping	0.655	0.660	0.992	Irs
Jinjiang	0.615	0.631	0.976	Drs
Nantian	0.643	0.654	0.984	Irs
Licai	0.639	0.662	0.966	Drs
Nanbing	0.595	0.597	0.996	Drs
Baoguo	0.773	0.775	0.998	Irs
Leguang	0.656	0.684	0.958	Drs
Shanrong	0.879	0.967	0.909	Drs
Guangba	0.772	0.808	0.956	Drs
Honglin	0.872	0.878	0.994	Irs
Baisha	0.870	0.876	0.993	Irs
Longjiang	0.909	1.000	0.909	Drs
Bangxi	0.936	0.971	0.963	Drs
Yangjiang	0.879	0.937	0.938	Drs
Wushi	0.686	0.721	0.952	Drs
Changzheng	0.730	0.836	0.873	Drs
Average	0.671	0.733	0.919	—

Note: data from model calculating results. Irs means increasing scale efficiency, Drs means decreasing scale efficiency.

The calculation results showed that the average overall production efficiency of the stated farms was 0.671 which was in the medium range. There were 6 farms whose overall production efficiency was completely effective and so was pure technical efficiency and scale efficiency. The average of scale efficiency was 0.919, but pure technical efficiency was only 0.711 and most farms achieved the economies of scale.

In terms of scale efficiency, it was at a high level, which was related to the nature of business. Respectively there were 19 farms with the increased and decreased returns of scale, which illustrated that the returns of scale of 43% farms needed to be improved. The output could be improved by expanding the production scale to improve production efficiency.

4 CONCLUSION

The results of the above analysis illustrated that returns to scale under family long-term contract made great contribution to the overall efficiency of rubber industry. The improvement of efficiency mainly depended on the expanding of scale and its effect was small, but pure technical efficiency has not fully played its role in overall efficiency improvement. So in order to improve the overall efficiency, the importance of technology should be fully paid attention to. Scientific and technological innovation should be promoted to achieve the application and transformation of scientific technological achievements.

REFERENCES

H.Q. Liu, 2011. *Chinese Agricultural Science Bulletin 11*(30) 111–115.

C.H. Kuang, 2003. *China State Farms Economy 03*(11) 25–26.

U. Goel, 2015. *Journal of Information and Optimization Sciences 36*(6) 595–616.

F.F. Liu, 2006. *Journal of Statistics and Management Systems 9*(2) 301–317.

Q. Wei, 1986. *Annual Meeting of Systems Engineering Society* 422–429.

A. Charnes, W.W. Cooper and E. Rhodes, 1978. *Eur J Oper Res 2*(6) 429–444.

Service and experience design

Smart Science, Design & Technology — Lam et al. (eds)
© 2020 Taylor & Francis Group, London, ISBN 978-0-367-17867-3

Research on the relationship between smart service design and consumer demand for qualitytaking a tourism factory as an example

Tsenyao Chang*
Department of Creative Design, National Yunlin University of Science and Technology, Yunlin, Taiwan

Chienwen Kao
Intelligent & Local Design Service Center, National Yunlin University of Science and Technology, Yunlin, Taiwan

ABSTRACT: In response to the advent of the experience economy and the intelligence era, smart services have become the main force to drive upgrade and transformation of enterprises. This study is a case study on the Tourism Factory and tries to explore the consumer experience benefits brought forward by smart services in the Tourism Factory from the perspective of Service Design. The factors for evaluation of relevant benefits can be obtained from analysis of Consumer Demand for Quality and Customer Satisfaction. Here, this study mainly investigates the consumer experience model shaped by the innovative.

1 INTRODUCTION

1.1 Research background

According to the statistics of the Ministry of Economic Affairs in 2018, the number of newly established tourism factories has increased in recent years. Among the 135 tourism factories, there are 58 food tourism factories, accounting for more than 40% of the total. There has been a significant upward trend in the establishment of tourism factories (Ministry of Economic Affairs, 2018). Tourism factories can enhance the delivery of brand image by introducing the concept of Service Design, as Service Design is also an effective way to achieve brand communication (Thamtheerasathian et al., 2018). In a nutshell, Service Design is an innovative integration model for companies to provide services for customers and enhance the corporate brand image (Li et al., 2008).

In addition to improving the service quality, the introduction of Smart Service Design into the enterprise can also provide services for consumers to experience and enjoy (Allmendinger and Lombreglia, 2005). Functions through smart networks and smart technology devices can enhance the customer's pleasure experience, so they can be used as substantial transformation strategies by tourism factories; for example, improving the effectiveness of corporate network services and providing service integration of smart services and service design in the Tourism Factory. On the one hand, Smart Service Design involves service design and evaluation of Consumer Demand for Quality and Customer Satisfaction in the overall service process; on the other hand, Smart Service Design also involves satisfying Consumer Demand for Quality and improving Customer Satisfaction with the smart service project. In other words, intelligence, services, and the relevant integrated designs based on their combinations are the key to Smart Service Design.

As indicated by the study of Niu (2018), Consumer Qualia Demand can stimulate enterprises to launch quality services and products for increasing revenue, and it also encourages enterprises to achieve organizational innovation benefits in service functions. Since Consumer Qualia Demand can also bring happiness and satisfaction to customers (Liu, 2003), meeting Consumer Demand for Quality can increase satisfaction and the added value of the service process, as well as consumers' evaluation and perception of corporate services and eventually Consumer Behavior (Yen et al., 2015). Therefore, this study focuses on "Smart Service Design, Consumer Demand for Quality and Customer Satisfaction" in tourism factories.

1.2 Research purpose

For the Tourism Factory, this study mainly explores actual application models in the industry through literature review and combines them with the relevant theories proposed by scholars to derive innovative services developed in line with specific business needs. The research topics and research structure related to the research purpose are set out as follows:

---

*Corresponding author: E-mail: changty8908@gmail.com

(1) Explore the application models of Smart Service Design and their main functions.
(2) The relevance of Consumer Demand for Quality to Smart Service Design.
(3) The relevance of Customer Satisfaction to Smart Service Design.
(4) Smart Service Design, Consumer Demand for Quality, and Customer Satisfaction are introduced into the design integration model of tourism factories.

2 LITERATURE REVIEW

2.1 *Tourism factory*

Tourism factories can bring the substantial effects of service and experience to customers through product manufacturing processes, links between products and local resources, and the brand story of Tourism Factory (Lin, 2018). The speciality of the tourism factories is that they value the interaction between the company and its customers, and they are the domain that integrates the business service design concept and provides the consumer experience model (Hung et al., 2017). In Taiwan, tourism factories are flourishing robustly. In order to advertise the food production process, these tourism factories allow customers to experience and taste their self-made foods, so that consumers are attracted to experience the overall service process of the tourism factory.

2.2 *Definition of smart service design*

Smart services refer to the use of smart technologies to help companies provide optimized service interfaces. Usually, smart service projects are designed to serve user functions, such as smart technology platforms, electronic wireless networks, mobile smart map apps, etc. Smart devices, wireless networks, smart business apps, and many more functions are also included (Liudmila, 2015). Through Smart Service Design, we can help develop a service model to help companies improve the quality of the service process and innovate service models. Smart Service Design focuses on customer experience and pursues the enhancement of Customer Satisfaction with smart services. This means that for companies, Smart Service Design can create a new business value model (Marquardt, 2017). That is to say, in the new business value model, Smart Service Design can provide customers with convenience of service and instantly serve customer needs.

2.3 *Definition of consumer demand for quality*

Consumer Demand for Quality refers to consumer perception of the nature of a product or service (Yen et al., 2015). For consumers, Qualia Demand consists of three major aspects. The first is aesthetics, which means that products and services can attract customers and give customers a good visual experience.

The second is creativity, which means that fresh and amazing feelings about products and services can improve customers' interest in relevant products and services. The third is delicateness, which allows customers to enjoy the fine sense of products and services. In addition, products and services can bring about a sense of comfort through ergonomics and human factors design and pursue customer satisfaction based on this (Small and Medium Enterprise Administration, Ministry of Economic Affairs, 2009).

2.4 *Relationship between smart service design and consumer demand for quality*

In Service Design, Consumer Demand for Quality can be displayed in both models of aesthetics and life, and the integration of these two models can help companies develop business innovation models (Yen & Lin, 2012).

In the intelligence era, Smart Service Design can even improve the service experience through the smart service project for enterprises, thereby enhancing market competitiveness and reducing risk to achieve intelligent service function benefits (Maurer & Schumacher, 2018).

In summary, by satisfying Consumer Demand for Quality, Smart Service Design can produce good service benefits, which mainly involves three factors: aesthetics, creativity, and delicateness.

2.5 *Definition of customer satisfaction*

Customer Satisfaction refers to the practicality of and satisfaction with the products and services perceived by the consumer. Influencing factors may include cost concession, product perceptions, service quality, safety of use, and technological techniques (Mohamed et al., 2018). In other words, the customer's perception of intimacy and service capabilities of an enterprise will increase consumer loyalty (Jain & Aggarwal, 2017). In addition, Customer Satisfaction or the level of acceptance of the consumer can also be assessed by measuring consumer preferences for services and products, including the customer's response to spatial planning and service quality (Li & Chen, 2009).

Even most customers will use the value-added benefits of consuming the products and services as a measure of satisfaction (Ismail and Yunan, 2016). From the perspective of customer relationship management, Customer Satisfaction can be the main consideration in evaluating the practicality and efficiency of products or services used or experienced by consumers (Wei et al., 2018).

3 RELATIONSHIP BETWEEN CUSTOMER SATISFACTION AND SMART SERVICE DESIGN

In terms of service functions, satisfying customer needs and improving customer experiences is an important key to earn revisits by customers. This way, companies can continuously update service technologies and enhance service functions. If the service functions can be redesigned based on the metrics measuring Customer Satisfaction, the consumer will focus on the service rate and service items in the service environment while evaluating the service capabilities of the enterprise, which may further affect the customer's satisfaction with services of the enterprise (Babin et al., 2005).

In Smart Service Design, planning a marketing strategy for the market as a whole can also be achieved through the actual business model of the enterprise, where the service process benefits are assessed from analysis of the smart phone function device interface and the big data analysis. Through the consumer satisfaction with use of the enterprise intelligence function interface and the service process, relevant results are obtained, and relevant analysis and decision-making are conducted based on such data. Then, service functions are updated and improved based on such decisions, thereby enhancing the company's service capabilities in the market. Smart services are updated based on the latest decisions, which will not only increase the number of customers, but also increase the competitive advantage of the enterprise in the market (Shin, 2015).

3.1 Integrated architecture of smart service design

Finally, this study uses the Tourism Factory as an example to propose the Smart Service Design and Integration Framework with five constructs. From this framework, the conceptual architecture/model of Smart Service Design is developed based on factors such as aesthetics, creativity, and delicateness in Consumer Demand for Quality and the service quality and technological techniques in Customer Satisfaction, which are further explained as follows:

Construct 1: For Aesthetic Design, use Smart Service Design to create a visual aesthetic experience.

Construct 2: Integrate the sense of creativity into Smart Service Design to provide a fresh and interesting service process.

Construct 3: In the smart service project, import the delicateness model to provide comfort.

Construct 4: Evaluate customer satisfaction with the service quality as a basis for the improvement of Smart Service Design in the Tourism Factory.

Construct 5: The technological techniques of the Tourism Factory will influence the customer's experience/satisfaction, so the technological techniques can be used to enhance the service process through Smart Service Design.

As shown in the above five constructs, Smart Service Design can satisfy the Consumer Demand for Quality in the Tourism Factory. This mainly involves three aspects: the visual aesthetic experience provided by the visual design of products and services,

Figure 1. Architecture of smart service design.

the creative experience shaped by relevant products and services, and the refined enjoyment of products and services.

The service quality and scientific technologies can satisfy Consumer Demand for Quality in terms of the three aspects, as the service quality and scientific technologies are the main items covered by Smart Service Design. In other words, for the Tourism Factory, Smart Service Design, Consumer Demand for Quality, and Customer Satisfaction are elements subject to mutual influence, as shown in Figure 1.

REFERENCES

Statistics Report for December of 2018, Department of Statistics, Ministry of Economic Affairs. Idei Nobuyuki, Quantum Leaps - Finding the Edge in A Changing World (Translator: Chin-Hsiu Liu). Taipei: Business Weekly Publications, Inc. (Original Edition, 2002).

T. Li, H. Ming, F. Kung, H. Wang and P. Wang, 2008. *A Preliminary Study on Service Design 24*(6), 6–10.

H. Yen, P. Lin and R. Lin, 2015a, *Journal of Design 20*(2), 1–24.

J. Lin, 2011. *Journal of Design 16*, 3–P5.

Y. Lin, 2018. *Journal of General Education 18* (7), 217–234.

C. Hung, C. Chen and J. Lin, 2017. *Journal of Design 22* (3), 69–92.

The Project of Qualia Advancing for SMEs in 2009. Small and Medium Enterprise Administration, Ministry of Economic Affairs.

H. Yen and R. Lin, 2012. *Journal of National Taiwan College of Arts 91*, 127–152.

C. Li and Y. Chen, 2009. *Journal of Customer Satisfaction 5*(1), 93–119.

L. Thamtheerasathian, S. Choemprayong, P. Teerathammongkol and S. Srisatriyanon, 2018. *Get Access 43*(4), 175–184.

A.R. Lombreglia, 2005. *Harv Bus. Rev. 83*(10), 131–4, 136.

F. Ye, Y. Qian and R.Q. Hu, 2019. *Journals & Magazines, IEEE Network 33*(1), 84–91.

Niu Han-Jen, 2018. Qualia: Touching the inner needs of consumers' hearts. *Australasian Marketing Journal.* P1–13.

Tatiana Gavrilova Liudmila, 2015. Smart Services Classification Framework. Position Papers of the Federated Conference on Computer Science and Information Systems, pp. 203–207.

K. Marquardt, 2017. Smart services – characteristics, challenges, opportunities and business models. DOI: 10.1515/picbe-2017-0084, pp. 789–801.

H.Y. Yen, P.H. Lin and R. Lin, 2015b. The Effect of Product Qualia Factors on Brand Image-Using Brand Love as the Mediator. 2015 62 Volume 3, p. 3_67–3_76.

F. Maurer and J. Schumacher, 2018. Organizational Robustness and Resilience as Catalyst to Boost Innovation in Smart Service Factories of the Future. *2018 IEEE International Conference on Engineering, Technology and Innovation (ICE/ITMC)*, Date of Conference: 17–20.

Samira Mohamed Abd Elmeguid Mohamed A. Ragheb Passent I. Tantawi Ahmed Moussa Elsamadicy, 2018. Customer satisfaction in sharing economy the case of ridesharing service in Alexandria, Egypt. International Conference on Restructuring of the Global Economy, P373–3381.

P. Jain, V.S. Aggarwal, 2017. *The Journal of Management Awareness 20*(2), 29–42.

A. Ismail, Y. Sufardi and M. Yunan, 2016. *LogForum 12*(4), 269–283.

K. Wei, F. Tseng and C. Hsieh, 2018. *Customer Relationship Management to Import Cloud Services*. Northeast Decision Sciences Institute 2018 Annual Conference, Providence, Rhode Island, USA.

H.Š. Erjavec, T. Dmitrović and P.P. Bržan, 2016. *Journal of Business Economics and Management 17*(5), 810–823.

B.J. Babin. M. Griffin and Y. Lee, 2005. *Journal of Services Marketing 19*(3), 133–139.

D. Shin, 2015. *Total Quality Management 26*(8), 91–93.

Smart Science, Design & Technology — Lam et al. (eds)
© 2020 Taylor & Francis Group, London, ISBN 978-0-367-17867-3

Brand creation and business model with service design implemented into micro agriculture: A case study of LuGuang Rice

Tsenyao Chang*
Department of Creative Design, National Yunlin University of Science and Technology, Yunlin, Taiwan

Jizhen Wu
Intelligent & Local Design Service Center, National Yunlin University of Science and Technology, Yunlin, Taiwan

ABSTRACT: Regarding managing and marketing of agricultural product, brand design and service design can create market differentiation and uniqueness and are the key to create product features and market position. Brand loyalty of customers is increased, and marketing performance is enhanced through the creation of brand and services. An industry–academia collaboration, LuGuang Rice, is used as the research case in this study. The collaboration relationship among the three parties, production end, design and service end, and agricultural product counselling end on agriculture brands, is discussed. Brand design, service experience, and information design are implemented into agriculture brands. The stages and strategies of developing a primary industry into tertiary sectors are analyzed. The opportunity points for strengthening agricultural production and marketing are discovered. Thus, the service resources of local industries can be connected and integrated, the possibilities of approaching consumers market can be expanded, and the strategies of agricultural products production and brand marketing can be readjusted. Focus group method and case study analysis are performed on our research case, LuGuang Rice. Production and marketing mechanism of micro agriculture is being counselled through the perspective of brand and service design to enhance the revenue of farmers, upgrade industries, and make micro agriculture become resource sharing and enhance co-prosperity of brand agriculture.

1 MARKET BOTTLENECK IN THE WEDDING INDUSTRY

Agriculture in Taiwan has faced the following problems: insufficient labor force, low willingness of becoming a farmer, ageing practitioners, and emigration of young adults in rural areas. Council of Agriculture Executive Yuan has promoted multiple projects and plans for young adults to become farmers and stay farmers regarding these issues continuously to encourage young adults to go back home and adjust the industrial structure. Agricultural producers focus more on the direction of interaction and communication with the customers in their business operations and simplify the cycles of sales process nowadays to reduce the commission during sales process. Not only can the loyalty of customers be increased, but also the profits of the products can be enhanced. However, on the other hand, operators have to devote more time and effort to manage customer experience and sales, create their own brand, and open up product identification and channels. Brands can make the consumers realize the differences of company products or services among many competitors and be willing to pay a higher price

based on its value (de Chernatony, 1993). Agricultural economics can be transformed into cultural economics through the construction of cultures and regions and then its economic benefits can be enhanced (Hsiao, 2006). Branding creates an add-on effect to agricultural products. Hence, agricultural products can be transformed and upgraded by brand designing, including the operation of all elements from value positioning, vision plans, design and execution, to service design, to open up consumer groups and sales channels and create more marketing and promotion highlights and product profit ranges.

Government policies focus more on the improvement of counselling techniques and the enhancement of operation and management capabilities, etc., but they lack the counselling part on the product marketing end after harvest (Tsai & Lai, 2012). One of the five most wanted counseling measures to be added or strengthened in the future for young farmers from the government is "Assist in the creation of brands or provide guidance in product image design," which is indicated in "Choice for Retention on Farming: An Empirical Study of Young Farmers in Taiwan" (Huang, 2017). This shows that the desire of young farmers in the two big issues, the service

*Corresponding author. E-mail: changty8908@gmail.com

gap in brand creation and the transformation and upgrading of agriculture, is to seek for innovative agricultural economics and opportunity points.

The study investigates how to collaborate with different units and integrate local resources and the industry–academia collaboration mechanism to create resource sharing and co-prosperity, build the image of an agricultural brand, and develop market position and brand marketing during the process of micro agriculture from production to brand creation using the example of LuGuang Rice. By analyzing a rice case, the study organizes the brand creation model for industry–government–academia collaboration of agriculture brands, offers a reference mechanism of domestic micro agriculture from production to branding, and provides actual cases and execution process referenced by domestic counselling centers to help the farmers to create brands and increase revenue.

2 LITERATURE REVIEW

2.1 Brand agriculture

The meaning of brand agriculture refers to add-on values created on different sides of agriculture (including primary, secondary and tertiary agricultural products), such as quality, safety, health, and ecological environment, by modern brand concepts and shaping methods (Huang, 2017). Lin (2013) pointed out three major points of brand agriculture: developing knowledge and skills based on agricultural sciences; capable of clearly recognized brands with local features cultivated by agricultural resources, which connect the production chain and service chain; and satisfying the needs of the market by excellent agricultural products and services. The three layers of agriculture, production and management, and marketing are innovated sequentially with its branding. Symbol of quality is promoted, and the industry integrated with marketing is counselled in agricultural products, driving all agriculture organization to create brands. In the marketing layer, networking, experiencing, and reputation are applied to respond to the consumption habits of the experience era nowadays; the internet is used to develop opportunity points for the products; and the management of brand image and relationship with customers are valued (Huang, 2017).

2.2 Service design

Service design focuses on the interaction between service and people. Not only can the existing services be remodeled through design, but also new service models can be added to create new service value (Parker & Heapy, 2006). Service design is a multidisciplinary field, which integrates different methods and tools in all kinds of professional fields to create a complete service experience through tangible or intangible media (Stickdorn & Schneider, 2013). Taiwan Design Center pointed out that

service design can offer six assistances to small and medium enterprises in their research report (2011) as follows: developing a service vision and strategy, focusing on customers, designing the new service, developing internal processes, creating better experiences, and creating and maintaining a brand. Design Council (2004) introduced the double diamond design process to assist the development of innovative services. Four major stages, discover, define, develop, and deliver, constitute the service design process to solve the problems faced by the targets systematically. Stickdorn and Schneider (2013) inducted five major principles of service design thinking for design service process: users as the center, execute in order, create jointly, reification of objects, and symbols and integrity. Brands can rethink their interaction with stakeholders and take on the customer-oriented collaboration form through multiple the diverse tools of service design, such as Persona, service blueprint and customer journey map, etc., to create values for each other. Service design integrates tangible and intangible media to create a smooth and satisfying service process and provide cheerful service experience to keep customers and create brand loyalty with the objective of providing users a complete service. Thus, a self-checking service process can create opportunities and create brand advantages.

2.3 Brand marketing

Brand marketing is the action of conveying business concepts and brand name to create brand value and maintain brand image for the brands to be recognized by consumers in the society (Wood, 2000). The promotion of agricultural products depends on products with quality and differentiation. Chan et al. (2011) think that the creation of local brands of agricultural products includes three aspects: origin condition, brand shaping, and marketing promotion. Market competitiveness of the quality of the products is made by origin conditions; customers understand local brands through brand shaping; and origin channels are integrated by marketing promotion. Successful brand shaping depends on the marketing strategy adopted. Tsai and Lai (2012) pointed out two key brand marketing strategies: creating brand position and value and implementing brand marketing activities. The creation of brand position and value relies on product quality as its core value, brand positioning with local features, the creation of special brand identification and participation in competition and government assessment, and selection to ensure the brand has an association advantage on the market. Regarding the implementation of brand marketing activities, six operational methods can be applied to enhance brand reputation: getting hold of all kinds of sales exhibition opportunities, experience marketing with tourism factory, developing brick-and-mortar retail channels, managing social media on the internet, having delicate advertising strategies, and collaborating with different industries.

2.4 *Experience design*

A series of special experiences are conveyed to customers through service design while experience marketing is known as marketing products by providing customers with complete services and experiences (Schmitt, 1999). The design of different service contact points can attract the sight and touch the inner feelings of customers and become one of the elements for customers to consider the brand as a primary selection among congenial brands. Customers can understand more of the brand image through experiences with the medium of product presentation, space and environment, and electronic media. Consumption is promoted by satisfying both sense and sensitivity needs of the customers. Tuan (2002) applied five perspectives on planning experiences mentioned in the book *The Experience Economy* to explain the procedure of applying experience design in leisure agricultural experiences, including establishing themes, shaping images, removing negative clues, adding souvenirs, and activating the stimulation of five senses, to provide customers special experiences. Chi (2018) pointed out that designers are required to pay attention to three major items during experience design of agricultural products: respecting the motivation of the farmers, considering the agricultural experiences as a part of the business strategies for the farmers, and developing the experience design based on the current end consumer groups of the farmers. The three major items are closely related to the element of "people." Considering the users as the center in the design is required. By considering the preferences and characteristics of different groups from the farmers to the consumers, the most suitable service design and experience can be provided.

3 METHODS

The study uses the agriculture brand LuGuang Rice as an example to investigate the key roles of the three parties, "farmers," "counselling centers," and "design team," and their relationships by the focus group method. The main theme of the interview process is service process. The stand and collaboration mode of each parties in the path of brand creation are investigated separately. The focus group method is that a trained moderator will interview a group of people with similar background, which is usually composed of 5–10 people and can be adjusted flexibly under the criteria of at least 4 and at most 12 people (Krueger and Casey, 2000), and by natural method to hold a 90-minute to 2-hour meeting (Cooper and Schindler, 2009). The main purpose of the method is to investigate

possible causes by limited samples; however, the main factors will still require quantitative research in order to make decisions eventually (Chiu, 2007). The method used can obtain extensive and detailed information in a short amount of time (Wang, 2002).

The focus group forum held for the study had invited four key roles during the brand operating process of the case for discussion, including the responsible person of the LuGuang Rice Farm, which is the producer and operator of the agricultural products in the primary industries, a brand designer, which is also in the academic research unit in the tertiary sectors, and the representatives of agricultural counselling centers and agriculture promotion consultants, who connect the gaps between the two ends. The four people discussed their brand creation process comprehensively and made suggestions and investigated the service process and needs of brand agriculture from an objective perspective.

4 ANALYSIS & CONCLUSION

The study analyzed the core values of operating a agricultural brand and the roles and experiences of "farmer," "counseling centers," and "design team" and proposed the resources sharing mode for micro agriculture branding.

4.1 *Cooperation between brands*

The research case was proposed by the counselling center with social practical objectives and accumulated technical skills to look for local farmers who are open-minded and have the potential to brand creation through sales exhibition and other relevant channels to activate counselling and collaborative service mechanism; provide agricultural techniques, such as guidance on cultivation techniques, management of pesticides and the promotion of friendly agricultural methods, etc., to the farmers; and enhance the quality and production volume of the agricultural products. After the production quality had been enhanced, the counselling center approached the design team to integrate brand value claims, create brand image design to add values to the agricultural products, integrate brand marketing, and expand the possibilities of consumption market through networking, experiencing, and reputation.

4.2 *Factor analysis of farmer for micro agriculture*

The crop of the farmer in the case is rice, which belongs to the micro business scale. The farmer in the case believed that the creation of brand agriculture has three major elements: the motivation of brand creation, the enhancement in production techniques, and brand marketing. The main motivation in the initial stages of brand creation is to enhance the add-on values of the agricultural products, create separation from the products from food wholesalers in the market, and serve the consumers directly for devoting into own brand. For the element of enhancing production techniques, a dominant variety is grown, cultivation techniques are enhanced to strengthen the advantage of product quality, and derived product series of the brand are developed from the services and suggestions provided by the counselling centers. Brand marketing is performed currently through referral, holding experience activities and operating social networking sites, getting hold of all kinds of exhibition opportunities to strengthen the promotion and collaborate with the counselling centers to expose the products through public activities and press conferences, and have co-prosperity with the linked collaboration units.

4.3 Factor analysis of farmer for micro agriculture

The core of the counselling center of the case is Satoyama Initiative, which is devoted to creating a joint lifestyle of forest community and agriculture ecosystem and taking on the objective of achieving social practices. The center utilizes its own existing resources to help developing features of local brands and performs resource integration to provide the counselling services. The main counselling items of the center are discovering and developing brand features, matching experts to enhance cultivation techniques, assisting product examination and certification, implementing professional teams to create the brand and brand marketing, develop brand-derived products, etc. The center builds the development base and strengthens the brand in all aspects of the brand agriculture based on scheduled milestones. The counselling center is also the core of the three parties in the case, which links experts of different fields and serves as the communication bridge between the three parties.

4.4 Factor analysis of design team

After the participation of the design teams through the match by the counselling center, they have implemented elements such as brands, services, experiences and designs, etc. to give energies to the brand design of micro agriculture. The designers serve multiple roles and tasks during the brand operating process of the case. First, they need to know the local culture of the brand, collect information, and interview in the field to understand the deep value of the agriculture brand. Later on, they will use this information to develop and design elements and highlights of the brand. They will need more interaction with the farmers to understand the actual production process of the agricultural products and then organize local features and

brand value claims by collaborating with other teams. Other than the visual design of the brand, the insights and analysis capabilities from the designers are also required to transform the observed features of the brand during the process, propose designs to convey brand value claims, and enhance the knowledge of consumers regarding brand image. The design teams bring the producers and consumers closer by information transparency to promote brand marketing. One-step brand marketing of transparent production and marketing of agricultural products is built by 360-degree panorama video filming in the field, establishing traceability, managing a cloud platform for agriculture, and launching products in catering association with traceability. The design team believe that the key factor that affects its success in the brand marketing collaboration is "the attitude of the farmers." The trust of the famers in the design teams affects the level of information transparency. Self-exposed, open techniques and long-term interviews have resulted in pressure on the farmers for brand marketing, which caused tense relationships between the two parties. In addition, there will be positive effects for brand marketing if the farmers can expose information on social media platforms actively and provide after-sales services to consumers.

4.5 Factor analysis of design team

The study investigated the collaboration mechanism of multiple parties on the micro agriculture branding business by going deep into a case and found the following: the counselling team was served as the core and matched experts in agricultural techniques to enhance product quality and create brand value for micro agriculture step-by-step and accumulate sufficient base and energy for the brand with schedule milestones; the matched design teams integrated resources to perform product branding based on its brand value and promote experience marketing and brand marketing by information optimization to create brand image. It has been pointed out that the attitude of farmers toward brand creation affects the feasibility of brand creation during the brand creation process. Hence, the attitude of the farmers towards brands should be taken into consideration when formulating relevant plans, and the counselling centers can be served as the connection bridge to maintain the joint collaboration relationship among the three parties. The professionalism of each unit and its own brand image affect one another and even overlap with one another in the collaboration mechanism. In our case, our school served as the counselling unit. Its professional image can integrate with the brand image of the agricultural products to create brand sharing and co-prosperity to accelerate micro agriculture branding.

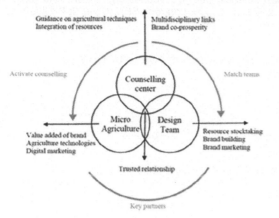

Guidance on agricultural techniques
Integration of resources

Multidisciplinary links
Brand co-prosperity

Activate counselling

Match teams

Counselling
center

Micro
Agriculture

Design
Team

Value added of brand
Agriculture technologies
Digital marketing

Resource stocktaking
Brand building
Brand marketing

Trusted relationship

Key partners

Structure chart of the key elements

Figure 1. Structure chart of the key elements.

REFERENCES

Wang, M.-L. 2002. Focus Group: Theory and Application. *Bulletin of Library and Information Science*, 40, 29–46.

Lin, J.-C. 2013. Comprehensive Research of Domestic and Foreign Brand Agriculture. *Acta Agriculture Jiangxi*, 25(7), 143–146.

Chiu, C.-S. 2007. Marketing Research: Practices and Theoretical Applications. Best-Wise Co., Ltd.

Chi, S.-T. 2018. A Study of the Experience Design of Agricultural Value. Master Thesis.

Tuan, C.-L. 2002. Experience Economy and Educational Farm. *Agricultural Extension Article Collects*, 60, 209–223.

Huang, P.-W. 2017a. New Era of Taiwan Brand Agriculture Created by the Revolution of Supply Side. *Agricultural Policy Review*, 2:2(2017), 33–42.

Huang, H.-M. 2017b. Analysis on Top Hundred Young Farmers' Satisfaction toward Current Government Measures. Master Thesis.

Yang, C.-F. & Huang, T.-C. 2011. Service Design Tools and Methods. Taipei City: Taiwan Design Center.

Tsai, M.-T. & Lai, C.-K. 2012. Research on Brand Marketing of Agricultural Products of Farmers' Association. *Agricultural Extension Article Collects*, 57, 59–79.

Chan, I-H., Tsai, P.-K., Chiang, H.-K. 2011. Research on the Process of Local Brand of Agricultural Products Implemented by Farmers' Organization. *Journal of the Agricultural Association of Taiwan*, 12(2), 126–150.

Hsiao, K.-S. 2006. Research on the Extension Strategy of Agricultural Industry and Culture. *Review of Agricultural Extension Science*, 21, 1–3.

Cooper, D.R. & Schindler, P.S. 2009. *Business Research Methods* (10th ed.). UK: McGraw-Hill.

de Chernatony, L. 1993. Categorizing Brands: Evolutionary Processes Underpinned by Two Key Dimensions. *Journal of Marketing Management*, 9(2), 173–188.

Krueger, R.A. & Casey, M.A. 2002. *Focus groups: A practical guide for applied research* (3rd ed). CA: Sage.

Parker, S., & Heapy, J. 2006. The journey to the interface: How public service design can connect users to reform. London: Demos.

Schmitt, B.H. 1999. *Experiential Marketing* (Translated by Wang, Y.-Y. & Liang, H.-Y.). Taiwan: Flag Technology Co., Ltd.

Stickdorn, M., & Schneider, J. 2013. This is service design thinking: Basics, tools, cases: Bis.

Wood, L. 2000. Brands and Brand Equity: Definition and Management. *Management Decision*, 38(9), 662–669.

Smart Science, Design & Technology — Lam et al. (eds)
© 2020 Taylor & Francis Group, London, ISBN 978-0-367-17867-3

Live happily ever after—exploring innovative service construction for wedding ceremony planning

Tsenyao Chang*
Department of Creative Design, National Yunlin University of Science and Technology, Yunlin, Taiwan

Shihchieh Chang
Graduate School of Design, National Yunlin University of Science and Technology, Yunlin, Taiwan

ABSTRACT: Taiwan's wedding industry is facing the bottleneck of market shrinkage. Scientifically reviewing service provision and searching for creative extraction and design for the Taiwanese wedding industry is a real need. In this study, 25 wedding consultants discussed service design, aiming at the pain points of ritual cumbersome, unclear, time-consuming, laborious, and interactive, and proposed a merged marriage ceremony called "Once a Lifetime," simplified customary materials, and dating events that will increase interaction. After two years and six months of experiment, the results show that 84% of newcomers and 53% of parents are satisfied. Another questionnaire conducted with 45 wedding consultants averaging 6–10 years experience in hotel, banquet hall, and restaurant wedding planning and practice, from the north, central, and south regions of Taiwan, showed that 81.5% consider the simplified wedding ceremony to be acceptable, 70.4% feel it is feasible, and the actual practice rate was 71.1%. As a result the booking rate has increased by 62.2%, and 92.9% of the people can feel a reduction in tediousness, which indicates that the introduction of service design can help design innovation.

1 MARKET BOTTLENECK IN THE WEDDING INDUSTRY

Taiwan's wedding industry is facing a bottleneck in the market's shrinking pattern. The current industry's solution trend is to expand and build new venues or propose new ideas through traditional brainstorming meetings (e.g., wedding experience, theme weddings, wedding fairs, wedding fairs) to attract more customers, hoping to increase their operating profit margin (OPM) and at the same time to reduce their research and development costs. However, the so-called new ideas tend to be mostly hype, and once the accessible market share is taken, the bottleneck situation would then resurface. Therefore, a scientific review of service provision and the continuing search for creative extraction and design remains relatively important for Taiwan's wedding industry.

This study takes service design as the academic foundation, and the wedding ceremony as the carrier, to extract the pain points through the service contact point, and to study the way of transforming pain points into sweet spots. The aim is to help the wedding industry generate innovative thinking by means of a scientific and logical discussion in order to improve the current situation.

2 SERVICE TOUCH POINT IN THE WEDDING SERVICE PROCESS

Tim Brown, who proposed design thinking, thinks that design is a comprehensive and complete way of thinking whose power will end old ideas and provide a path to innovation, and that the need to grapple with service can also transform the world in the future (Wu, 2010). Moreover, the cross-disciplinary discussion and extraction of service design is in line with the value of design thinking, and the feasibility of specific practices is gradually increasing. Therefore, this study was conducted in the wedding industry to explore the feasibility of the introduction of design thinking.

2.1 *The interpretation of the principles of service design*

Marc Stickdorn and Jakob Schneider propose five principles for service design thinking: (1) User-Centred, (2) Co-Creative, (3) Sequencing, (4) Materialized Items & Evidencing, and (5) Holistic (Chi, 2013). There may be a feasible chance of introduction of service design for wedding services providers, but although there is occasional discussion in the wedding industry, there have been no significant results

---

*Corresponding author: E-mail: changty8908@gmail.com

with any substantive deepening. The service environment in different fields will have different ways of operation. Birgit Mager likewise proposed eight principles of service design: (1) The value of service must be strategic thinking, (2) The value transfer of shareholders to stakeholders, (3) Listening to consumers, understanding the system, and not solely relying on survey data, (4) Co-creation and production, (5) Attention from multi-oriented experienced professionals, (6) Creating service organizations and roles, (7) Service staff has the initiative, and (8) Embrace creativity and continuous innovation and improvement of services (Xu, 2012). Both studies show their central value focus on "human thinking." In other words, when the wedding service can think in terms of contemporary human nature, service design has a real chance for introduction.

2.2 The touch point of the wedding ceremony service

The wedding industry has a wide range of fields. There are preparation, implementation, and results stages, that is, a three-stage journey of service design: before, during, and after the service. This study is based on ritual services as the core of experiment and discussion and tries to analyze whether the various service-oriented services of Taiwan's wedding industry have a reference basis. As far as the ceremony is concerned, the service touch point occurs at the time of the wedding ceremony, which is the most common pain point. On the single point of contact, the current engagement processes are defined in the following order: starting, welcoming, introducing the relatives and friends of both parties, offering engagement gift, servicing tea, pressing tea, wearing a ring, meeting the ceremony, returning the bride's dowry, and the wedding ceremony. The processes of the wedding are as follows: setting out, firing fireworks, welcoming the bridegroom, approving, appreciating parents, boarding the car, paying the car, worshiping the ancestors, serving tea, and entering the room. Both of these sets of 10 processes symbolize 10-10 perfection, and within these 20 processes, each of the small touch points can be subdivided; but this study uses only large touch points as a basis for discussion.

Because the wedding ceremony service symbolizes "the perfection of the perfection," as the saying goes, it is the most painful point in the practice according to the field researchers. Therefore, this study uses wedding ceremony services that are introduced with service design based on "human thinking" and "empathy" as the basis for the service thinking of the overall wedding industry. Can it be practiced? Subsequent actual case verification will be carried out.

3 RESEARCH METHODS AND IMPLEMENTATION PROCESSES

This study attempts to extract the value of the newly married couples and key participants in the wedding preparation, and then integrates the wedding plan, asks the wedding consultant to use the same method of value extraction and construct the module. It then actually applies the innovative process planning service design to the wedding case, and finally proceeds to conduct effective surveys to verify that innovative thinking through service design can be truly practiced.

3.1 Wedding ceremony - the new couple's value extraction

The 10 groups will be married to the key participants in the wedding and W01-W10 (W stands for Wedding), and a total of 10 wedding ceremony preparation meetings will be held in each session. The story of the ceremony will be recorded. Three experts (wedding consultant, wedding design, the wedding researcher) use the narrow KJ method to create the "one mind and one card" based on the 10-10 perfection process, and then classify and summarize the card to achieve the final value extraction.

3.2 Wedding ceremony - the service provider's value extraction

Twenty-five wedding consultants are invited to code M01-M25 (M stands for Maker), and also take the value extraction of the step 1 wedding ceremony service journey. After the extraction, the result is compared with W01-W10. This study is aimed at the common pain points that occur in couples' value extraction as well as in service providers', to propose innovative ways and build modules to conduct experiments to clarify the benefits of service design introduction.

3.3 Innovative wedding ceremony service design experiment impact survey method

The twenty-five wedding consultants seek opportunities in their practice to conduct practical case application of innovative wedding ceremony service design, accumulate and summarize the experimental results, and conduct an experiment to verify whether the pain points are improved. The questionnaire design is designed to improve pain points.

4 DISCUSSION ON THE APPLICATION OF SERVICE DESIGN IN WEDDING SERVICES

In response to the value extraction of process 1 and 2, we have drawn four common pain points: the cumbersome ceremony, the unclear meaningfulness, the time-consuming and laborious procedure, and lack of intimacy in the interaction. After the research, an innovative service process was designed, named "Once a Lifetime," which was used to conduct case experiments by the 25 consultants. The results of the experiment showed that 84% of married couples and 53% of parents were satisfied. The result of the survey on process 3 increased the wedding service booking rate by 62.2%. The actual reduction is cumbersome by 92.9%, and interaction between the two parties by 42.3%. The results of each step are listed as follows.

4.1 Wedding ceremony - the new couple's value extraction

In the 10 groups, the new couples and their key participants (Table 1) were recognized by the "10-10 Perfection" ritual, the three experts used a narrow KJ method, and the three-stage extraction was carried out through the story journey, followed by "Feelings," "Meaning," and "Value." "Feelings" with "One Mind and One Card" was used as the first card division (named "Meaning"): meaningless waste, unhelpful wedding customs, and beautiful. There were five items, such as happiness, unnecessary face problems, and meaning of observance. The second time the card was divided into islands (named "value"), and the three experts agreed that there was a "contradictory and helpless superficial happiness."

4.2 Wedding ceremony - the service provider's value extraction

The 25 wedding consultants are all female, aged 20–46 years old, with 5–15 years of experience, 4 practicing in the northern area, 15 in the central region, and 6 in the south. The value extraction of the wedding ceremony service journey was conducted in the same way as the first step (as shown in Table 2); the final definition of "value" is also carried out by three experts, and "forced happiness" is obtained.

According to the preliminary research results from the customers (new couple) and the service providers (consultants), the "value" of the wedding ceremony is mostly "negative." The interpretation of the three experts is as follows: Expert 1 believes that the general cognition of the wedding ceremony is still under the impression of the traditional sense; expert 2 believes that the wedding ceremony process is not innovative enough and still of the old style; expert 3 believes that the current wedding ceremony customs have not been universally recognized and the good aspects of traditional values have not been highlighted. Therefore, in this study, we attempted to design an innovative wedding ceremony excluding pain points and to conduct case

Table 1. Wedding ceremony value extraction table for the new couples and their key participants.

No.	Key participants	Area	Feelings	Meaning	Value
W01	The new couple, and their parents	northern	cumbersome & troublesome		
W02	The new couple, and their parents	central	annoying doubts		
W03	The new couple, their parents, and agents	central	feel happiness		
W04	The new couple, their parents, and grandparents	southern	blessing of inheritance	meaningless waste helpless wedding customs	
W05	couples' parents and their uncles	southern	blessed	inheriting good happiness	contradiction
W06	The new couple, their parents, and agents	southern	ancestor's wisdom	unnecessary proud influence	helpless happiness
W07	The new couple, and their parents	central	a waste of money	not knowing the meaning	
W08	The new couple, and their parents	northern	the easier the better	for obeying	
W09	The new couple, and their parents	central	everything is meaningful		
W10	The new couple, and their parents	central	afraid of criticizing from elders		

Table 2. Wedding ceremony - wedding consultant's value extraction.

Event #	Consultant #	Area	Avg. Age	Yrs of experience	Feelings	Meaning	Value
1	M01-M04	northern	30.3	7.4 yrs	Mostly Simplified		
2	M05-M19	central	33.5	10.7 yrs	What a waste	simplicity	forced happiness
3	M20-M25	southern	35.7	5.2 yrs	Obeying the elders	obeying	

experiments to verify whether the innovation of service design thinking can be improved.

4.3 The modeling of an innovative wedding ceremony service

According to the three experts' comprehensive conclusions of new couples and service providers' value extractions, the current wedding ceremony services will result in "difficult, contradictory and helpless superficial happiness" and four pain points: cumbersome, unclear, time-consuming and laborious, and interaction is not intimate. Therefore, the modeling is based on the common meaning of the new couples and the service providers.

The "proud influence" is interpreted as signifying that the parents' interaction is insufficient and it is difficult for them to understand each other; "cumbersome and time-consuming" is interpreted as needing to be simplified; "meaningless waste" is interpreted to mean that the ritual preparation can be omitted; and "happiness inheritance" is interpreted as the joyful happiness of the wedding ceremony. Reinterpreting all the above processes and redesign results in "Family rendezvous party" → "Wedding ceremony and banquet" → "The gathering of two families." The interaction between the two families increased to three times, conducted before, during, and after the wedding. At the same time, the engagement and wedding banquet, which used to be held separately, are now handled together. All relevant ceremonies are performed at the wedding banquet hall.

The wedding banquet holding day is not necessarily held on the so-called good day, but on any holiday. Trilogy (family meeting and gathering; ceremonies; banquet) is to be completed in one major event, once for three; this study takes its title from the "Once a Lifetime" wedding ceremony module (Figure 1).

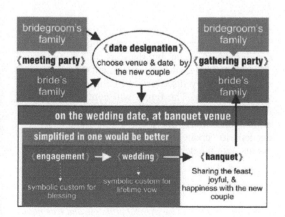

Figure 1. Flow chart of the "Once a Lifetime" wedding ceremony module.

This study upholds the principle of "not one marrying another, but for the two people who love each other and are going to get married, there is no need for a good day or good hour; as long as there is true love, true happiness, there is a good day." It outlines a trinity model of the wedding ceremony of "Once a Lifetime": meetings and gatherings, ceremonies, and banquets. Meetings and family gatherings are to be improved by the habits of the agents (matchmakers) during the wedding preparations, in which the two families are less familiar (one out of the four pain points); this experiment hopes, through pre-marriage family meetings and gatherings after the marriage, to promote the relationship of the two families and enhance intimacy. In the ceremonial part, most of the simplifications were carried out. In addition to the whole ceremony of the wedding ceremony, the six gifts and/or the twelve gifts were changed to the "Blessing Gift Monetary" written on the gift book. The book also had the custom of blessings. After the introduction of the program, the two families will keep one copy and take a group photo; that is, one of the four pain points—cumbersome and meaningless in the wedding ceremony, such as welcoming, levying, meeting, and returning the bride—is simplified. Other complexities could be kept, such as family member introductions, serving tea, and pressing tea in the innovation ceremony, since they might be able to increase interaction between the two and are not excessively time-consuming. In addition, since the wedding ceremony and the banquet are all handled at the banquet venue, the departure, the pickup, the fireworks, the limousine, the car, and the entering room are exempted. The ancestor worship may vary, depending on the customs. Overall the fourth pain point can be improved.

4.4 The experimental innovative wedding ceremony service model

Improvement was verified from the 10-10 perfection process to the "Once a Lifetime." The "Once a Lifetime" wedding ceremony module is designed to be used in actual wedding cases, not necessarily on a so-called "good day," and is briefly planned by simplifying traditional Taiwanese wedding customs. It uses conclusions from 25 wedding consultants from June 2016 until December 2018. Over a period of two years and six months, the 25 wedding consultants had the opportunity to conduct 19 case trials, 5 in the north, 10 in the central, and 4 in the south. Compared with the total number of regions in each session, the northern willingness is 83%, the central willingness is 67%, and the southern willingness is 100%. According to the feedback of the 25 wedding consultants after the interview, the new couples in the south are eager to break through the traditions and are willing to try something new. In the north, the new couples are not happy with the traditional customs, while the new couples in the central region can accept the traditions and want to try something new. Among the 19 newlyweds who agreed to try the experiment, 16 new couples were 84% satisfied, 10 parents were 53% satisfied, and 7 key participants were satisfied at only the 37% level, showing that the wedding ceremony design promoted by the service design method still has a lot of room for development outside the married couple.

In order to verify the practical feasibility of the service design during the wedding ceremony, a study was conducted through 20 questionnaires with 45 wedding consultants, who have an average of 6–10 years of experience in planning and practice in the hotel, banquet hall, and restaurant business in the northern, central, and southern regions of Taiwan. The 20 questions used by the wedding consultants are consistent with the questionnaires and interviews of the four wedding ceremony modules (as shown in Table 3). The simplified wedding ceremony is considered to be acceptable by 81.5%, feasible by 70.4%, and has an actual practice rate of 71.1%; the increase in the subscription rate is 62.2%; and it is considered that 92.9% of the people can feel a reduction in tediousness, but the family interaction between the two parties is only 42.3%.

To sum up, this study introduces the experimental results of the wedding ceremony design service to infer the statistical point of view. The positive rate of positive sense formation is 62.4% (the total average of all high and very high ratios in the questionnaire), and the actual reduction is cumbersome. The pain and improvement of the subscription rate is the initial positive result of the service design application. However, the increase in the interaction between the parents of both parties is limited, and it is worthwhile to study and continue to innovate and improve the service principles.

Table 3. Questionnaire - results of the "Once a Lifetime" wedding ceremony module.

Wedding Ceremony Simplified Consumer Conformity at the Banquet Venue/Average Acceptance 81.5%		
What do you think of the new couples' acceptance against conducting engagement ceremony at the banquet?	8. What do you think of the new couples' acceptance against conducting wedding ceremony at the banquet?	11. What do you think of the new couples' acceptance against combining engagement & wedding ceremony at the banquet?
H57.8% +EH37.8%	H33.3% +EH24.4%	H48.9% +EH42.2%

Wedding Ceremony Simplified Consumer Concise/Average Feasibility at the Banquet Venue 70.4%		
What do you think of the feasibility of engagement ceremony conducted at banquet with increased booking rate?	7. What do you think of the feasibility of providing wedding ceremony service at banquet?	10. What do you think of the probability of simplifying by combining engagement & wedding ceremony?
H40% +EH33.3%	H31.1% +EH31.1%	H35.6% +EH40%

The wedding ceremony simplifies the consumer perception held at the banquet venue/the actual rate of the event is 71.1%		
What do you think of the new couples' level of satisfaction against conducting engagement ceremony at the banquet?	9. What do you think of the rate of current practices conducting wedding ceremony at banquet?	12. What's your observation of the current actual situations of simplifying by combining engagement & wedding ceremony?
H48.9% +EH37.8%	H33.3% +EH20%	H40% +EH33.3%

P.S: H = High, EH = Extremely High.

5 VISIONS OF INTRODUCTION OF SERVICE DESIGN INTO THE WEDDING INDUSTRY

When consumers are not satisfied with the provided services and then raise complaints or dislikes, a doubt of that service being in place is naturally generated. However, when dealing with difficulties in response to problems, it is often impossible to get the energy to solve the problem adequately. Although service design is an emerging study of design, there are many examples of methodology that can be used as examples. This study uses years of wedding experience in the bottleneck of the industrial pain points and hopes to employ scientific logic to transform the pain points to selling points in order to discover innovative

new business opportunities in Taiwan's wedding industry. This service design application has achieved positive results so far, but there are still many explorations to be done, such as experimental samples, data analysis, sensory discovery, meaning discussion, and value extraction. The future will continue to expand the maternal parameters for more service touch points, and more fine-grained information will be needed to create innovative Taiwanese wedding results with the development of services.

REFERENCES

D. Huang, 1995. *Taiwan-style KJ principle and techniques*. Taipei City: China Productivity Center.

H. Xu, 2012. Eight operational principles of service design. *Brain Magazine*. Available: https://www.brain.com.tw/news

L. Wu, 2010. *Design thinking to transform the world*, original author: Tim Brown. Taipei City: Linking Publishing.

S. Chang, 2019. Cultural Constraints - A Probe into the Application of Taiwan Wedding Culture under the Perspective of Social Design, *Social Work and Social Design Seminar*. Nanhua University, Chiayi.

X. Chi, 2013. *This is the service design thinking*, original author: Jakob Schineider, Marc Stickdorn. New Taipei City: China Productivity Center.

Smart Science, Design & Technology — Lam et al. (eds)
© *2020 Taylor & Francis Group, London, ISBN 978-0-367-17867-3*

From traditional pastry to experiential tourism: Investigating the development of business model of baking tourism factories

Tsenyao Chang*
Department of Creative Design, National Yunlin University of Science and Technology, Yunlin, Taiwan

Pinfeng Liu
Intelligent & Local Design Service Center, National Yunlin University of Science and Technology, Yunlin, Taiwan

ABSTRACT: According to the promotion of Taiwan's media and tourism industry, pineapple cake is one of Taiwan's traditional cakes, and it not only has an international reputation, but also brings economic effects to Taiwan's baking industry and tourism industry. The market of Taiwan's baking industry is diversified. Under the impact of the exotic baking snack market, traditional cakes other than pineapple cakes have a lower market share in baking snacks. In the competitive environment of the baking market, traditional pastry manufacturers optimize their own enterprises through the transformation of operation or sales model; and the Taiwan government knows that the global industrial structure has changed. In order to preserve and enhance the competitiveness of traditional industries in the market, the enterprises that have features of "local industry characterization" and "local characteristics industrialization" are encouraged to propose "sightseeing factory projects" to transform their business management model. Three dimensions of operation of the baking tourism factories are analyzed: "Culture," "Education," and "Marketing." Driven by changes in industrial structure and policy promotion, baker industries have planned and built sightseeing factories to diversify their own business. This study selected the two major brands, the earliest and the latest from Taiwan's four traditional cake industry sightseeing factories, to conduct field surveys and draw their business model charts, as well as further evaluate and analyze the service experience model of their operations, and analyze the business model of Taiwan's traditional pastry factory, i.e., how to add its brand image in the market so that the company can continue to grow well.

1 INTRODUCTION

The lifestyle and consumption pattern of people are no longer subjected to only the aspects of food, clothing, housing, and transportation due to the global change in economic structure and lifestyle nowadays. The demands of education and entertainment have increased; hence, people are willing to devote themselves to consumption in recreation and entertainment. The government not only is aware of the economic effect brought by the recreation and tourism industry but also pays attention to the impact suffered by many traditional industries, which have declined gradually. The government has proposed the policies of "tourism factory plans" to improve the competitiveness of traditional industry in the market. Diversification operations of enterprises not only can enhance competitiveness but also is a way of attracting customers. Enterprises cannot depend on operation optimization alone and should have operation strategies of brand in order to have sustainable operation when facing an environment with constantly changing market demands. The study assesses and analyzes the operated service experience model of the two tourism factories of the four traditional pastry industries registered by the government in Taiwan: Kuo Yuan Ye, which is the earliest starting one, and Vigor Kobo, which is the latest starting one, through field observation and interview by performing field research and the business model canvas, and investigates the key factors for their success in the competitive market.

2 LITERATURE REVIEW

2.1 *Historical development of traditional pastries tourism factory in Taiwan*

Pastries were considered as luxuries in the years before 1895 (Ming Zheng Dynasty and Qing Dynasty) and were purchased only on special days, such as weddings, funerals, and festivals, since the people were relatively poor. Thus, pastries in the early stages did not have much variety in terms of types and their

*Corresponding author: E-mail: changty8908@gmail.com

flavors are much simpler. During the period of Japanese rule (1895–1945), Japanese catering culture as well as pastries, such as "wagashi," "western bread," and "western cakes," were brought in and altered the perspective of people toward the pastry industry. Hence, the pastry industry in Taiwan began developing rapidly during this period, and the terms "Chinese pastry store" and "wagashi store" were used to distinguish the types of pastries (Lu, 2007; Hsu, 2000). Mass amounts of people from China settled down in Taiwan after Kuomintang established the government of R.O.C. in Taiwan in 1949. The baking industry since then has grown vigorously, and the traditional pastry has evolved into various types since Taiwan has been through the governance of different races and influenced by the catering culture of different provinces in China. The baking industry in Taiwan has become more and more diversified, and people can choose different cultural baking products in Taiwan. That affects the market of the traditional pastry.

Under the effects of the industrial structure, professionals of local and industrial culture suggested that factories transform their operation; that is, not only can the production process of traditional industries be preserved, but also the idle space and hinterland can be reprofiled by the method of creating recreation and entertainment places and, in addition, the tourists can feel the situation personally so as to experience and inherit the historical culture of the industries (Tai, 2018). On the other hand, the governments of many developed countries, such as the United States, England, Netherlands, France, and Japan, assessed that tourism and leisure industries are considered as one of the indispensable elements to drive the traditional industries under the economic policies. Therefore, the Industrial Development Bureau, Ministry of Economic Affairs and Central Region Office, and Ministry of Economic Affairs promoted a "Tourism Factory Plan" with "local industry characterization" and "local characteristics industrialization" jointly in 2003 (Official Tourism Factory website 2018) to link the local industries with the tourism routes to activate manufacture industry, think outside the box of old operations, and create tourism zones of various themes with the value of technique inheritance.

2.2 Transformation development of industries from the perspective of service design

Service design is connected to the senses of customers and service quality and takes on the aim of providing a complete and satisfying service process. Different types of industries provide different service processes to customers. Therefore, Stickdorn and Schneider (2012) proposed five basic principles of service design:

a. User-centered: Service design focuses on "people." Not only is the service experience of customers considered as the core, but also the needs of the service provider are valued to achieve a balance status for all the people participating in the service process so that the service can have sustainable development.

b. Sequencing: The service process should have logic within it, and the progress and tempo of the service experience should be controlled. Service process with a fast or slow tempo affects the inner feeling of people experiencing it.

c. Co-creation: Service design has evolved from "Design to People" to "Design for People" to "Design with People" and then revolutionized into "Design by People" nowadays. The evolution process of service design shows three points: first, customers are the ones who experience the service process from beginning to end; second, the mood of the service provider affects the service atmosphere created; third, the service experience process provides direct contact between the service providers and customers. Thus, the service process of "Design by People" would provide a proper service quality.

d. Evidencing: Many service processes, such as the production of desserts and pastries, are performed in the back stage. Customers cannot experience the effort put in by the dealers or service providers participating in the service process. The carefulness of the dealers in service design can be felt by the visual stimulation of objects and symbols such as delicate product appearance.

e. Holistic: Service design process requires comprehensive considerations, from the creation of atmosphere, configuration of the route to operators, users and service providers, etc. The needs of all parties should be considered with a broad view. Customers will experience the service by their five senses—visual sensation, olfactory sensation, taste sensation, hearing sensation, and touch sensation—to increase the feeling of the customers and further affect service quality.

Tourism factories of the baking industry integrate two different industries, including baking industry and tourism industry. The objects participating in the whole service process increased and are even more diverse. New industries should follow the five principles of the service design—user-centered, sequencing, co-creation, evidencing, and holistic—to create a proper service experience process to strengthen the impression of the brand in the heart of the customers.

3 RESEARCH METHODS

3.1 Business model canvas

Business model canvas is a kind of systematic blueprint for organization strategies and can be used as an accessory appliance by the brands to assess their own strategy organization and process. Osterwalder et al. (2012) proposed nine perspectives to assess the operation pattern of the enterprises, which can be further made into the business model canvas of the enterprises. Operators should position the core concepts of the brand clearly during the creation and decision processes of the brands and use them as the basis of business decisions sequentially. The business model can be expanded in the conditions without affecting and changing the operating objectives of the brands, while the spirit of the brand can be conveyed to consumers effectively. The organization of the assessed contents of the business model canvas is shown as follows:

a. Key Partnership (KP): resources from collaboration when holding activities.
b. Key Activities (KA): activities that can be held in the long term and are helpful to the marketing and operation of the enterprises.
c. Key Resources (KR): the resources provided to the enterprises to convey their business concepts and values.
d. Value Propositions (P): business concepts of the enterprises and beliefs to be conveyed while satisfying the needs of the customers at the same time.
e. Channels (CH): channels for customers to know their brands and perform consumption.
f. Customer Relationships (CR): enterprises are connected in their relationships with customers through different business concepts and methods.
g. Customer Segments (CS): features of the customer group appealed to mainly by the enterprises.
h. Cost Structure (CS): key for enterprises to focus on when putting effort in operation.
i. Revenue Streams (RS): main revenue source of business operations.

3.2 Analysis of the research objects

The government of Taiwan has promoted the policies of tourism factory since 2003. Five brands of traditional pastry operate the tourism factories, including Kuo Yuan Ye Museum of Cake and Pastry, Ah-Tsung-Shih Taro Culture Museum, I-Mei Foods Apprentices Training Center, Invention Center of I-Lan Cake, Pineapple Cake Dreamworks of Vigor Kobo, etc. This study analyzes the five brands of relevance and their differences (as shown in Figure 1).

From the figure above, it is known that Kuo Yuan Ye is the earliest established traditional brand of desserts and pastries; however, its pastries are not represented locally. The latest established brand is I-Lan Cake, and the impressions of their products are linked to the local region so that tourists going sightseeing in Ilan will think of I-Lan Cake as a souvenir. Hence, the features of these two products are different. Vigor Kobo is the second latest

Tourism Factory / Items	Date of Establishment of the Brand	Date of Establishment of the Tourism Factory	Locally Representative Products	Diverse Products
Kuo Yuan Ye Museum of Cake and Pastry	1867	2002	✕	○
I-Mei Foods Apprentices Training Center	1934	2006	✕	○
Ah-Tsung-Shih Taro Culture Museum	1967	2012	○	○
Pineapple Cake Dreamworks of Vigor Kobo	1992	2012	✕	○
Kuo Yuan Ye Museum of Cake and PastryInvention Center of I-Lan Cake	2000	2009	○	○

Figure 1. Comparison chart for the tourism factories of traditional desserts and pastry brands (Osterwalder, A. et al. 2012).

established brand, and its representative products can be purchased all over Taiwan. The date of establishment of its tourism factory differs from that of Kuo Yuan Ye by ten years. Thus, the study selects "Kuo Yuan Ye Museum of Cake and Pastry and "Pineapple Cake Dreamworks of Vigor Kobo" for comparison of their business models and discusses the key to their successful business and how they can hold a portion in the competitive environment nowadays.

4 RESULTS AND DISCUSSION

4.1 Kuo Yuan Ye Museum of Cake and Pastry

Kuo Yuan Ye Museum of Cake and Pastry was established and activated in 2002. Its usage is separated by floors. There are two floors in total. The first floor is mainly for sales exhibition room and DIY experience space, while the second floor is for the introduction of the history of Kuo Yuan Ye and its meaning to Taiwan etiquette by preserving cakes so that people of the younger generations and children can understand the meaning of the cakes (Official website of Kuo Yuan Ye Co., Ltd. 2018). The concepts of its business model canvas are as follows (Figure 2):

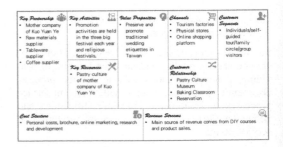

Figure 2. Business model canvas of Kuo Yuan Ye Museum of Cake and Pastry.

64

Key Partnership: Other than Kuo Yuan Ye itself, its raw material suppliers, tableware suppliers, and coffee suppliers are all its key partners. Key Activities: Promotion activities are held mainly in three traditional festivals of Taiwan and, in addition, they are also held in other festivals (such as thankful seasons, anti-drooling ceremony and anniversary, etc.). Key Resources: Pastry cultures of the mother company of Kuo Yuan Ye. Value Propositions: Preserve and promote the traditional wedding etiquettes in Taiwan. Channels: Sell mainly through physical stores. The tourism factory superimposes the impression of the brand and, together with online order and home delivery, is the secondary sales method. Customer Relationships: Pastry Culture Museum and Creative Baking Classroom are separated by floors. They not only convey the concepts and evolution of traditional festival activities but also plan handmade, environmental education and festival activities to create experience services to deepen the impression of brands on customers. A reservation system is adopted to ensure service quality. Customer Segments: Other than the basic group visitors, they also value individuals, self-guided tour, and family markets. Cost Structure: They focus mainly on the aspects of product marketing, research, and development. Revenue Streams: The main source of revenue comes from DIY courses and goods.

4.2 Pineapple Cake Dreamworks of Vigor Kobo

Pineapple Cake Dreamworks of Vigor Kobo was activated in 2012. The appearance of pineapple cake is used as a visual impression. Interactive playground and videos of pineapple growth are created through technologies. In addition, there are also transparent production process of the pineapple cake and a DIY experience zone. Customers walk through a series of experience processes and then enter the sales zone (Official website of Pineapple Cake Dreamworks of Vigor Kobo. 2018). The concept of its business model is as follows (Figure 3):

Key Partnership: Other than Vigor Kobo itself, its raw material suppliers, tableware suppliers, and coffee suppliers are all its key partners. Key Activities: Main

promotion activities of the enterprises are cross-industry alliance (such as tourism and travel industry) and participation in important pastry competition. Key Resources: Participated in pastry competition and obtained food certification. Value Propositions: Devoted effort in the inheritance of the delicate pastry culture in China. Channels: Customers can purchase goods through the visit to a tourism factory and physical stores as well as online order and home delivery. Customer Relationships: Growth process of the raw material, pineapple, and the conveyed image are shown, and technologies are applied to achieve the purpose of implementing education in entertainment to enhance variety and richness in the factory zone. Experience activities of live teaching of pastry production are held to enhance interactive experiences with customers. Customer Segments: *Other than the basic group visitors, they are also valued individuals, self-guided tour, and family markets.* Cost Structure: Focus mainly on personal costs, make alliance with travel agencies, and operate by profit-distributed model. Revenue Streams: Main source of revenue comes from DIY courses and goods.

5 CONCLUSION

Industrial structure is affected by the change in the lifestyle of our citizens based on the statistics made by Ministry of Economic Affairs. In recent years, the tourism and travel industry has driven the baking industry. Keeping hundred-year brands in line with the times and enabling new brands to create market in a market environment with vigorous competition are the keys to their sustainable operations. Three key elements for strengthening the brands are found in the study, which are culture, marketing and education (as shown in Figure 4), by organizing the business models for hundred-year brands and new brands of traditional pastry in Taiwan, and they are described as follows:

a. Culture: Pastries were sent to share the joy during the celebration activities in early stages, such as wedding, anti-drooling ceremony or

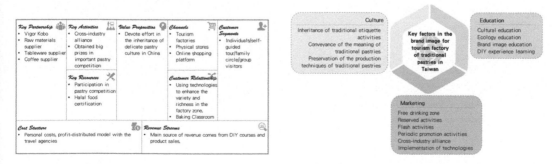

Figure 3. Business model canvas of Pineapple Cake Dreamworks of Vigor Kobo.

Figure 4. Key factors in the brand image for tourism factory of traditional pastries in Taiwan.

anniversary, etc. Traditional etiquette activities were changed or faded gradually along with the change in social and catering structure. Tourism factories of pastries in Taiwan can maintain the historical culture of Taiwan by the inheritance of the traditional etiquette activities, conveyance of the meaning of pastries, preservation of traditional pastry techniques, etc. Brand image can also be strengthened through the association of traditional pastries and etiquettes.

b. Education: Culture inheritance is conveyed through the visual design inside the factory as well as the interactive teaching of tour guidance to strengthen cultural impression on customers; other than this, there are also teaching topics, such as food safety and brand image, the growth and production of the raw materials, the historical development of catering culture related to the products, etc. Taking food safety and brand image as interactive learning topics can increase the richness of the museum; DIY experience courses can even increase the experience of taste sensation, olfactory sensation, and touch sensation. Creating experience courses for the five senses can deepen customer experience of the brand.

c. Marketing: Different business models resulted in different marketing decisions. Traditional pastry brands in Taiwan should consider the inheritance of the culture in Taiwan and also assess business models of other tourism factories when building tourism factories. Free drinking zone, reserved activities, flash activities or periodic promotion activities, etc. can be provided to attract consumers; cross-industry alliance can be applied to expand consumption circle and boost consumption; culture of the brand can be conveyed, or interactive education can be performed by technology (such as AR, VR, etc.) to increase experience service process and enhance the brand impression on customers.

The development trend in the market of traditional pastry industries had been drawn an equal sign with local souvenir. Not only has the souvenir market of domestic and foreign travels increased, but also the brand image to domestic and foreign customers has been enhanced through integration with the tourism industrial chain. Hence, traditional pastry brands expand their business model through the creation of tourism factories. Operators should consider the three aspects, culture, education, and marketing, in their business plans and apply the five senses, visual sensation, hearing sensation, taste sensation, olfactory sensation, and touch sensation, in their consumption experience model and link their brand image for the benefit of the sustainable operations of their enterprises.

REFERENCES

Lu, M.Y. "The Study on Commercial Space of Baking Industry in Fengyuan-A case of Jungjeng Road," M.S. thesis, Dept. Hist. Geo., Nat. Chaiyi Univ., Chaiyi City, Taiwan, 2007.

Hsu, H.C. "Review and Outlook (Part 1)," *Baking Industry*, vol. 91, pp. 51, 2000.

Tai, Y.H. "A Research on Inheritance and Innovation for the Industrial Culture of Tourism Factory: A Case Study of Yilan Tourism Factory," M.S. thesis, Dept. Cultural Assets & Reinventio, Fo Guang Univ., Yilan County, Taiwan, 2018.

Official Tourism Factory. https://taiwanplace21.org.tw/map.php

Stickdorn, M. and Schneider, J. *This is Service Design Thinking*, 1st ed. New Jersey: John Wiley & Sons Inc., 2012.

Osterwalder, A. Pigneur, Y. Smith, A. Clark, T. van der Pijl, P. *Business Model Generation*, (translated by Yu, C.L.), 1st ed. Taipei: Goodmorning Press Co., Ltd. 2012.

Official website of Kuo Yuan Ye Co., Ltd. Available: http://www.kuos.com/

Official website of Pineapple Cake Dreamworks of Vigor Kobo. Available: http://www.dream-vigor.com/

Smart Science, Design & Technology — Lam et al. (eds)
© 2020 Taylor & Francis Group, London, ISBN 978-0-367-17867-3

ANIMAP- A location-based social networking services

Shun Yao, Winglun Siu & Yuren Chen
Department of Information Engineering and Computer Science, Feng Chia University, Taichung, Taiwan

Tzuping Lai & Yichung Chen*
Department of Industrial Engineering and Management, National Yunlin University of Science and Technology, Yunlin, Taiwan

Chioujye Huang
School of Electrical Engineering and Automation, Jiangxi University of Science and Technology, Ganzhou, PR China

Pinghuan Kuo
Computer and Intelligent Robot Program for Bachelor Degree, National Pingtung University, Pingtung, Taiwan

ABSTRACT: The rapid development of the internet in recent years has prompted the birth of various social networking services (SNSs) such as Facebook, Twitter, and Instagram. A valuable feature of these SNSs is that they enable people to stay in contact with friends and family who are far away. However, real communities are built within physical boundaries. Conventional SNSs do not consider the geographical relationships which determine the majority of communities. We therefore developed a new SNS app called ANIMAP that takes into account the geographical relationships as well as the friendships between people. The aim of this app is to enable users to share information and opinions not only with friends and family but also their physical neighbors.

Keywords: Social networks, LBSN, iOS

1 INTRODUCTION

The internet has become an essential tool of modern life. One of its advantages is that it gives people the ability to share their lives with others, and people of all ages seem to enjoy communicating with others through online media. Whether it is text, images, or as currently trending, videos, online sharing is taking up a growing proportion of our lives. However, existing social networking services (SNSs) are mostly based on friendship rather than the geographical relationships between people [1]. Many people have set up clubs or fan pages on SNSs for the purpose of sharing local information and strengthening community solidarity. The information may include local news, small daily joys, and even local politics. These clubs have contributed much to local prosperity, but unfortunately, they are often difficult to find and lack consolidation. Furthermore, most platforms do not offer comprehensive functions for local clubs and therefore do not allow local communities to display their value. The objective of this study was thus to establish a comprehensive local platform

called ANIMAP on iOS that enables people who care about local matters to share local information, brings life back from the internet to the land we live on, and draws communities closer together.

We aimed to create not merely a platform for sharing local news but also an online activity center where communities can congregate to discuss local issues. We hope this kind of sharing and discussion can resonate through everyday life to bring people closer to one another via exchange and mutual assistance. This platform can also serve as a source of information regarding local life, transportation, and tourism for non-locals. Therefore, in our app we included the following features, (1) Display of local information using maps and current statuses, (2) Filtering function to enhance platform readability, (3) Anonymous reviews, (4) Distribution analysis of video viewing locations, (5) Map of personal memories, (6) Video stickers, (7) Simple easy-to-use interface, and (8) Commercial model.

The chapters of this paper are organized as follows. We discuss in Chapter 2 the differences of the proposed platform with existing apps. Chapter 3 is

---

*Corresponding author. E-mail: mitsukoshi901@gmail.com

the framework design. The proposed main functions and technology are presented in Chapter 4, and the app operations are presented in Chapter 5. Finally, we present our conclusions in Chapter 6.

2 COMPARISON WITH EXISTING APPS

Below we introduce three of the most popular social networking apps.

Facebook [2][3]: Facebook is one of the most popular SNSs at present. The accessibility of information shared on this platform depends on user privacy settings, so information that is only shared with the friends of users is difficult to find. Furthermore, there many clubs or fan pages that exist for the purpose of sharing local information such as weather, traffic, and news. Such information is continually updated, so users can obtain valid local information. However, the platform contains substantial amounts of advertisements, which can severely affect readability, and the report is slow to take effect.

Instagram [4][5]: Instagram is another popular SNS that enables swift information sharing. The app is easy to use and has no complex procedures. It uses hashtags to help users search for information, but there is little interaction among users on the platform. Users mainly use it to share information with friends, but it is difficult to determine the accuracy of the information based on posted discussions. There are also few local fan pages on this platform, which makes it difficult for users to obtain local information.

Twitter [6][7]: Twitter is less common in Taiwan because it has a more complex interface and complicated functions and procedures. It also has fewer users, which makes it relatively difficult to obtain up-to-date local information on this platform. It also focuses on people rather than places. It has few advertisements, which increases readability. However, its report is poor.

Comparison of aforementioned apps and proposed app: We compare the aforementioned apps with the proposed app in Table 1. As can be seen, the proposed app has a user-friendly interface, shares local information, provides up-to-date local information in an easy-to-access way, and has good readability and

Table 1. Comparison of proposed app with other social networking apps.

	User-friendly interface	Shares local info.	Provides up-to-date local info.	Good readability	Report function
Facebook	O	O	O		
Instagram	O	O		O	
Twitter				O	
Animap	O	O	O	O	O

a swift report function, which make it significantly superior to the other apps for our purposes.

3 FRAMEWORK DESIGN

This system was mainly written using XCode and Swift3. It enables users to share various types of local information in the form of current statuses. A unique aspect of the system is that it displays videos depending on where the user is. This can give users a better understanding of local conditions and also allow users to query the latest location information more quickly. Each video has a message board. Users can find replies here, and it can greatly increase interactions on the platform and bring people closer together.

4 MAIN FUNCTIONS AND TECHNOLOGY

Below we introduce nine main functions and technology of our app.

Local current status: The primary function of ANIMAP is to share local information. Users can upload short 5-15 sec video clips that they have filmed and some text to share information or opinions about things around them, which are then placed on a map using the GPS on their mobile phone. Users can watch the video clips shared by different people on the app and give comments or support. To maintain video quality and quantity, videos with more support can remain on the map for a longer period of time. For better readability, users can also choose the categories of information that they wish to view. Local information is displayed in lists to make it more user-friendly.

Video message board: To promote interactions among users on the platform, ANIMAP provides a message board for each video. Users can choose to discuss and rate videos semi-anonymously to prevent personal factors from affecting the authenticity of reviews. It is also hoped that the option of semi-anonymity will encourage interactions on the platform. In addition, the platform allows users to cite comments from other users in reviews and ratings. This is also aimed at encouraging interaction and engagement via discussion to strengthen community ties. The messages and ratings of the videos will help other users determine the accuracy of the information and avoid the posting of false information.

Notification settings: The notification settings allow users to choose what types of information they wish to receive notifications about from the system. In addition, to prevent users from being overwhelmed by notifications, users can set notifications so that they are only notified of videos with more than a certain number of likes. They can also set friend notifications and continuous notifications. For sudden incidents or emergencies such as earthquakes, typhoons, or severe incidents, the system will automatically turn on the notification function. This can be coordinated with

Apple Watch for faster notifications so that users are prepared in the case of emergencies.

Map of personal memories: ANIMAP hopes to give users a method to record and view their happy memories. It therefore provides users with personal maps that display their past videos based on where the videos were filmed. This enables users to return to the same sites and record more videos, and they can also choose the videos they want to view based on category and time. We hope that this can bring back memories of interesting things in their past and help them construct their own personal map that they can share with those around them.

Personal sticker archive: ANIMAP also provides different stickers based on weather, place, and significant landmarks so that whenever a user goes somewhere, he or she will have unique local stickers to use when editing videos. ANIMAP also provides a personal sticker archive that records which stickers users have applied and serves as a collection of achievements to encourage users to share more videos from different places.

Video distribution analysis: To help users improve video quality, the system provides a distribution analysis function that gives users an understanding of where their viewers watch their videos. The accounts, numbers, and locations of viewers will help them understand the preferences and needs of other users so that they can continuously improve the quality of their videos, attract more viewers, and enhance the overall quality of videos on the platform.

Simple easy-to-use interface: The interface is crucial to a user's first impression of a software program. However, many programmers overlook the importance of the interface in their attempt to provide more functions and services, which ultimately causes users to gain a poor first impression and delete the program. To avoid this problem and increase the usability of our app, we designed a simple interface for users. Optimizing the user's learning of how to use the software program is also very important, so illustrated instructions are given to first-time users to strengthen their impression of the product.

Commercial model: The platform allows companies and stores to upload promotional videos. The purpose of this is to enable stores to highlight local specialties through the platform and give locals a better understanding of the place they live in. To prevent the platform from being flooded with advertisements, specific interfaces are given to companies, and they are limited to uploading one video a day so as not to overwhelm users.

5 APP OPERATIONS

Below we explain how ANIMAP is used. Its functions are divided into three parts.

Viewing messages from others on personal pages: Figure 1 exhibits the main page of ANIMAP,

Figure 1. The main page of ANIMAP.

Figure 2. The first format of message display on ANIMAP.

Figure 3. The second format of message display on ANIMAP.

Figure 4. The view of watching video.

which shows the current statuses shared by users on a map. The number at the upper right corner of the videos indicates the number of videos that have been uploaded at that location. Figure 2 shows the first format of message display on ANIMAP. Users can tap the icon in the upper left corner and choose the categories of information that they wish to see. Figure 3 shows the second format of message display on ANIMAP. Users can view the information they need in a list instead of on a map. Figure 4 presents the view of watching video. The upper right corner displays the category of the video and how long it is in seconds. Figure 5 exhibits the message board, where users can rate or review the video based on the comments or cite other comments for their review. For instance, Viewer #3 cited the comment left by Viewer #2. Figure 6 displays popular videos. Users can use the filtering function in the upper right corner to choose the videos they want to see for other videos or users.

Personal page: Figure 7 displays one of the video editing pages, where users can use weather

Figure 5. The message board.

Figure 6. Popular videos.

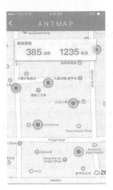

Figure 11. Viewer distribution function. (in a small map).

Figure 12. Viewer distribution function. (in a large map).

Figure 7. Video editing pages.

Figure 8. The interface that user can view their video history I.

Tapping the map icon in the middle brings them to a map of past videos. Tapping the book icon on the right shows their personal sticker archive.

Video analysis page: Figure 11 displays the viewer distribution function. Users can see the numbers of likes and views on this page and thus understand where most viewers watch their videos. As shown in Figure 12, zooming out from the map gives a clearer picture of where a video is popular. Users can analyze different areas to help them improve the quality of their videos.

6 CONCLUSION

SNSs have mushroomed in recent years, but most only consider the friendships rather than the geographical relationships between people, which does little to foster physical communities. In view of this, we developed a new SNS app called ANIMAP that takes into account the geographical relationships as well as the friendships between people. The aim of this app is to enable users to share information and opinions not only with friends and family but also their physical neighbors.

Figure 9. The interface that user can view their video history II.

Figure 10. The interface that user can view their video history III.

ACKNOWLEDGEMENT

This work was supported in part by the Ministry of Science and Technology of Taiwan, R.O.C., under Contracts MOST 107-2119-M-224-003-MY3 and MOST 107-2625-M-224- 003.

REFERENCES

D.N. Yang, C.Y. Shen, W.C. Lee, and M.S. Chen, "On socio-spatial group query for location-based social networks," Proceedings on ACM international conference on Knowledge discovery and data mining (SIGKDD), pp. 949–957, 2012.

stickers in their videos. Figures 8, 9, and 10 present three ways in which users can view their video history. In the first, users can view the current statuses that they filmed in the past on their personalpage.

L. Humski, D. Pintar, and M. Vranić, "Analysis of Facebook Interaction as Basis for Synthetic Expanded Social Graph Generation," IEEE Access, vol. 7, pp. 6622–6636, 2018.

A.A. Galán, J.G. Cabañas, Á. Cuevas, M. Calderón, and R.C. Rumin, "Large-Scale Analysis of User Exposure to Online Advertising on Facebook," IEEE Access, vol. 7, pp. 11959–11971, 2019.

K. Han, H. Jung, J.Y. Jang, and D. Lee, "Understanding Users' Privacy Attitudes through Subjective and Objective Assessments: An Instagram Case Study," Computer, vol. 51, no. 6, pp. 18–28, 2018.

D. Vilenchik, "Simple statistics are sometime too simple: A case study in social media data," IEEE Transactions on Knowledge and Data Engineering, (Early Access), 2019.

Y.C. Chen, M.Y. Tsai, and C. Lee, "Recommending topics in dialogue," World Wide Web, vol. 21, no. 5, pp. 1165–1185, 2018.

X. Zheng, J. Han, A. Sun, "A Survey of Location Prediction on Twitter," IEEE Transactions on Knowledge and Data Engineering, vol. 30, no. 9, pp. 1652–1671, 2018.

Xuanzang theory

Smart Science, Design & Technology — Lam et al. (eds)
© 2020 Taylor & Francis Group, London, ISBN 978-0-367-17867-3

The benefit of creative activities in community rehabilitation center

Tsai-Chieh Chien*
Ph.D Program of Mechanical and Energy Engineering, Kun Shan University, Tainan, Taiwan

Huann-Ming Chou
Department of Mechanical Engineering, Kun Shan University, Tainan, Taiwan

ABSTRACT: Creative activities are a crucial method of intervention in occupational therapy and are commonly applied to inspire creative thinking and problem-solving abilities in patients. Various activities such as art, theater, dancing, and music can be adopted as creative exercises. Such activities exert a wide range of positive effects such as improving patients' motivation, learning outcomes, self-appraisal, and self-expression. Although creative activities are considered a type of therapy, the application of such activities in Taiwan's community rehabilitation centers remains limited. In this study, a method for preparing handmade soap was developed based on the selection of this activity by a group of patients. The preparation activity was implemented over a 6-week period and a satisfaction survey was conducted after completion of the activity to determine the activity outcomes. The results revealed that with respect to activity benefits, activity content, creativity exerted during the activity, hope for the future, and interpersonal interaction, at least 80% of the patients were extremely satisfied, suggesting that handmade soap preparation is not only a cultural and creative activity but also an effective measure for improving the therapeutic outcomes of patients in community rehabilitation centers.

Keywords: Creative activity, Community rehabilitation center, Occupational therapy

1   BACKGROUND

Occupational therapy is a profession that regularly employs creative activities. All activities provided by therapists can serve as therapeutic exercises (T. Schmid, 2004). Moreover, occupational therapeutic activities such as arts and crafts can serve as sources of inspiration for creative activities (R. E. Levine, 1987). Creek noted that creative activities constitute a type of psychological therapy that can relieve pain in humans without the use of language. Adults and children can regard such activities as games to trigger feelings of fun and self-motivation; by engaging in such activities, patients' emotional recovery can be accelerated and their health can be improved. Numerous common activities such as cooking, knitting, domestic arrangement, gardening, dancing, and playing music require creativity; therefore, such activities can be learned to develop new skills or practiced to sharpen existing skills (J. Creek, 2005). In addition, creative activities require participants to exhibit a positive attitude, the intention to participate, and basic creative potential. Creativity affects various aspects of daily life from personal care and productivity to social relationships and leisure activities (H. du Toit, 1991). Because creative activities can promote peoples' self-motivation, strengthen their learning abilities, and improve their self-expression, encouraging patients in community rehabilitation centers to engage in such activities can enhance their creativity and greatly facilitate their recovery.

The greatest challenge for occupational therapists is to identify patients' creative potential and inhibitors to their participation in activities. To overcome these challenges, a therapeutic environment must be developed. For example, a topic must first be selected; said topic must then be the subject of brainstorming to stimulate patients' in generating ideas related to the topic. Required materials and equipment must be prepared to facilitate creative exercises and relevant information and instructions regarding the topic's creative aspects must be provided to enable new skills and knowledge to be taught in an appropriate manner. Finally, therapists should inform patients that the importance of an activity lies in the process rather than the end product to prevent excessive praise of a patient's work making others feel that their work is inferior, which in turn could suppress creativity. In addition, time pressure is not necessary in activities conducted by therapists. Occupational therapists should respect their clients' decisions to complete or not complete work and should give attention to all finished products. Finally, activities require a quiet and undisturbed environment. The process for handmade soap is presented in Table 1.

---

* Corresponding author: ball0818@seed.net.tw

Table 1.   Preparation process for handmade soap.

Mix fats by placing them in a stainless steel pot and heating them using a water bath or induction cooker.

Dissolve sodium hydroxide in an acid and alkali resistant container by constantly stirring the substance until it has fully dissolved.

Measure the temperature of the fat mixture after placing it in the water bath.

Measure the temperature of the lye.

Mix the fat and lye when both are a similar temperature; constant stirring is required for the first 20 minutes.

Stir until the soap liquid has become thick; the soap liquid is sufficiently thick when the lines drawn on it by a blender remain.

Pour the liquid into soap boxes purchased at a chemical materials store

Release the soap products from the molds.

Package the end product.

Table 2.   6-week handmade soap course.

Week	Content
Week 1	Introduce the method for preparing eco-friendly handmade soap
Week 2	Purchase the equipment required for soap preparation
Week 3	Begin soap preparation
Week 4	Cool soap
Week 5	Pack soap by using ecofriendly materials
Week 6	Deliver soap bars to neighbors to enhance community integration

Table 3.   Activity satisfaction survey.

Strongly agree ———— Strongly disagree

Item	1	2	3	4	5	6
Overall, I am highly satisfied with this activity.	80%	20%	0%	0%	0%	0%
The content of this activity was helpful to me.	90%	10%	0%	0%	0%	0%
I gained substantial rewards from this activity.	60%	30%	10%	0%	0%	0%
I look forward to participating in similar activities.	60%	20%	20%	0%	0%	0%
I think this activity was creative.	80%	20%	0%	0%	0%	0%
My creativity could be adequately applied.	50%	30%	20%	0%	0%	0%
Participating in this creative activity improved my mood.	70%	30%	0%	0%	0%	0%
Participating in this creative activity improved my self-confidence.	80%	20%	0%	0%	0%	0%
Participating in this creative activity generated hope in me.	80%	20%	0%	0%	0%	0%
Participating in this creative activity enhanced my interactions with others.	90%	10%	0%	0%	0%	0%

2   PURPOSE

In a meeting of patients in a community rehabilitation center, several cultural and creative activities were named from which the patients were asked to choose one; handmade soap was selected. A 6-week activity was subsequently implemented and a satisfaction survey was conducted to analyze the activity outcomes.

3   SAMPLE/METHOD

A community rehabilitation center in central Taiwan was chosen as the research site. Ten patients in that center attended a patient autonomy meeting on a Friday to discuss the creative activity to be run. Handmade soap was selected as the activity topic after voting. Occupational therapists subsequently analyzed and gained an understanding of the preparation method for handmade soap. The cold process was adopted in this study to produce cold process soap, which is made by mixing fat, sodium hydroxide, and water. After saponification, which takes 1 to 2 hours, the soap must sit for at least 2 weeks for its alkalinity to drop, after which time it is ripe and ready to use. When the soap preparation process was understood by the occupational therapists, a 6-week activity schedule was developed in Table 2.

4   RESULT

A satisfaction survey was conducted after completion of the 6-week activity. The survey was developed by the researcher and the results are detailed in Table 3.

The patients' opinions offered as responses to open-ended questions are presented in Table 4.

5   DISCUSSION

Tables 3 and 4 show that 90% of the patients were extremely satisfied with the increased interaction with others and the benefits they received from participating in the activity. Generally, the patients felt that through the activity arrangement and process of

Table 4. Open-ended questions in the activity satisfaction survey.

Item	Reflection
The aspect of this activity that interested me the most	Most patients considered the process of preparing hand-made soap and the moment of removing the cooled product from the container the most interesting aspects of the activity. Some patients expressed that the sense of achievement was felt most strongly when distributing trial products to neighbors. Most patients felt that the most rewarding aspect was learning about the use of soap and its preparation pro-
The aspect of this activity that I consider to have been the most rewarding	cess. In addition, many patients rated this activity to be the most meaningful among numerous other activities in which they had participated at the community rehabilitation center.
Content that could be added to this activity	Most patients thought that the current content was appropriate and consistent with their ability level. They hoped that similar activities would be held in the future.
Topic suggestions for the next activity	Most patients did not offer specific opinions regarding this item
Other opinions and comments	Most patients did not offer specific opinions regarding this item, except one, who expressed that the stages of the soap activity were complicated and hoped that the process could be simplified slightly.

group preparation, interpersonal interactions were enhanced; moreover, the patients subjectively perceived physical, mental, and spiritual benefits from this activity. Regarding activity satisfaction, 80% of the patients were extremely satisfied with the effects of this creative activity in increasing their creativity, self-confidence, and hope for the future. The patients claimed that because all participants were deeply engaged in the learning process of this activity and a sense of achievement was felt when the finished

product had been produced, the level of overall satisfaction with the activity was acceptable. In addition, through molding, the finished soap products exhibited multiple appearances; creativity was applied to product packaging. The presentation of the final products instilled self-confidence in the patients and enabled them to feel capable of changing themselves through creative activities, thereby reducing their sense of uncertainty about the future. Regarding treatment effects, 70% of the patients were extremely satisfied with the activity outcomes, indicating that this activity exerted a calming effect on the patients and enabled them to temporarily disregard their depression because the process required concentration. Regarding the richness of rewards and expectation of participating in similar activities, 60% of the patients were highly satisfied. The relatively low satisfaction levels of some patients may have been due to the products being intended for sale for charity or other purposes, leading to a sense of subjective deprivation in some patients. Furthermore, such activities generally require substantial time, physical strength, and focus; patients with physical or mental conditions often experience side effects due to failure to take medicine regularly, and thus sometimes experience difficulty focusing and reduced physical strength. Finally, the least satisfying aspect among the patients was creativity exertion, possibly because the patients had relatively low levels of creativity to begin with and were relatively incapable of conducting effective thinking because of apathy or negative symptoms. Therefore, the patients often echoed the ideas of the opinion leader in group settings and passively acted in accordance with these ideas; one example was the final decision to purchase molds at a chemical materials shop, which limited the patients' creative application, and thus lowered the creative exertion score.

REFERENCES

T. Schmid, 2004. Meanings of creativity within occupational therapy practice, *Australian Occupational Therapy Journal*, vol. 51, pp. 80–88.

R.E. Levine, 1987. The influence of the arts-and-crafts movement on the professional status of occupational therapy, *American Journal of Occupational Therapy*, vol. 41, pp. 248–254.

J. Creek, The therapeutic benefits of creativity, 2005. *Promoting health through creativity*, pp. 74–89.

H. du Toit, Initiative in occupational therapy, 1991. *Patient volition and action in occupational therapy. Vona and Marie du Toit Foundation, Hillbrow, South Africa.*

Smart Science, Design & Technology — Lam et al. (eds)
© 2020 Taylor & Francis Group, London, ISBN 978-0-367-17867-3

Preliminary research into the ancestral culture of Taiwan's indigenous tribes

Lichin Chang*
Ph.D. Program of Mechanical and Energy Engineering, Kun Shan University, Tainan, Taiwan

Huannming Chou
Department of Mechanical Engineering, Kun Shan University, Tainan, Taiwan

ABSTRACT: Research conducted by linguists into the evolution of Austronesian languages in the Pacific area, DNA-based research by international genetic anthropologists, as well as the spread of the sweet potato and paper mulberry into other countries provide evidence that the Austronesian languages originated from Taiwan. Taiwan's indigenous tribes each have different culture and language, demonstrating the richness of its cultural assets; hence Taiwan can be said to be the source of the ancestral origins of the Asia Pacific tribes. Taiwan also plays an important role in preserving the cultural heritage of the Austronesian languages. Subsequently, the onslaught of the industrial revolution, modernization, effects from foreign influences, as well as religious beliefs resulted in the predicament of cultural preservation and disappearance. This not only impacted the maintenance of ancestral spirit worship but also led to the gradual disappearance of substantive traditional techniques and crafts such as art, architecture, fishing, and hunting skills. This paper commences with a study of the reasons for the Taiwan indigenous tribes' homage to ancestral spirits and how this worship is an extension of the culture of filial piety. The paper also includes a study into the characteristics of the worship of ancestral spirits and how this heritage can be enhanced and optimized to fit into a contemporary setting yet preserve the heritage and characteristics of this fine traditional culture. The paper will further investigate the possibility of developing this ancestral heritage to become the origin of a cultural and creative-based vehicle in order to inject economic power into this area of development and enhance the long-term preservation of this cultural heritage.

Keywords: ancestral culture, culture of filial piety, optimization of ancestral culture, cultural and creative aspects of ancestral culture

1 A PRELIMINARY STUDY ON THE REASONS FOR TAIWAN'S INDIGENOUS TRIBES' ANCESTRAL WORSHIP

1.1 *Parents and elders as the objects of lifelong learning in a difficult living environment*

Most Austronesian Taiwanese live near mountains or beside the sea and hence depend heavily on nature for their living and face high risk against survival. Beasts and vipers in the mountains and waves at the beach or fishing pose risks to their survival. Hence they become adept at using natural resources and dealing with an arduous living environment, and have a deep sense of harmony with and dependence on nature. In view of this, they develop a reverence for the mountains and seas, and believe in a dominant force that can bless or punish humans, which naturally results in the worship of heaven and earth and faith in animism (Indigenous Religious Rituals. Taiwan's Indigenous Peoples Portal, 2019). From an early age, they must abide by the teachings and heritage of their parents, elders, and chiefs, and learn the skills of survival, which is a lifelong learning career. Study subjects include people and nature, and skills include not only crafts, dance, shipbuilding, hunting, and fishing, but also astronomy, geography, architecture, farming, weaving, worship, and daily living. The diversity and difficulty level of these skills require diligence and lifelong learning, hence the peoples' respect for fathers, brothers, and tribal elders. Parents not only fulfill parental roles

---

*Corresponding author: liqin@cht.com.tw

for their children, but also function as their lifelong teachers. Therefore, they are extremely respectful of ancestral wisdom and experience, resulting in ancestral worship among Austronesian Taiwanese. As such, ancestral worship culture is in fact an extension of filial piety culture. Although the ancestral worship culture among the various ethnic groups may differ in its expression of affection and respect for ancestors, it mainly comprises veneration of ancestors and all past souls. Some may even include animal spirits, mountain gods, water gods, spirits and ghosts, and totems.

1.2 Resisting outside tribes

To protect their survival resources and economic territory as well as to demonstrate their bravery, most Austronesians have a head hunting custom (Li, 2011), making it difficult to grow the population in the tribes. Indigenous tribes are mutually helpful within their own group but at the same time are strongly mindful of intergroup conflict, thereby creating independent and diverse Austronesian Taiwanese communities. The threat of being killed by enemies and mortal fears make it easier for them to seek ancestral spirits to comfort their minds. In their unique perception of the principle of cause and effect, they also worship the heads of their decapitated enemies in the hope of taming the beheaded souls and balancing their negative karma. Faced with the uncertainties of life, they believe in an unknown divine power that determines fate and events. This gives rise to a certain degree of awe toward nature and ancestral spirits.

1.3 Nontextual heritage

With a lack of text to pass on skills, all survival knowledge and experience must be taught from generation to generation by the elders, fathers, brothers, mothers, and sisters in a tribe. Hence, familial and tribal affection and closeness and great respect for the old are further highlighted. Emotional cohesion within ethnic groups is strong, and invasion by outside tribal culture or force is not tolerated. As such, the language and culture of each ethnic group are preserved. Ancestral spirits are also defended in full force; as a result, the ancestral culture not only represents the extension of filial piety, but is also substantially significant in its reverence toward nature, respect for all things, and preservation of ecology, and is not general superstition and faith.

Given the above reasons, it is evident that filial piety among Taiwan's indigenous tribes is more extensive and profound. It not only pertains to the old and elders within a tribe, but also involves a better recognition of the meaning of filial piety because of the tribe's relationship with the environment.

2 CHARACTERISTICS OF ANCESTRAL RELIGIOUS VIEWS AMONG TAIWAN'S INDIGENOUS TRIBES

2.1 Rich legendary mythologies

Although each ethnic group has a certain cognition and connotation of the ancestral spirits, most are based on legendary mythologies. For example, the Rukai tribe believes that their ancestors came from the hundred-pace snake, hence making the hundred-pace snake a symbol of their ancestral spirits. The Taroko tribe believes that their ancestors originated from the stone god, while the Thau tribe has the ula-laluan and the Saisiat tribe celebrates the PaSta'ay festival (Indigenous Religious Rituals. Taiwan's Indigenous Peoples Portal, 2019), and so forth. Compared to the concept of God in monotheism and reincarnation in Buddhism, Austronesian Taiwanese beliefs in ancestral spirits have less specific or strict origins.

2.2 Collective concept of ancestral spirits

For Taiwan's indigenous tribes, ancestral spirits are mainly a collective concept. Rather than the spirit of a single entity or individual, ancestral spirits are mostly a pluralistic combination. The spirits are typically not worshipped in a home, but collectively by the tribe or ethnic group on a specific festival (Indigenous Religious Rituals. Taiwan's Indigenous Peoples Portal, 2019). The ancestor worship is significantly different compared to the Han culture. The collective worship of ancestral spirits and the heavens and earth is not only an expression of gratitude for and acceptance of blessings from ancestral spirits, but also a strengthening of tribal cohesion and demonstration of tribal hierarchy and communication. The shared meal brings joy and friendship within the tribe and also serves as an opportunity to learn and express mutual respect.

2.3 Taboos and precepts are cumbersome

Taiwan's indigenous tribes have integrated their respect for ancestral spirits into their lives, and many taboos and precepts can be found with regard to food, clothing, housing, traveling, sacrifice, and hunting. A fear of angering their ancestral spirits and bringing misfortune to their descendants further fuels the mystique. For example, whether farming, fishing, hunting, visiting, or head hunting, they must listen carefully to the birds and strictly observe the taboos and remain silent when hunting. During worship, women are forbidden from touching anything or doing anything. Unquestionably, these cumbersome tribal taboos (C. R. Li, 2010, M. C. Wang, 2001) stem from the fear of bad luck caused by angering ghosts, gods, or ancestral spirits.

2.4 Concept of causality

For most of Taiwan's indigenous tribes, their understanding of ancestral spirits is based on the causality concept of "reaping what you sow" and the distinction between good and evil spirits. Generally, they perceive ancestral spirits as good spirits, and violation of ancestral teachings or evil deeds will bring about punishment by ancestral spirits. In other words, they believe that selfishness will eventually result in bad consequences, and thus arises their care for people and respect for the universe. However, unlike the doctrine of karma and reincarnation in Buddhism, their concept of causality does not include the original entity that stores the seeds of good and evil deeds.

In addition to the above brief descriptions, the philosophy, worldview, and spiritual power of ancestral spirits also characterize Austronesian Taiwanese ancestral worship.

3 OPTIMIZING ANCESTRAL WORSHIP CULTURE

People are the most important carriers of culture. Therefore, the quality of people is key to cultural heritage. To ensure the transmission of cultural origins, there must be good content and carriers. For example, although the Chinese culture and Jewish culture were once threatened with annihilation, they persevered unshaken. These two cultures are similar in their great emphasis on education, making it evident that educating talent is the essence of culture. Through cultivating the people of the tribes and keeping abreast with modern development, the ancestral culture can be further advanced.

3.1 Exalting the concept of causality in the culture of ancestral worship

Most Taiwan's indigenous tribes believe that only good spirits can be regarded as ancestral spirits, thereby suggesting the value of encouraging good, sacrificing self for the good of the community, disregarding personal safety, and promoting good while suppressing evil. Moreover, the belief that cause and effect is the only truth is a form of balance for survival. This concept of balance is applicable to all humanities, geography, and ecology, such as in ecological balance, not overhunting and respecting all lives, and so on.

3.2 Should legends and taboos be preserved?

Progress of time and scientific advances bring about the question of whether the beauty of legends can be preserved. By pursuing the truths behind legends and their conformity to scientific knowledge, understanding the difficulties of ancestral background, and reinterpreting their

significance, the culture of ancestral worship can be reborn and its mystique unveiled to reinvent ancestral spirits as symbols of closeness and wisdom.

3.3 Reflecting and tempering the culture of killing and drinking (H. T. Chuan, 2006, M. R. Hsu, 2004)

With the progress of time, the heroic actions of indigenous tribes should be manifested in the people themselves and their conduct, with perseverance and an unyielding attitude towards the truth. Cherishing life based on the concept of cause and effect can ensure a long life, and eliminating drinking and bad habits can also increase the life expectancy of indigenous tribes (Health Inequality in Indigenous Villages—Improvement Strategies and Action Plan (2018–2020), 2018). However, only through early education can bad habits and culture be reversed to enhance health and tribal strengths among Austronesian Taiwanese.

3.4 Promote filial piety

History reveals that Austronesian Taiwanese have always shown a broad understanding of filial piety. This positive tradition and consciousness of filial piety should therefore be embraced, and the promotion of ancestor worship culture is the first step toward the return of filial piety.

In addition, we should praise the knowledge and skills of the ancestors in all aspects of life, ecology, art, architecture, and sculpture, and their reverence for nature, protection of nature, and self-contentment should be admired. The culture of ancestral worship is not about protection, but about development and identification. With respect and tolerance as the starting point, it preserves the dignity and cultural diversity of Taiwan's indigenous tribes, and eliminates mutual prejudice and discrimination. Only through more open-mindedness and ideas can Austronesian Taiwanese culture be optimized. In addition to being respected and recognized by outsiders, self-realization and the courage to confront conflict are even more important because only a culture that has been tempered can withstand the test of trials. The Austronesian Taiwanese must first identify with and refine their own culture and at the same time remain conscious of themselves while remembering their ancestors. The people of today will naturally become the ancestral spirits of tomorrow and therefore play a role in the future culture of ancestral worship.

4 CONCLUSION

Ancestral worship culture can be regarded as an extension of filial piety culture. The more successful the filial piety heritage is, the more flourishing the ethnic reproduction and generational alternation. The promotion of ancestral worship culture by Taiwan's indigenous tribes can be regarded as a clarion call for Austronesians around the globe to trace their roots (2001 Special Austronesian Cultural Events, 2002). Taiwan has an innate historical advantage. Regardless of the meanings of ancestral spirits among Taiwan's indigenous tribes, most represent their reverence toward the driving power of nature, their predecessors and ancestors, and the nether world. Ancestral worship is both a cultural heritage of filial piety and an expression of self-humility. Ancestral spirits are spiritual sustenance for Austronesians, and a religious and life assurance. The fusion and disappearance of cultures often occur subtly over time rather than due to a powerful invasion of force. Therefore, cultural conformity to truth and humanities can also be studied. Optimizing the culture of ancestral worship may enable Austronesian Taiwanese to regain their persistence and enthusiasm for their ancestral spirits as well as to learn about their ancestors' contributions to Taiwan. Despite changes through the passage of time, when future generations seek blessings from their ancestors, they can discover the timeless essence of their inheritance and continue to glorify their ancestral spirits.

REFERENCES

R.D. Gray, A.J. Drummond, S.J. Greenhill, 2009. *Science 323* (5913), 479–483.

Y.M. Chen, 2004. *Historical Monthly 199*, 34–40.

J.M. Diamond, 2000. *Nature 403*, 709–710.

Y.C. Chen, 2015. Publisher: Ink Literary Monthly Co., Ltd., 5th printing.

J.A.V. Tilburg, 1994. *Easter Island: Archaeology, Ecology and Culture*. Washington DC: Smithsonian Institution Press.

L. Robert, 2001. *The Journal of Pacific History 36*(1).

B.C. Wu, 2015. *The Seed that Rewrote History: From the Austronesians in Taiwan*. Liberty Times.

C.K. Li, 2011. *The Tribes and Migration of Austronesian Taiwanese*. Indigenous Religious Rituals. Taiwan's Indigenous Peoples Portal, Wikipedia-Cutting the Grass. (Head hunting).

C.R. Li, 2010. *Taiwan's Indigenous Taboos and Mythologies*. Shu I Elementary School.

M.C. Wang, 2001. Taipei: Indigenous Peoples Cultural Foundation, pp. 136.

H.T. Chuan, 2006. Drinking and Original Sin, Taiwan Academy of Ecology, Issue 10.

M.R. Hsu, 2004. Exploring Drinking Behavior Among the Indigenous People. Graduate Student, Family Education Research Center, National Chiayi University.

Health Inequality in Indigenous Villages—Improvement Strategies and Action Plan (2018–2020), 2018. Ministry of Health and Welfare, Republic of China.

2001 Special Austronesian Cultural Events, 2002. Council for Cultural Affairs, Executive Yuan, Taiwan.

Smart Science, Design & Technology — Lam et al. (eds)

A discussion on the Chan concept underlying the Japanese way of tea based on the essence of Buddhist realization: The tea ceremony of Sen Sōtan

Wei-Hsuan Fan*
Ph.D. Program of Mechanical and Energy Engineering, Kun Shan University, Tainan, Taiwan

Huannming Chou
Department of Mechanical Engineering, Kun Shan University, Tainan, Taiwan

ABSTRACT: The Japanese tea ceremony is world-renowned for its concept of "Tea and Zen are of one flavor." As this saying suggests, whether there is Chan in the tea ceremony (Way of Tea) or whether the tea masters are enlightened Chan practitioners or not is of significant relevance. In order to understand the Chan concepts behind the Japanese tea ceremony stretching back over the past four hundred years, this research studies the essence of attaining awakening—the eighth consciousness, which is in alignment with the Mahāyāna's Consciousness-Only school (S. Yogācāra) and the Chan tradition—to explore Sen Sōtan's concept of "Tea and Zen are of one flavor." As a result of the findings, this paper suggests that Sen Sōtan's way of tea and "Zen" comes nowhere near the self-abiding state of the eighth consciousness, but is merely a frame of mind without language and discourse within movements and discursive thoughts.

Keywords: Japanese chadō, Sen Sōtan, the essence of enlightenment, Chan tradition, consciousness-only school

1 PREFACE

Due to the prevalent tea-drinking culture of the Chan temples, chadō (tea ceremony; way of tea) was introduced to Japan when Buddhism was brought into the East during the Tang dynasty, and was developed to become a completely systematic culture. Japanese chadō was established by Murata Jukō during the 15th century, until Sen Rikyū further epitomised the practice (Ito Kokan, 2005; Kuwata Tadachika, 2016); Sen Sōtan, common ancestor of the mainstream Japanese chadō's three Sen schools, was Sen Rikyū's grandson. Ito Kokan, a renowned modern Japanese tea monk, said, "There is nothing like tea, for which the Japanese culture has such a remarkable affinity. Tea is what ordinary people can't live without in their daily life … It is not an overstatement to call the tea ceremony as the bloom of Japanese artistic legacies" (Ito Kokan, 2005).

As the Japanese chadō has always prided itself on the Zen of its tea culture, Sen Rikyū mentioned in The Southern Record that "The top priority of cha-noyu zashiki, also known as grass hermitage tea ceremony, is to attain enlightenment by practicing Buddhism" (Nanbô Sôkei, 2016; Teng, 1992; Lin, 1991). Also, as stated in The Record of Yamanoue Soji, "The tea ceremony comes from Zen Buddhism, which simultaneously serves as a foundation of the tea ceremony" (Ito Kokan, 2005; Teng, 1992; Lin, 1991; Isao Kumakura, 2006). These formulated the basis of the Japanese "Tea and Zen are of one flavor" ideology. As the original purpose of organising tea ceremonies was to allow enlightened masters to help the participants in achieving enlightenment, in order to find out if Chan does exist in the Japanese chadō, we should study if the people involved in tea ceremonies are indeed the enlightened ones by referring to the standard of enlightenment of Mahāyāna Buddhism. As the thoughts of the three Sen schools were inherited from Sen Sōtan, we are able to uncover the Zen of Japanese chadō for the past four hundred years from the connotations of Sen Sōtan's chadō. Hence, our research will first explain the essence of enlightenment of the Chan tradition, before using them to assess if Sen Sōtan's chadō matches the true nature of Chan. None of the existing research has employed

*Corresponding author: E-mail: pretty2004216@gmail.com

this method to study chadō; thus the results of this paper should be able to contribute to the sustainability and development of Japanese chadō.

2 ESSENCE OF ENLIGHTENMENT OF MAHĀYĀNA BUDDHISM

Xuanzang travelled to the West to acquire the original Buddhist sutras and translated them accordingly to prove that the fundamental of Buddhist Dharma lies in the eighth consciousness, which is the Mahāyāna's absolute truth that believes "All three realms are manifestations of the citta, and all dharmas are generated by the vijñāna." As for the realizing the "true suchness," this expression refers to the personal realization of the eighth consciousness of every sentient being. Thus, the purpose of the absolute truth or the eighth consciousness revealed by Buddha Śākyamuni was made known (Yu and Chou, 2017). In the third volume of Discourse on the Perfection of Consciousness-only, Xuanzang stated, "It is also called ālaya because impure dharmas and it embrace and harbor each other, and sentient beings grasp it as a self. The hosts of bodhisattvas who have entered the path of insight and obtained actual contemplation of reality (tattva-abhisamaya) are called 'superior ones' [in the verse]. They have the ability to realize and comprehend the ālaya consciousness, and for that reason our World-Honored One properly revealed it. On the other hand, all bodhisattvas are called 'superior ones.' Even though prior to the path of insight they are unable to realize and comprehend the ālaya consciousness, they are still capable of believing in, understanding, and seeking the transmutation of the support, and the Buddha was also referring to them. Other evolving consciousnesses do not have this meaning." The statement on bodhisattvas who have entered the path of insight and obtained the actual contemplation of reality means that they have realized the Alayavijnana (the eighth consciousness). If a practitioner can "realise the Alayavijnana," and actually observe directly that Alayavijnana is indeed a true mind-entity that is neither arising nor ceasing, he then becomes a saint or sage who has entered the stage of vision of the path in Mahāyāna Dharma. Previous papers have referred to the example of the Sixth Patriarch of the Chan tradition, Huineng's Platform Sutra of the Sixth Patriarch, to illustrate: "The self-nature is able to contain all dharmas; it is the 'store-enveloping consciousness.'" This shows that the Chan tradition of enlightenment of self-nature refers to the storehouse consciousness, which is also known as ālayavijñāna. This is in line with the core nature of entering the path of insight by Xuanzang's Consciousness-Only school (Fan and Chou, 2017). In the following sections, we will apply this standard to clarify the Zen concepts in chadō by assessing Sen Sōtan's establishment on chadō based on the

characteristics of the eighth consciousness and the mental consciousness.

3 SEN SŌTAN'S IDEOLOGY IN *THE TEA AND ZEN ARE OF ONE FLAVOR*

Sen Sōtan (A.D. 1578–1658) was the son of Sen Shōan, who was the second son of Sen Rikyū. Sōtan was his tea name, and his birth name was Genpaku, thus he was also known as Genpaku Sōtan. Regarding the way of tea drinking, Sen Sōtan placed his focus on the plain side of chadō established by his grandfather Sen Rikyū. He assembled the thoughts on chadō by the Sen family and elaborated them to bring about significant impact on the prevalence of Rikyū's chadō. After Sōtan passed away, his three sons (Kōshin Sōsa, Sensō Sōshitsu, and Ichiō Sōshu) inherited his chadō and established the schools of Omotesenke, Kyoto Urasenke, and Mushakōjisenke at Fushin-an, Konnichi-an, and Kankyū-an respectively. These schools are collectively referred to as the three Sen schools.

As Sen Sōtan's chadō was known as "Wabi-style tea," Sōtan became famous because of this, and among his writings, the most popular one was *The Tea and Zen are of One Flavor*, also named *The Record of Tea and Zen*. The *Tea and Zen are of One Flavor* was divided into five chapters, where Sensō and Ittō compiled and elaborated Sen Sōtan's written wills. The following discussion on Sen Sōtan's chadō is based on the essence of this book, and the Chinese translation is from the book *Tea and Zen* (Ito Kokan, 2005).

3.1 *The tea ceremony lies in realizing Zen*

In this chapter, Sen Sōtan said, "The act of drinking tea is centered on realising Zen. Only when one concentrates on practicing the chadō will one embody the purpose of the tea ceremony. Dian cha (whisking tea) is all about Zen dharma, and one's effort is put to know one's self-nature … By the act of dian cha, one can actualise the observation of one's own mind and there is no difference between this kind of instruction and the teachings of Buddhas." From here, he made it clear that the purpose of Japanese chadō is to attain enlightenment of the self-nature, fundamental mind, which is the same as the Buddhist teachings. Then, he went on to state that the tea ritual is merely a side issue and the practice of chadō should not be attending to such trifles; thus the practice of the Zen of tea should place its focus on the essence of chadō instead of the ritual. As Sen Sōtan had expressed mournfully, the chadō during that time was more about the tea ritual, with a lack of actual enlightenment of the Zen mind. Hence, there is even more reason for us to review his explanations through the eighth consciousness, which is the essence of Buddhist enlightenment, so that we can find out the authentic connotations of his chadō.

3.2 The tea ceremony is a way of practice

Regarding the way of practice through the tea ceremony, Sōtan said, "Those who practice Zen by doing things related to the tea ceremony are utterly concentrated. They pick the tea wares with exclusively one mind, entering samadhi. When picking a tea ladle, they pay attention to it alone without any discursive thoughts. Their mind focuses on the act of picking. When putting down the ladle, they maintain the same concentration and focus on the act of putting it down. Their mind operates in the same way as mentioned above when they pick or put down any wares in addition to the ladle. Their mind concentrates on when their hands let go of objects; their mind naturally lets go when they pick anything in their hands. Their conscious mind remains mindful wherever. Accordingly, their chi during dian cha flows incessantly and only streams into the Samadhi of Tea. It is not necessary to spend many months and years understanding them; they can be well understood from their mind. They just fully dedicate themselves to diligently doing the practice and to making progress in the Samadhi of Tea."

Sen Sōtan believed that the Samadhi of Tea encompasses the action of whisking tea while maintaining a concentrated mind, which is to focus on the picking and placing of tea wares, to move one's hands but not one's mind, and to allow continuous flow of steam by whisking the tea without keeping discursive thoughts. Such a view, however, falls within the scopes of "abiding in a mind without any distracting thoughts" and "thoughtless pristine awareness" of the conscious mind, because by realising the picking and placing of tea wares, one has entered the realms of form objects and tangible objects. Yet, by maintaining a concentrated mind without discursive thoughts, such mental consciousness corresponds to the mental object of the samadhi state. These are exactly the states of the conscious mind corresponding to the six sense objects. The existing samadhi states in Mahāyāna Dharma, including the Dharma-Flower Samadhi, the Treasure-King Samadhi, the Diamond Samadhi, the Buddha-Mindfulness Samadhi, the One-Characteristic Samadhi, and the One-Practice Samadhi, are all Samadhis pertaining to the prajna wisdom that have attained realisation of the eighth consciousness, the fundamental mind, and do not belong to the practice of meditative concentration of the mental consciousness. The nirvanic quiescence and cessation nature of the true mind, the eighth consciousness, does not arise through cultivating the meditative concentration of thoughtlessness, as it is being intrinsically free from thought and without any thought since eons without beginning. Therefore, there is no need to practice sitting meditation to clear thoughts or subdue any affliction, and there is also no need to remove any deluded thought through the action of whisking tea. The realms of sitting meditation in stillness, or whisking tea in movements

through "abiding in a mind without any distracting thoughts" or "thoughtless pristine awareness," are all attained after practice. As Sen Sōtan focused on whisking tea while abandoning all afflictions and maintaining a mind without discursive thoughts to obtain the correct concentration, this was merely the meditative concentration samadhi of one-pointed absorption within phenomena, and it was the concentration state of the conscious mind that stopped at one realm because it had yet to enter the wisdom samadhi pertaining to prajna of the one-pointed absorption within principle. It would be a mistake for us to insist that the removal of discrimination and deluded thoughts from the conscious mind is indeed the combination of chadō and the mind of Zen, or to regard it as the attainment of the Samadhi of Tea, because this is only a practice according to the phenomenon, and has only the samadhi (meditative concentration) but not the wisdom, and thus is essentially a divergence from Chan tradition of attaining awakening.

Additionally, Sen Sōtan believed that the way of practicing chadō should not adopt sitting meditation to contemplate phenomena, as sitting meditation for a prolonged period can easily lead to numerous afflictions and discursive thoughts. If one can concentrate on whisking tea and the picking and placing of tea wares, deluded thoughts will not arise easily, and the path of completion can be entered without second thoughts. Hence the saying goes, "To observe the fundamental nature (fundamental mind) from the picking and placing of tea wares, to teach people the method of sitting meditation, as it is not simply sitting still in silence." To observe the fundamental nature, however, the fundamental mind should be realized and attained first, or else how can we observe the fundamental nature? If the fundamental nature can be observed without attaining enlightenment, this would fall into the mind of either with thoughtless pristine awareness or pristine awareness with thoughts. Thus, if one concentrates on the picking and placing of tea wares, or even maintains a mind without any affliction, language, text, and deluded thought during daily routines, which are, in fact, the perception and observation minds that are interacting with the six sense objects, it is still within the state of the mental consciousness.

3.3 The concept of tea

Sen Sōtan said, "The alleged concept of tea is with only one intention. Some people take this intention as the concept of tea with true Zen and their countenance shows the façade of enlightenment. They are in a presumptuous mind, accusing others of not knowing the concept of tea. Other people believe the concept of tea cannot be spoken out, and the format of the tea ceremony cannot be taught. The essence of tea can only be learned by one's own experience of the tea ceremony, which is then considered a separate transmission outside the scriptures, but in fact breeds partial erroneous views. These views lead to negative karma." The phenomenon of Japanese chadō observed by Sen Sōtan was certainly not pointless, but it is also noteworthy to emphasize the fact that Buddhist enlightenment always places its essence solely on the eighth consciousness, and this eighth consciousness is intrinsically staying away from the five destinies and six paths, with a state without language or mentation. This is indeed the mind of liberation that is freed from the three realms; thus, as the saying goes, "Upon entering the gate, it is necessary to identify the host and distinguish between the sacred and the mundane face to face." We should crosscheck diligently and repeatedly to see if it matches the insights of the separate transmission outside the scriptures and the core concept of Buddhist masters from the West.

Sen Sōtan believed that "In terms of the Dharma, it sees rise of thought as the first violation of precepts. The top priority of Zen is to maintain a state of mindfulness. Holding a purposeful mind is what someone who practices Zen of tea should abhor most. Thus, when one's thoughts rise up and do the things related to the tea ceremony, this is a reverse of aptitude in Zen." Here, Sōtan interpreted meditative concentration as Zen because he viewed the crux to samadhi as training the conscious mind to become an unstirred mind. However, Chan refers to prajna but not meditative concentration, and the Chan allegory is not an unstirred mind of samadhi but the innately immovable principle entity, the eighth consciousness. Even if the conscious mind is practiced and developed into the immovable mind of the formless realm, it is still subjected to destinies and has not departed from the mental object of the samadhi state; thus Sen Sōtan's statement here is not in line with the essence of Chan.

3.4 Tea wares

Sen Sōtan said, "Take the round vacant clean, pure mind as the container. The tea within such a clean pure mind as its container is the tea with aptitude in Zen. Whether the ware is good or bad, one can obtain a true clean, pure container from one's mind as long as one removes the deviant erroneous thoughts of good and evil... Once one's clouds of afflictions obstruct the light of true suchness, the mind is polluted by five sense objects and breeds desires. Three poisons, greed, aversion and ignorance, are then gathered, thereby turning the clean pure mind into a container of three poisons."

Indeed, if we hope to attain awakening through the way of tea, we should not focus on the price and aesthetics of the tea wares, but on attaining and realizing the profound, pure treasure ware, the principle entity of eighth consciousness. However, "a true clean, pure container" mentioned by Sōtan here may turn into "a container of three poisons" due to pollution; thus what he meant by "a true clean, pure

container" was actually the conscious mind. Yet, the nature of the conscious mind will certainly correspond to the mental factors such as primary afflictions, secondary afflictions, and five specific concomitants. Even if all afflictions are subdued and eradicated under great efforts, when the thoughtless pristine awareness still exists, it will still correspond to the five specific concomitant mental factors. Hence, the pure fundamental mind, namely the eighth consciousness, cannot be attained or transformed using the mental consciousness to eliminate all thoughts of good and evil, the three poisons, and abide in a mind without any distracting thoughts. This paragraph shows the fact that Sen Sōtan's chadō falls within the scope of mental consciousness but not the realm of eighth consciousness, which is the essence of enlightenment realized by Chan patriarchs.

Besides, Sen Sōtan presented his view as "A cause must lead to an outcome. Bad causes lead to bad residence while good causes contribute to good residence. One should brace up and strive with a bold omnipresent mind, diligently work on all things related to Zen of tea, so that one can exempt oneself from a lord's prison, close the doors that lead to three evil paths, ascend to heavenly world and attain enlightenment. This should not be doubted. Such an accomplishment corresponds to the oneness of heaven and earth. This precious round, clean and pure container is also the aptitude in Zen of tea."

Such an expression by Sen Sōtan is far from correct! According to Buddhist Dharma, ascending to heaven is not equivalent to attaining enlightenment, because one is still subjected to the cyclic existence of birth and death. After one has devoted all efforts to practice the Zen of tea, he should focus on attaining realization of the fundamental mind, the eighth consciousness, so as to achieve liberation by aligning oneself with the characteristics of the eighth consciousness that intrinsically transcends the three realms and six paths of rebirth.

3.5 Dew-laden ground

Sen Sōtan said, "Remove all afflictions, and manifest your fundamental nature of true suchness, which is called abhyavakasika (uncovered area). If misleading views are removed, but the thought still exists, this is not called abhyavakasika. Once the thought of three realms is gone, the mind can then be called abhyavakasika. In addition, the site of Buddhist practice center has the same meaning as abhyavakasika. It is a clean pure sphere. This can also be called a tea room, a symbol of one's own mind. Buddha says, 'A world is a non-world. One should bring forth the non-abiding Mind.'"

As Ito Kokan explained, the "dew-laden ground" here was borrowed from the Simile and Parable chapter in the Lotus Sutra [1]. Reference to the original text indicates, "Then the elder, seeing that his children had safely escaped and were all sitting in the open square." Thus the term "dew-laden ground" actually refers to "open-air ground," but the Japanese chadō has widely employed the term to denote the garden path before entering a tea room. Sōtan offered a unique explanation of this, as according to the definition in this sentence, "dew-laden ground" refers to the self-abiding state of the eighth consciousness. However, from the perspective of Mahāyāna Buddhism, the fundamental nature of true suchness has been clearly revealed at all times even before the practitioners are freed from all afflictions, and this is what the Sixth Patriarch meant by "The Buddhadharma is here in the world; Enlightenment is not apart from the world" and "one can realize bodhi without eradicating afflictions." Hence, as stated in the Diamond Sutra, "One should bring forth the non-abiding Mind." This means that the eighth consciousness is innately non-abiding in the six sense objects but is able to generate the mind at all times; thus it is not necessary to eliminate all afflictions before the fundamental nature can be revealed, and Sen Sōtan's words in this case show that he was unclear on the different characteristics between the mental consciousness and the eighth consciousness.

Summing up, we believe that even though Sen Sōtan imposed high standards on chadō, he himself was unable to live up to these standards, as he did not comprehend the differences between mental consciousness and the eighth consciousness, and he mistook the state of mental consciousness that is free from language, text, discursive thoughts, and abides in a mind without any distracting thoughts as the state of the eighth consciousness, the essence of enlightenment to Chan tradition. Therefore, even though his chadō possesses artistic conception, there is a lack of Chan because he had not gained authentic insight of reality into Chan tradition.

4 CONCLUSION

Since ancient times, the Japanese chadō has been known for the Zen within its tea culture, and as Sen Sōtan is the common ancestor of the three mainstream Sen schools in today's Japanese chadō, research on his chadō is helpful in clarifying the connotations of the so-called Zen in Japanese chadō stretching back over the past four hundred years. As the existence of Zen in chadō is related to whether the people involved in tea ceremonies are enlightened Chan masters, this paper employs the standards of attaining awakening by the Consciousness-only school and Chan tradition to assess the Zen of Japanese chadō. Through initial observations on Sen Sōtan's *The Tea and Zen are of One Flavor*, we discover that Sōtan regarded the conscious mind without language and discursive thoughts within movements as the essence of chadō. This corresponds to the state of mental consciousness but not the eighth consciousness; thus Sen Sōtan's chadō does not contain any concept of Zen, even though it is free from luxurious indulgence. Hence, if modern Japanese chadō really wishes to achieve the Zen of tea, it has to rediscover its origin and opportunities in Chinese Chan tradition.

REFERENCES

Ito Kokan Trans, 2005. *Tea and Zen*. Trans Dong Zhi, Tianjin: Baihua Literature and Art Press, 2005.

Kuwata Tadachika, 2016. 600 *Years History of Tea Ceremony*. Trans. Li Wei, Beijing: Beijing Shiyue Literature and Art Press, 2016.

Nanbô Sôkei, 2016. *The Southern Record*. Annotated by Nishiyama Matsunosuke, Volume 1: Memorandum. Tokyo: Iwanami Shoten.

J. Teng, 1992. *An Introduction on the Japanese Chadō Culture*. Beijing: Oriental Press.

R. Lin, 1991. *The Origin of Japanese Chadō: Notes on the Southern Record*. Taipei: Lu-Yu Tea Culture Institute.

Isao Kumakura, 2006. The Record of Yamanoue Soji: Some Ideas of Sado (Tea Ceremony). Tokyo: Iwanami Shoten, Publishers (Japan).

J.W. Yu and H.M. Chou, 2017. *Proceedings of International Conference on Innovation Applied System Innovation*, Paper ID 0067, 2017.

W.H. Fan and H.M. Chou, 2017. *Proceedings of International Conference on Innovation Applied System Innovation*, Paper ID 0144, 2017.

Smart Science, Design & Technology — Lam et al. (eds)

A preliminary study of the characteristics of the psychological dissection techniques applied on the minimally conscious state

Lichen Tsai*

Ph. D. Program of Mechanical and Energy Engineering, Kun Shan University, Tainan, Taiwan

ABSTRACT: The study of an extremely minimal conscious state becomes an important topic in contemporary research into the structure of human being's consciousness. This paper introduces the results and physiological characteristics derived from the various stages of practicing the advanced psychological dissection technique (APDT). During the process of practicing APDT, the sensory and perceptive functions of taste, smell, vision, hearing and touch will gradually diminish. As a result, the scope of sensory awareness of our consciousness will also be limited. This paper is of the opinion that the modern research of the minimal conscious state will bring vital inspirations if it is provided with the characteristics of the physiological and psychological aspects described in the psychological dissection techniques of the human mind.

Keywords: Minimally conscious state, Levels of consciousness, Psychological dissection technique, VS/UWS

1   INTRODUCTION

To assist with trauma diagnosis, Giacino et al. (2002) published the definition of minimally conscious state (MCS) in medical research on brain injury (Giacino et al. 2002). MCS is mainly used to distinguish between coma, vegetative state (VS) and full-consciousness in the locked-in syndrome. In Giacino's article, coma refers to a person who is unconscious with no spontaneous eye opening and unable to be awakened under strong stimulation. A person in a vegetative state is characterized by no identifiable behavioral evidence for self or environmental awareness, but is capable of spontaneous or stimulus-induced arousal, evidenced by sleep–wake cycles. A person with locked-in syndrome has preservation of cognition but is almost unable to express himself. Individuals in MCS are in a fluctuating state of consciousness. Although MCS is the result of serious neuropathology, behavioral evidence of consciousness is still discernible except that these behaviors occur inconsistently. As such, this person does not meet the criteria of vegetative state, and is therefore regarded as being in the minimally conscious state (Giacino et al. 2002).

However, it is inconclusive whether a vegetative state is definitely an unconscious state and hence distinct from MCS. The above article acknowledges that assessment of a patient's level of consciousness is based fully on observable behaviors, and factors such

as sensor damage, effector impairment or decreased motivation may lead to an underestimation of a patient's consciousness and cognitive function (Giacino et al. 2002). Furthermore, Kotchoubey et al. (2014) criticized that in the above definition of the MCS, there is semantic ambiguity as to whether consciousness is unitary (all-or-none) or non-unitary (gradual or continuous) (Kotchoubey et al. 2014). It could be thought that defining MCS as an unstable state of fluctuating consciousness alludes to the unitary nature of consciousness. Yet many clinical evidence indicate the non-unitary nature of consciousness. In addition, between coma (unconsciousness) and locked-in syndrome (relatively full consciousness), the existence of different states such as VS and MCS seem to suggest that consciousness is gradual rather than unitary, thereby rendering the definition of MCS contradictory.

In diagnosing coma, VS, MCS and locked-in syndrome, because the definition of MCS in trauma medicine is still controversial, therefore finding a new research approach is a key topic for this article.

2   LEVELS OF CONSCIOUSNESS AND PSYCHOLOGICAL DISSECTION TECHNIQUE

The trauma medicine provides an important contribution to the study of consciousness level, and the definition of full consciousness gives a broad premise for

---

*Corresponding author: lejintswa@gmail.com

the definition of MCS. A lack of understanding and consensus on what defines full consciousness would render it difficult to define the boundaries of partial consciousness, VS and MCS. This article regards full consciousness in terms of the five senses of a normal awake person, namely visual, auditory, olfactory, taste and tactile perceptions, and their underlying innate conscious functions and scope of self-cognition and action. Any loss of consciousness and function would render the function and scope of consciousness incomplete, and therefore constitute a lack of full consciousness. For example, when human vision is damaged alone (whether due to damage to the eye structure or visual area of the brain), the function and scope of full consciousness will be limited by the lack visual information input. By the same token, consciousness of locked-in syndrome would involve the severe loss of sensory "control over the body's ability to act." In other words, the consciousness is greatly limited in its tactile function and scope. In most cases, locked-in syndrome is caused by damage to the brain stem and death is extremely likely. Therefore, although cognitive function may remain complete in locked-in syndrome, consciousness is only partial rather than full. Trauma can lead to partial functional loss in the consciousness, thereby indicating that full consciousness comprises the division of labor and coordination among partial functions. Essentially, full consciousness comprises consciousness related to five senses and cognition that is not directly related to the five senses. In the above independent visual impairment, the defined function and scope represent the overall dissection of full consciousness functions (Tsai et al. 2016), and the same is true for the overall dissection of the other sensory functions.

Although trauma medicine provides an important contribution to research on consciousness level, there are certain limitations. Full consciousness functions rely on a sound neural network and physiological base. However, the location and severity of neurological and physiological damages contain elements of uncertainty. Furthermore, conducting controlled experiments on neural and physiological damage violates ethical norms, thereby limiting trauma medicine in its ability to fully determine how consciousness functions may be divided and organized. The criticism by Giacino et al. on the definition and diagnostic ambiguity of VS and MCS is caused by these limitations in trauma medicine research.

This article asserts that human psychological functions include the ability to ignore and forget, resulting in a dissection effect similar to that in trauma and therefore provide an alternative method and approach to research. For example, concentrating on a problem at hand to the extent of completely ignoring surrounding conversations is a dissection effect where auditory processing is temporarily cut off. Through attention training, the average person can achieve the ability to ignore the five senses and dissect psychological functions and scope, a skill called psychological dissection technique (PDT) (Pai

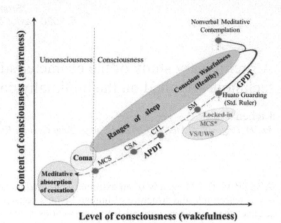

Figure 1. Psychological dissection technique chart.

et al. 2018). Figure 1 shows a PDT map where the x and y-axes represent the level of wakefulness and awareness for analyzing consciousness, respectively. The z-axis represents active willpower and the associated technique but is omitted to simplify the diagram. A complete PDT curve consists of two parts: one is the solid line, which represents general psychological dissection technique (GPDT), and the dotted line represents advanced PDT (APDT). For further explanation of the relationship among conscious wakefulness, senses and techniques, and Laureys (Laureys, 2005) research outcome, please refer to the previous article (Tsai et al. 2018).

The GPDT curve describes the use of focused self-awareness training to achieve a wakefulness that temporarily ignores the five senses and their changes. All states within the GPDT curve occur during full consciousness. In contrast, in locked-in syndrome, the loss of sensory function and scope in the body's ability to control movement should mainly pertain to the impairment of tactile perception. Therefore, given the same level of wakefulness, perceptual function and scope in a locked-in state would be lower than at any stage in the GPDT. Hence the horizontal position of the ellipse representing locked-in syndrome state should be below the GPDT curve and intersects with VS.

The MCS* is found within the intersection of the locked-in syndrome state and VS, representing the minimal conscious state, whose definition is controversial. This article believes that the correct MCS position should be on the APDT curve and intersect the vertical line that distinguishes consciousness from unconsciousness. The overall structure of the psychological dissection map shows that the ellipse covering the range between sleep and wakefulness tilts at 45 degrees. This indicates that under normal physiological functioning, full consciousness enters unconscious state during sleep mode, or returns from unconsciousness to full consciousness when leaving sleep mode, thereby manifesting a positive correlation between wakefulness and awareness. In the lower right of the diagram, the overlap between the VS and the locked-in state indicates that damage causes consciousness to go from normal full conscious perception to a swift downward displacement of ΔY on the y-axis. This ΔY value represents the loss of full consciousness control over body movement in terms of sensation and scope, and the abrupt downward displacement into a damaged state. In between normal physiological state and damaged/unhealthy state, controllable psychological functions and skills that exist in healthy physiological state lead to gradual pass from full consciousness into MCS, and finally into the state of absorption of cessation. The following will describe the physiological characteristics that are manifested in the gradual process from full consciousness to MCS.

3 PHYSIOLOGICAL CHARACTERISTICS OF PDT AND MINIMALLY CONSCIOUS STATE

Attention training is a commonly used technique. For example, the well-known mindfulness training uses attention training to manage emotion by changing the emotional associations of cerebral nerves (Williams, 2011). This attention training technique is not only helpful for emotional management, but also applicable to psychological dissection. In Table 1, Row 1 contains cognitive, visual and other perceptions, and psychological functions such as breathing

Table 1. 6-week handmade soap course.

Row	State	Psychological Function	Key Characteristics and Stage
1	Normal	Cognition, vision, hearing, smell, taste, touch, breathing and heartbeat	General psychological dissection technique (GPDT)
2	1st Dhyana	Cognition, vision, hearing, touch, breathing and heartbeat	Samatha mode (SM) and Advanced psychological dissection technique (APDT) activation
3	2nd & 3rd Dhyanas	Cognition, breathing and heartbeat	Controllable tactile loss (CTL)
4	4th Dhyana	Cognition	Controllable suspended animation (CSA)
5	4 Formless Concentrations	Minimal consciousness	MCS
6	State of absorption of cessation	None	

and heartbeat present under normal condition. In GPDT, active willpower and techniques are used to train attention to produce the dissection of the five senses while maintaining full consciousness.

Row 2 onwards shows the scope of advanced psychological dissection technique. The contents in Table 1 summarize the Buddhist concept of Four Dhyanas and Eight Samadhis. This article describes the Four Dhyanas and Eight Samadhis in terms of advanced psychological dissection techniques because they are consistent with the basic principles of anatomy. However, the process from GPDT into APDT is beyond the scope of this article and is therefore omitted. Nevertheless, humans have a psychological function that allows the brain to move from wakefulness into sleep, and enter the unconsciousness of deep sleep. Such operability can also be found in the process of activating the samatha mode (SM) during wakefulness, then moving into APDT stage, and finally into the unconscious state of absorption of cessation. The key to switching the modes is in the active willpower and techniques in the different levels of attention training.

In Row 2, psychological function in First Dhyana shows the loss of both food associated smell and taste. The APDT activated SM mode is similar to but distinct from sleep mode. Cognition and the five senses are gradually lost during sleep, and even if the five senses are ignored under GPDT, their order of loss is extremely difficult to determine. In contrast, under the APDT activated SM mode, the loss of smell and taste occurs in a fixed order and pattern. However, the loss of smell and taste is not an easily observable outward manifestation, and hence this article does not include it as an identifiable physiological characteristic.

In Row 2, psychological function in First Dhyana shows the loss of both food associated smell and taste. The APDT activated SM mode is similar to but distinct from sleep mode. Cognition and the five senses are gradually lost during sleep, and even if the five senses are ignored under GPDT, their order of loss is extremely difficult to determine. In contrast, under the APDT activated SM mode, the loss of smell and taste occurs in a fixed order and pattern. However, the loss of smell and taste is not an easily observable outward manifestation, and hence this article does not include it as an identifiable physiological characteristic.

In the Second Dhyana and Third Dhyana in Row 3, the loss of visual, hearing and tactile begins in Second Dhyana, while cognitive, breathing and heartbeat remain. Hence, this article regards controllable tactile loss (CTL) as a physiological characteristic of this stage. Visual, hearing and tactile loss means that individuals who have acquired this technique are able to refuse external light, sound or vibration stimuli. In particular, the impact of CTL is significant because in tactile loss, practicing individuals not only lose their mobility but also their sense of balance. Therefore to maintain body balance, practicing individuals usually assume a cross-legged sitting position. Otherwise, like patients with severe cerebral nerve damage who are generally bedridden, they will manifest a vegetative state with complete loss of bodily perception and movement.

In Figure 1, the horizontal position of the VS is above the CTL, indicating that sensory level in VS is much higher than in CTL. This is because in CTL, sensory loss is complete while in VS, the body maintains an immediate reflection to stimulus or exhibits startle reaction to sound and light (Giacino et al. 2002), thereby suggesting that the five senses are not completely lost in VS. Thus unresponsive wakefulness syndrome (UWS) would be a more accurate term for VS (Laureys et al. 2010). Since CTL is more typical of vegetative state, CTL could also be called controllable vegetative state (CVS). In the SM mode, CTL is indicative of functional loss from the total depression of the body's sensors and effectors, and is therefore even more typical of VS.

In the SM mode, CTL is similar to movement suppression in sleep mode. However, the key difference between SM mode and sleep mode is that in the sleep mode, movement suppression is a part of the overall mechanism of going from full conscious wakefulness to unconsciousness. In contrast, in CTL, the five senses are completely suppressed by the use of active willpower and technique. Hence under CTL, consciousness comprises purely cognitive functions that are unrelated to the five senses.

Row 4 shows that the life control center for breathing and heart beat is suppressed and the body goes into a controllable suspended animation state (CSA). In accident trauma, general suspended animation is usually manifested as unconsciousness, and with emergency treatment, consciousness might be restored. However, there are occasional cases of individuals coming back to life from suspended animation after death has been declared. In humans, basic physiological functions that are supported by breathing and heartbeat can be regarded as operating exclusively for the full consciousness. When the five senses in full consciousness are suppressed and specific levels of cognitive function and scope are further limited, the burden of basic physiological functioning is released and the body enters into CSA. Therefore, through APDT, autonomous controllable suspended animation can be achieved.

After CTL and CSA, MCS is reached in the meditative state of Four Formless Concentrations in Row 5. It actually occurs at the end of the Four Formless Concentrations, and is called Neither Perception Nor Non-Perception because further reduction of cognitive function would lead to the state of absorption of cessation which does not have mental consciousness in Row 6. Therefore, true MCS cannot occur in accident trauma, and only those trained in APDT can achieve MCS after having experienced the physiological characteristics of CTL and CSA.

4 CONCLUSION

The definition of MCS involves the basic problem of whether consciousness is unitary or non-unitary, and the structural problem of consciousness construction. Scientists use anatomical methods to study the structure of the human body; therefore psychological studies should use psychological dissection to study psychological structure. This article believes that the structure of consciousness that is revealed by the use of GPDT and APDT exhibits gradual, layered non-unitary nature. However, the effect of attention presents in unitary manner and can be used to suppress (cut off) the five senses and cognition, even to the point of MCS. Hence, consciousness can be both unitary and non-unitary.

This article believes that MCS should occur on the APDT curve, which describes a series of techniques that can be gradually superimposed to suppress sensory and cognitive functions. To attain MCS, the APDT curve must undergo various states such as SM, CTL and CSA. However, due to the superimposed effects, MCS would occur in a state where tactile perception is lost and breathing and heartbeat have stopped.

Although this article introduces APDT in terms of the Buddhist concepts of Four Dhyanas and Eight Samadhis, it is not cast in the mystery of religion. The characteristic physiological changes of each stage described by the APDT curve are similar to normal sleep mode. Hence the APDT curve is quite similar to the sleep mode, but further includes controllable suspended animation. Therefore in the study of MCS, sleep mode and suspended animation, psychological dissection technique provides an approach that is distinct from trauma medicine and other research methods, and is an alternative research approach and inspiration.

REFERENCES

J.T. Giacino, S. Ashwal, N. Childs, et al., 2002. *Neurology* 58 349–353.
B. Kotchoubey, et al., 2014. *Brain Injury 28*(9) 1156–1163.
L.C. Tsai and H.M. Chou, 2016. *A Preliminary Investigation into the Principle of Psychological Dissection Revealed in the Writings of Xuanzang*. International Conference on Innovation, Communication and Engineering, Paper No. C160249.
C.W. Pai, W.H. Fan, L.C. Tsai, 2018. *A Preliminary Study on the Psychological Classification and Rectification*. International Conference on Innovation, Communication and Engineering, Paper No. C180151.
S. Laureys, 2005. *TRENDS in Cognitive Sciences* 9(12).
L.C. Tsai, W.H. Fan, C.W. Pai, 2018. *A Preliminary Exploration of Psychological Classification and A Standard Ruler of the Conscious Mind*. International Conference on Innovation, Communication and Engineering, Paper No. C180150.
M. Williams, 2011. *Danny Penman, Mindfulness: an eight-week plan for finding peace in a Frantic World*. New York: Rodale.
S. Laureys, G.G. Celesia, et al., 2010. *Unresponsive wakefulness syndrome: a new name for the vegetative state or apallic syndrome*. BMC Medicine, 8:68.

Smart Science, Design & Technology — Lam et al. (eds)
© 2020 Taylor & Francis Group, London, ISBN 978-0-367-17867-3

Correlation analysis between renewable power generation and the seasonal climate in Taiwan

Juiwen Yu* & Lichen Tsai
Adjunct Lecturer of Mechanical Engineering and General Education Center, Kun Shan University, Tainan, Taiwan

ABSTRACT: The Taiwan government plans to increase renewable power generation capacity from now to 2025, while in 2017 the monthly generation ratio (M.G. ratio) for the solar photovoltaic generation was 0.54–1.42, and the maximum ratio was 1.42, showing that the maximum power generation occurred in the summer season from July to September. On the other hand, the M.G. ratio for wind power generation is 0.24–2.03, and the minimum ratio is 0.24, meaning that the minimum power generation also occurs in summer season from July to September. The ratio incline in the winter exhibited in the opposite direction. These results showed that renewable power electricity generation is much affected by the seasonal climate.

1 INTRODUCTION

In 2017, the capacity of solar photovoltaic and wind power generation (land area) equipment in Taiwan was 1,733 and 693 MW respectively. The Taiwan government has been actively developing renewable power generation in the past several years. The overall target is to achieve a solar photovoltaic capacity of 1.52 GW (KMW) within two years, and a cumulative capacity of 20 GW (17 GW on the ground and 3 GW on the roof) by 2025, and an accumulated wind power generation of 6.7 GW (land area 1.2 GW + offshore 5.5 GW) by 2025 (Executive Yuan of R.O.C., 2019). The output power generated by a photovoltaic module and its life span depends on many aspects. Some of these factors include the type of PV material, solar radiation intensity received, cell temperature, parasitic resistances, cloud and other shading effects, inverter efficiency, dust, module orientation, weather conditions, geographical location, and cable thickness (K. V. Vidyanandan, 2017). In wind power generation the important parameters are wind speed, turbine swept area, and air density. Selection of wind turbine should be based on the climate condition of the particular site. The power output is directly proportional to the swept area of the blades. The capacity of the wind turbine depends on the swept area. The maximum output is obtained at the maximum wind speed (C. Marimuthu et al. 2014). That is, the generation of renewable energy power is related to environmental factors such as temperature, solar radiation intensity, wind direction and velocity, etc. The subtropical climate in Taiwan, alternating the environmental factors season by season, would affect the generation of electricity from solar photovoltaic cells and wind power turbines. This paper reports and discusses the characteristics of the generation coefficient of solar photovoltaic and wind power generation for the application of renewable energy generation.

2 COLLECTION AND ANALYSIS OF GENERATION CAPACITY AND ELECTRICITY

The initial data of generation capacity and electricity generation is collected from the network website (Taiwan Power Company, 2019), which also provided the related capacity and electricity of generation from a private commercial electricity company (private company). The collected data was statistically analyzed and plotted using the Excel program to show the extreme or the incline of the electricity generation ratio. The electricity generation coefficient (E.G. coefficient) was defined as the ratio of electricity generation per year to the generation capacity (MKWh/KW.y or GWh/KW.y), which shows the electricity generation per unit generation capacity per year. The theoretical maximum E.G. coefficient of 8,760 KWh/KW.y indicated that 1 KW generation capacity could continuously operate 24 hours per day and generate electricity of 8,760 KWh within one year. The ratio of the E. G. coefficient was the actual E.G. coefficient to 8,760 KWh/KW.y. The monthly generation ratio (M.G. ratio) was defined as the ratio of the monthly electricity generation (Xi) to the average of the monthly electricity generation in the whole year, i.e., M.G. ratio $= Xi/((\sum_1^{12} Xi)/12) = Xi/\bar{x}$, which exhibited the variation amplitude of the monthly generation to the average monthly generation.

---

*Corresponding author: lejintswa@gmail.com

3 RESULTS AND DISCUSSION

In 2017, the total capacity for solar photovoltaic and wind power generation (land area) equipment in Taiwan was 1,733 and 693 MW respectively, while the power generation was 1,691 and 1,707 GWh respectively, which corresponded to an E.G. coefficient of 976 for solar photovoltaic and 2,463 KWh/KW.y for wind power generation. Both were much lower than the theoretical maximum E.G. coefficient of 8,760 KWh/KW, and were also lower than that of the actual ratio of 7,202 KWh/KW for firepower generation for Taiwan power company in 2017. The corresponding E.G. coefficient and the ratio of E.G. coefficient for Taiwan power generation in 2017 are listed in Table 1. The ratios of E.G. coefficient of 0.11 for solar photovoltaic and 0.28 for wind power were lower than the ratio of 0.82 for the Taiwan firepower generation in 2017, which indicated that the electricity generation efficiency for the renewable generation operation in Taiwan is low.

Figure 1 shows the variance of the monthly M.G. ratio for TPC solar photovoltaic generation and wind power, while Figure 2 exhibits that for private company solar photovoltaic generation and wind power generation. The monthly generation ratio (M.G. ratio) for the solar photovoltaic generation is 0.54–1.42, and the maximum ratio is 1.42, showing

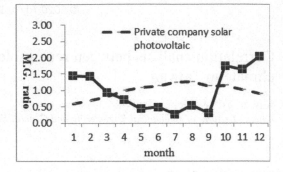

Figure 2. The variation of monthly generation ratio for private power company of Taiwan in 2017.

that the maximum power generation occurred in the summer season from July to September. On the other hand, the M.G. ratio for wind power generation is 0.24–2.03, and the minimum ratio is 0.24, meaning that the minimum power generation also occurs in the summer season from July to September. The ratio incline in the winter exhibited in the opposite direction. These results showed that the renewable power electricity generation was much affected by the seasonal climate. In Taiwan, the ambient temperature might be as high as 36–37°C and accompanied by the prevailing weaker southeast wind in the summer, which was advantageous for the solar photovoltaic generation but not the wind power generation. On the contrary, in the winter the ambient temperature was low in the range of 10–20°C and had a strong wind from the west-northern to the east-northern direction, which was disadvantageous for solar photovoltaic generation but profited wind power generation.

It is apparent that in Taiwan solar power generation and wind power generation are complementary in the same season, and are suggested to be installed at the same time. Furthermore, Taiwan is hot and needs to use more electricity for air-conditioning in summer, having less generation from solar power generation during the nights or rainy days or from wind power generation in typhoon season, and so has to rely on other power generation devices such as firepower or nuclear power to generate electricity. Therefore, it is necessary that green power generation in Taiwan should be prepared and set with sufficient firepower or nuclear power to provide sufficient demand.

On the other hand, Taiwan's total power generation capacity in 2017 was 49,953 MW, including 4,652 MW capacity of hydropower, 1,733 MW of solar energy, 693 MW of wind power, and total capacity of 42,675 MW for firepower and nuclear power generation. It indicates that the planned capacity of 26,700 MW by 2025, including capacity of 20,000 MW for solar energy and 6,700 MW for wind power generation, shall correspond to as high as 62.6% of the total capacity of 42,675 MW from

Table 1. The E.G. coefficient and the ratio of E.G. coefficient for Taiwan power generation in 2017.

Renewable power	Solar photovoltaic	Wind power	TPC Fire power
Generation capacity, MW	1,733	693	21,641
Electricity generation, GWh	1,691	1,707	155,864
E.G. coefficient	976	2,463	7,202
Ratio of E.G. coefficient	0.11	0.28	0.82

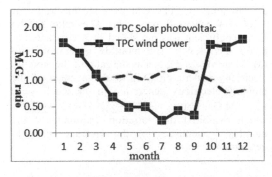

Figure 1. The variance of monthly generation ratio for Taiwan power company in 2017.

firepower and nuclear power generation, but only donate 9.6% of the potential electricity from firepower and nuclear power generation. Economic benefit assessment is suggested to be considered in order to avoid the low-efficiency investment.

Figure 3 shows the air temperature (month average from 1981–2010) variation in the southern, central, and northern regions of Taiwan. The air temperature was higher in the summer from June to September, and was lower in the other seasons or months. Figure 4 exhibits the variation of the hours of sunshine (monthly average from 1981–2010) in the southern, central, and northern regions of Taiwan. The air temperature and hours of sunshine were higher in the summer from June to September, and is lower in the other seasons or months. The air temperature and hours of sunshine in the southern region were higher than those in the central and northern regions of Taiwan. Both phenomena corresponded to the higher M.G. ratio value for solar photovoltaic generation in Figure 1 and Figure 2.

Figure 5 shows the variation of the wind velocity (monthly average from 1981–2010) in the southern, central, and northern regions of Taiwan. The wind velocity in the northern region was larger than 2.5 m/sec,

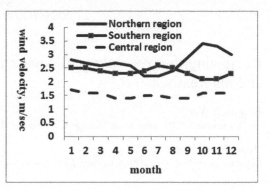

Figure 5. The variation of the wind velocity in the southern, central, and northern regions in Taiwan.

especially in the winter season, which was advantageous to the M.G. ratio for wind power generation as shown in Figures 1 and 2. On the contrary, almost all the wind velocity in the central region was lower than 2 m/sec, which was considered to be the minimum wind velocity for wind power generation.

4 SUMMARY

Taiwan is a subtropical country and has wide variation in air temperature, wind direction, and velocity in the different seasons, which apparently affects electricity generation efficiency from renewable power generation. Furthermore, no sunshine during the night and on rainy days is disadvantageous to solar photovoltaic generation, and less wind in the summer provides less electricity generation. It is suggested to assess the economic benefit for the installation distribution between renewable power generation, firepower, and nuclear power electricity generation in Taiwan.

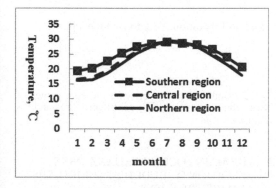

Figure 3. The air temperature variation in the southern, central, and northern regions in Taiwan.

REFERENCES

Executive Yuan of R.O.C., 2019. Forward-looking Infrastructure Design and Painting – Renewable Energy Construction, Policy and Plan, Important Policy.
K.V. Vidyanandan, 2017. An Overview of Factors Affecting the Performance of Solar PV Systems, Energy Scan, A house journal of Corporate Planning, NTPC Ltd., issue 27, pp. 2–8, New Delhi.
C. Marimuthu, and V. Kirubakaran, 2014. A Critical Review of Factors Affecting Wind Turbine and Solar Cell System Power Production, *Int. J. Adv. Engg. Res. Studies/III/II/*, pp. 143–147.
Taiwan Power Company, 2019. Information disclosure, Taiwan Power Company Website.

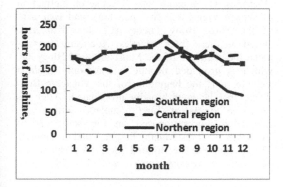

Figure 4. The variation of the hours of sunshine in the southern, central, and northern regions in Taiwan.

Smart Science, Design & Technology — Lam et al. (eds)
© 2020 Taylor & Francis Group, London, ISBN 978-0-367-17867-3

Reviewing the core values of Xuanzang's rendition of Buddhist scriptures and his treatises

Jungjung Wen*
Ph.D. Program of Mechanical and Energy Engineering, Kun Shan University, Tainan, Taiwan

Huannming Chou
Department of Mechanical Engineering, Kun Shan University, Tainan, Taiwan

ABSTRACT: The essence of the Buddha Dharma consists of two parts: the wisdom about all the dharmas on the Path to Liberation, and the wisdom encompassing the True Suchness and Buddha-nature on the Path to Buddhahood. Moreover, the wisdom of the True Suchness and Buddha-nature, further categorized into the wisdom of prajñā-emptiness and the wisdom of the knowledge-of-all-aspects, is also referred to as the Hundred Dharmas, Thousand Dharmas, or Million Dharmas of the Yogācāra Buddhism by the Consciousness-Only doctrine. Founded upon the central concept of Consciousness-Only, the Buddhist scriptures and treatises translated and authored by Chinese Dharma Master Xuanzang in the Tang Dynasty demonstrate the basis of the Buddha Dharma—the eighth consciousness. The ideology of Consciousness-Only includes both the theoretical value and the realistic value. The valid realistic value is grounded in the attaining of actual awakening to the eighth consciousness. The eighth consciousness, in essence, encompasses the entirety of the Buddha Dharma—the Path to Liberation and the Path to Buddhahood.

Keywords: Xuanzang, Consciousness-Only doctrine, Path to Liberation, Path to Buddahood, the eighth consciousness, true suchness

1 RESEARCH BACKGROUND AND MOTIVATION

Xuanzang became famous for his seventeen-year (628–645 AD) overland journey to India, seeking the Buddha Dharma and covering fifty-six countries, and returned to Chang'an in January of the year nineteen of Zhenguan, the Tang Dynasty; he brought back various Mahāyāna and Hīnayāna scriptures, treatises, and the disciplinary rules in Sanskrit texts, a total of 657 titles bound in 520 bundles. Emperor Taizong of Tang greeted him in Luoyang and accepted his wish to stay and translate Buddhist texts in Hongfu Temple, Chang'an, over a period of nineteen devoted years (645–663 AD) before his death. He completed translation of seventy-six titles of sūtra and treatises, totaling over 1,330 scrolls (A Biography of the Tripiṭaka Master of the Great Ci'en Monastery of the Great Tang Dynasty, 2014). This study reviews the core values of Xuanzang's rendition of Buddhist scriptures and his treatises, and thereupon traces his thought context to explore the relationship between the three turnings of the Dharma wheel and the doctrines of the Three-

Vehicle Bodhi, thereby verifying the core values of Xuanzang's rendition of Buddhist scriptures and his treatises.

2 THE MAIN FOCUS OF XUANZANG'S RENDITION OF BUDDHIST SCRIPTURES AND HIS TREATISES

In India, Xuanzang authored treatises in Sanskrit texts, including Treatise for Reconciling Various Doctrines (C. Huizong lun), Treatise on Refutation of Wrong Views (C. Zhi ejian lun), and Exponent of the Three Body Theory (C. San shen lun). Based on these three treatises, Xuanzang consolidated the principle of the Mahāyāna and Hīnayāna, which is grounded on the one-and-only Buddha Vehicle; it is the beginning of the Yogācāra Buddhism that he later broadly propagated. After his return to the Great Tang homeland, Xuanzang's translation works focused on the scriptures of Yogâcāra, Sarvāstivāda, Prajñā, and Vaipulya. These texts can be divided into three categories: part one of Yogâcāra, part two of Abhidharma,

---

*Corresponding author: E-mail: rrwen@ms45.hinet.net

Table 1. Xuanzang's translation works of Mahāyāna scriptures, and Vinaya's sūtras, vinaya, and treatises.

Table 1. (Cont.)

Mahāyāna scriptures	Huayen	Ārya-tathāgatānāṃ-buddhakṣetra-guṇôkta-dharma-paryāya (Sūtra Revealing the Qualities of the Infinite Buddha-Lands); C. Xian wubian fotu gongde jing, 1 scroll.
	Prajñā	Mahāprajñāpāramitāsūtra (Sūtra on the Great Perfection ofWisdom); C. Dabore boluomiduo jing, 600 scrolls. Prajñāpāramitāhṛ dayasūtra (Heart of the Perfection of Wisdom Sūtra), C. Bore boluomiduo xin jing, 1 scroll. Vajracchedikāprajñāpāramitāsūtra (Diamond-Cutter Perfection of Wisdom Sūtra), C. Jingang boluomi-duo jing, 1 scroll.
	Vaipulya	Kṣitigarbha-sūtra (Ten Cakras of Kṣiti-garbha, Mahāyāna Great Collection Sūtra); C. Dacheng daji dizang shi lun jing, 10 scrolls. Saṃdhinirmocana-sūtra (The Scripture on the Explication of Underlying Meaning); C. Jie shen mi jing, 5 scrolls. Vikalpa-pratītya-samutpāda-dharmôttara-praveśa sūtra (Sūtra on The Primacy of the Dharma Gate Distinguishing Conditioned Arising); C. Fenbie yuanqi chu sheng famen jing, 2 scrolls. Bodhisattva piṭaka-sūtra (The Bodhisattva Basket); C. Da pusa zang jing, 20 scrolls. Yogācārabhūmiśāstra (Treatise on the Stages of Yogic Practice); C. Yuqieshidi lun, 100 scrolls. Ārya-śāsana-prakaraṇa (Exposition of the Ārya Teachings); C. Xianyang shengjiao lun, 20 scrolls. Prakaraṇāryavākā (Verses on Exposition of the Ārya Teachings); C. Xianyang shengjiao lun sung, 1 scroll. Abhidharma-samuccaya (Compendium of Abhidharma); C. Dasheng Apida-moji lun, 7 scrolls. Mahāyānâbhidharma-samuccaya-vyākhyā (Exegesis on the Collection of Mahāyāna Abhidharma); C. Dasheng Apidamo za ji lun, 16 scrolls. Madhyānta-Vibhāga-Sāstra (Analysis of the Middle and the Extremes); C. Bian zhong bian lun, 3 scrolls. Madhyānta Vibhāga kārikā (Verses on Analysis of the Middle and the Extremes); C. Bian zhong bian lun song, 1 scroll. Mahāyānasaṅgraha (Summary of the Great Vehicle); C. She dasheng lun pen, 3 scrolls.

Vinaya	Mahāyāna Treatise	Mahāyāna-saṃgraha-bhāṣya or Mahā-yāna-saṃgrahôpanibandhana (Commentaries on the Summary of the Great Vehicle); C. She dasheng lun shi, 10 scrolls. Viṃśatikā-vṛtti (Twenty Verses on Vij-ñapti-mātra Treatise); C. Weishi ershi lun, 1 scroll. Triṃśikā (Thirty Verses on Vijñapti-mātra Treatise); C. Weishi sanshi lun song, 1 scroll. Karma-siddhi-prakaraṇa (Investigation Establishing [the Correct Understanding] of Karman); C. Dasheng chengye lun, 1 scroll. Pañcaskandhaka-prakaraṇa (Explanation of the Five Aggregates); C. Dasheng wuyun lun, 1 scroll. Nyāyapraveśa (Introduction to Logic); C. Yin ming ru zhengli lun, 1 scroll. Nyāyamukha (Gateway to Logic); C. Yin ming chengli men lun ben, 1 scroll. Mahāyāna śatadharma-prakāśamukha śāstra (Lucid Introduction to the One Hundred Dharmas); C. Dasheng baifa mingmen lun, 1 scroll. Explanation of the Lucid Introduction to the One Hundred Dharmas; C. Dshengg baifa mingmen lun jie, 2 scrolls. Ālambana parikṣa (Treatise on Contemplating Objective Conditions); C. Guan suoyuanyuan lun, 1 scroll. Catuḥśataka (Four Hundred [Stanzas]), C. Guang bai lun ben, 10 scrolls. Commentaries on the Four Hundred [Stanzas]; C. Dasheng guang bai lun shi lun, 10 scrolls. Karatala-ratna (The Jewel in Hand); C. Dasheng zhangzhen lun, 2 scrolls. Vaiśeṣika-daśapadârtha-śāstra (Treatise on the Ten Padârthas); C. Sheng zong shi ju yi lun, 1 scroll.
	Bodhisat-tva precept	Bodhisattva-śīla sūtra (On Conferring Bodhisattva Vinaya); C. Pusa jie ben, 1 scroll. Elaboration of On Conferring Bodhisattva Vinaya; C. Pusa jiejiemo wen, 1 scroll.

and part three of Mahāprajñā (Tables 1, 2, and 3). Reflecting his attitude of "not forsaking any bit of time" at work, Xuanzang's scriptural translation can be divided into three time periods:

(1) Zhenguan nineteen to twenty-three (645–649 AD), Xuanzang mainly worked on Yogâ-cāra sūtra and treatises, including 100 scrolls of Yogācārabhūmiśāstra (C. Yuqieshidi lun), five scrolls

(continued)

Table 2. Xuanzang's translation works of Hīnayāna scriptures.

Hīnayāna scriptures	Hīnayāna sūtra	Itivṛtaha sūtra (Sūtra on Original Occurrence); C. Ben shi jing, 7 scrolls.
	Hīnayāna treatises	Abhidharma-saṅgīti-paryāya pāda śāstra (Treatise on Pronouncements); C. A pidamo jiyimen zu lun, 20 scrolls. Abhidharma-dharmaskandha pāda śāstra (Treatise on Aggregation of Factors); C. A pidamo fa yun zu lun, 12 scrolls. Abhidharma Jñānaprasthāna śāstra (Foundations of Knowledge); C. A pidamo fazhi lun, 20 scrolls. Abhidharma Mahāvibhāṣa (Treatise of the Great Commentary on the Abhidharma); C. A pidamo dapipo sha lun, 200 scrolls. Abhidharmakośa-bhāṣya (A Treasury of Abhidharma, with Commentary); C. A pidamo jushe lun, 30 scrolls. Abhidharma-Nyāyānusāra śāstra (Conformity with Correct Principle); C. A pidamo shun zhengli lun, 80 scrolls. Abhidharma-samayapradīpika or Abhidharmakośa-śāstra-kārikā-vibhāṣya (Exposition of Accepted Doctrine); C. A pidamo zang Xianzong lun, 40 scrolls. Abhidharma-vijñāna-kāya-pāda-śāstra (Collection Consciousness); C. A pidamo shishen zu lun, 16 scrolls. Abhidharma Dhātu-kāya pāda śāstra (Treatise on Collection of Elements); C. A pidamo jieshen zu lun, 3 scrolls. Abhidharma-prakaraṇa-pāda śāstra (Treatise of Exposition); C. A pidamo jieshen zu lun, 18 scrolls.

Table 3. Treatises authored by Xuanzang.

Treatise on Refutation of Wrong Views; C. Zhi ejian lun, 1,600 verses, (found in Sanskrit treatise)
Exponent of the Three-Body Theory; C. San shen lun (found in Sanskrit treatise)
Treatise for Reconciling Various Doctrines; C. Huizong lun, 3,000 verses, (found in Sanskrit treatise)
Vijñapti-mātra-siddhi śāstra (Treatise on the Demonstration of Consciousness-Only); C. Cheng weishi lun, 10 scrolls.
Verses Delineating the Eight Consciousnesses; C. Bashi guiju song

of Saṃdhinirmocana-sūtra (C. Jie shen mi jing), twenty scrolls of Ārya-śāsana-prakaraṇa (C. Xianyang shengjiao lun), and one scroll of Prakaraṇāryavākā (C. Xianyang shengjiao lun sung). They are all important basis discourses for the Yogâcāra theory.

(2) Yonghui year one to Xianqing year four (650–659 AD), Xuanzang mainly worked on the Hīnayāna treatises, including thirty scrolls of Abhidharmakośa-bhāṣya (C. A pidamo jushe lun) and 200 scrolls of Abhidharma Mahāvibhāṣā Śāstra (C. A pidamo dapipo sha lun). At the same time he carried on textual research on Trimśikā vijñaptimātratā (C. Sanshi weishi lun song), clarified the explanatory notes of the other nine famed commentators, and authored the work of Vijñapti-mātra-siddhi śāstra (C. Cheng weishi lun), ten scrolls; (3) Xian Qing year five to Linde year one (660–664 AD), Xuanzang essentially worked on the 600 scrolls of Mahāprajñā-pāramitāsūtra as well as three scrolls of Madhyānta-Vibhāga-Sāstra (C. Bian zhong bian lun) and one scroll of Samayabhedo Paracanacaka Sastra (C. Yibuzong lun lun), among others (Pan, 1999). From the aforementioned, Xuanzang's intention and rendition of his scriptural translation works, which are grounded on the principle of Yogācāra and Consciousness-Only and consolidated the Three-Vehicle Bodhi as the entirety of Buddhist teachings, are distinctly discernible.

3 THE THREE TURNINGS OF THE DHARMA WHEEL ENCOMPASS THE ESSENCE OF THE THREE-VEHICLE BODHI

3.1 The first turning of the Dharma wheel – A brief introduction to the Āgamas

During the first turning of the Dharma wheel, Buddha Sakyamuni mainly emphasised tenets of the Two Lesser Vehicles, the Path to Liberation, expounding on how to put an end to segmented existence (C. fenduan shengsi). These were later compiled as the teachings of the Four Āgamas. The instructions do not touch on the actual realization of the ultimate reality of the dharma realm. As a result, the Path to Liberation of the Two Lesser Vehicles has no direct equivalence to the Buddha Bodhi that leads to Buddhahood. The primary practice method for the Sound-Hearer Vehicle [Path to Liberation] focuses on the phenomena of the five aggregates [skandha], twelve sense-fields [āyatana], and eighteen elements [dhātu]. Moreover, the Buddha taught in the lesser vehicles that the eighteen elements within the three realms are all illusory – the "characteristics of emptiness" and not the "nature of emptiness" – for all dharmas of the eighteen elements are impermanent and destructible, not eternal and everlasting. Therefore they are said to be void and empty due to their impermanent nature. Hence all conditioned phenomena are impermanent and unavoidably lead to suffering, emptiness, and selflessness; nevertheless, among the eighteen elements, the seventh consciousness [manas] is the only dharma that can continue on to future lives while the rest of the seventeen elements will perish since they can exist for only one single lifetime. Yet even manas is destroyable. A practitioner of the Sound-Hearer Vehicle fully observes the five aggregates, twelve sense-fields, and eighteen elements during walking, standing, sitting, or lying down to contemplate that all dharmas are impermanent and empty, and realize the impermanent nature of the physical body being void, no self, and no belongings of self. At the same time, the practitioner also truly understands the fact that his seeing, hearing, cognition, and knowing mind are also impermanent, empty, changeable dharmas; thereby he is able to "eliminate the view of self" and further eradicate self-attachment to attain arhathood. The complete cultivation of the Sound-Hearer Vehicle is to attain arhathood after abandoning this current karmic body; thereby his cycle of rebirth ceases and he will not incur any future existence. Consequently, there will be no more five aggregates, twelve sense-fields, and eighteen elements appearing within the three realms (Madhyama Āgama, 2014).

To attain the fruition of liberation of Solitary-Realizer Vehicles (S. pratyekabuddha), a practitioner needs to complete the cultivation on the method of "dependent origination without a fundamental cause" through direct daily observation. In other words, he must practice the dharma door of the twelve links of dependent origination, in the following sequence: ignorance as conditions, volitional action [come to be]; with volitional action as condition, consciousness [come to be]; consciousness as condition name-and-form [come to be]; name-and-form as condition, six sense bases [come to be]; six sense bases as condition, contact [come to be]; contact as condition, sensations [come to be]; sensations as condition, craving [come to be]; craving as condition, clinging [come to be]; clinging as condition, process of becoming [come to be]; process of becoming as condition birth [come to be]; birth as condition, aging, sickness, and death [come to be]. If a practitioner can directly observe the aforesaid sequence and eradicate ignorance and cease mentation from appearing again, he will not incur any future existence; as a result, there will be no more future name-and-form, nor physical body or six consciousnesses. That is, this ceases and therefore that ceases. The same goes for birth, aging, illness, and death that will all end, including sorrow, lamentation, grief, and despair, as all phenomena arise in dependence on other constituents are eradicated; this is called the dharma of "dependent origination without a fundamental cause." All sentient beings undergo incessant transmigration due to entanglement in these twelve links of dependent origination. When a practitioner realizes the true principle with a direct observation that "this exists and therefore that exists; this ceases and therefore that ceases," that is, the rule of dependent origination without a fundamental cause, he will then enter nirvana after abandoning this current karmic body and no longer incur any future existence. That is the way to achieve the cultivation Path to Liberation for the solitary realizer (Saṃyukta Āgama, 2014a; Ekottara Āgama, 2014).

3.2 The second turning of the Dharma wheel – A brief introduction to Prajñā

The major sūtras of the second turning of the Dharma wheel include Mahāprajñāpāramitāsūtra (C. Dabore jing), Aṣṭasāhasrikāprajñāpāramitā (C. Xiaopin bore jing), Vajracchedikāprajñāpāramitāsūtra (C. Jingang boluomiduo jing), and Prajñāpāramitāhṛdayasūtra (C. Xin Jing). These are sūtras for those who have realized the "true-suchness" to further acquire the knowledge-of-specific-aspects. As stated in the Mahāprajñāpāramitā-sūtra: "The true-suchness of all dharmas realized within the profound prajñāpāramitā is not unreal, not changing, extremely profound, and hard to see and perceive" (Mahāprajñāpāramitā Sūtra, 2014). Therefore, The Mahāyāna Path to Buddhahood is firstly realized by attaining direct perception of the origin entity of all beings – the eighth consciousness, i.e., true-suchness, the tathāgatagarbha, ultimate reality, the ālaya consciousness (Saṃyukta Āgama, 2014b; Cheng weishi lun, 2014). Thus, it is called the Path to Buddhahood – the wisdom of the dharmadhātu that is generated by attaining direct realization of the ultimate reality of the nature of the dharma realm entity. "True suchness" represents that the nature of ālaya consciousness [store vijñāna] is real and immovable. "True" represents its everlasting, permanent, and indestructible nature, and is a real existing mind entity; "suchness" represents its perpetually immovable pure nature that is eternally detached from all sensations generated by either the seven consciousness minds self or self-belongings of the six sense-objects. Having realized the true suchness, a bodhisattva's wisdom of emptiness prajñā would be brought forth by directly observing both the illusory nature of the aggregates, sense-fields, and elements, as well as observing the essential nature of sentient beings' mind entity. The wisdom pertaining to prajñā that the Path to Buddhahood needs to cultivate consists of three parts: first, the knowledge-of-general-aspect; second, the knowledge-of-specific-aspects; and third, the knowledge-of-all-aspects. The knowledge-of-general-aspect refers to the overall prajñā knowledge, i.e., the fundamental wisdom of non-discrimination, the wisdom of the essential nature of the dharma-realm, that is, having acquired the true suchness, the bodhisattva is able to directly observe, experience, and understand its general aspect. Based on this knowledge, a practitioner further acquires the knowledge-of-specific-aspects pertaining to prajñā, referring to the specific knowledge stated in all the prajñā scriptures, also known as the intrinsic nature of true suchness, including its functions, characteristics, the mundane dharma, supramundane dharma, etc. These are all known in Buddhist terms as the subsequently acquired non-discriminating wisdom. The mastery of the knowledge-of-specific-aspects enables a practitioner to acquire the cultivation of higher wisdom concerning the knowledge-of-all-aspects, provided that the practitioner has truly experienced and realized the prajñā pertaining to the knowledge-of-specific-aspects. The perfect realization of the knowledge-of-all-aspects entails the achievement of the ultimate Buddhahood.

3.3 The third turning of the Dharma wheel – A brief introduction to Yogācāra

The Buddha Bodhi is the third part of cultivation that falls within the scope of the knowledge-of-all-aspects pertaining to prajñā and is the core wisdom of attaining the ultimate Buddhahood; therefore, a practitioner must further practice the expanded teachings of these Consciousness-Only scriptures. Of all the sūtras, the most outstanding are the Śrīmālādevī Siṃhanāda Sūtra (C. Sheng man jing), the Laṅkāvatāra Sūtra (C. Leng qie jing), and the Saṃdhinirmocana-sūtra (C. Jie shen mi jing). They all elaborate on the distinct functional potentialities of the seeds stored in the ālaya-consciousness, on the Mahāyāna unsurpassed definitive and ultimate meanings, and on the Buddha Dharmas that enable a bodhisattva to eventually attain Buddhahood by following those cultivation steps and stages. As the Yogâcāra viewpoint states that "the three realms are mind only and all phenomena are mere consciousness," it clearly demonstrates the true reality of life and the essential nature of all phenomena. Since a bodhisattva has realized the intrinsic nature of the dharma realm and verified that the ālaya-consciousness possesses both the emptiness nature of nirvana and the existing nature of saṃsāra, he understands that the ālaya-consciousness is able to generate the seven consciousness minds, the fifty-one mental concomitants, and all the phenomena, causing sentient beings to undergo cyclic existence. Based on the wisdom of the dharmadhātu that derived from realizing the intrinsic nature of the dharma realm, a bodhisattva further gradually fosters the knowledge-of-all-aspects through the eight consciousnesses. A bodhisattva is able to extinguish the transformational cyclic existence upon fully completing the cultivation of the knowledge-of-all-aspects. Thereafter, the tathāgatagarbha no longer undergoes any fostering and is renamed "stainless

consciousness [amalavijñāna]"; this denotes the achievement of Buddhahood.

4 DISCUSSION AND CONCLUSION

The Buddhist sūtras of the three turnings of the Dharma wheel expound the Three-Vehicle Bodhi, namely the Sound-Hearer Bodhi, the Solitary-Realizer Bodhi, and the Buddha Bodhi. The foundational practice of these three vehicles consists of realizing and verifying the characteristics of emptiness, impermanence, sufferings, and selflessness of the aggregates, sense-fields, and elements. Based on this realization, the practitioner of the Sound-Hearer Bodhi is able to extinguish self-attachment through confronting external situations and attains the fruition of liberation – arhathood. The practitioner of the Solitary-Realizer Bodhi further observes the method of dependent origination without a fundamental cause, mostly focusing on the twelve links of dependent origination to attain the fruition of a pratyekabuddha. In both of these cultivation paths, practitioners extinguish their eighteen elements after abandoning their current karmic bodies and will not incur any future existence. In the Path to Buddahood, a bodhisattva must, based on the direct realization of the intrinsic entity of sentient beings – the tathāgatagarbha – gradually learn the following: the knowledge-of-general-aspect and the knowledge-of-specific-aspects, both pertaining to prajñā, in the three stages of worthiness, followed by the knowledge-of-all-aspects on the First Ground and above. These contents consist of the knowledge-of-all-aspects attributed to the Consciousness-Only Hundred, Thousand, and Thousands Dharmas of the eight consciousnesses. Upon fully completing the cultivation of the knowledge-of-all-aspects, a bodhisattva is said to have attained the ultimate Buddhahood. The scriptures of the three turnings of the Dharma wheel just elaborate on the various aspects of the Three-Vehicle Bodhi and of the profound, shallow, extensive, and brief doctrines. The Buddha Dharma is not complete without the entirety of the sūtras taught during the three turnings of the Dharma wheel. The cultivation and realizations based on the Buddha Dharma lead to the achievement of Buddhahood.

Xuanzang established the Yogācāra principle of "True Consciousness-Only Realisation" to demonstrate the "True Tathāgatagarbha Realisation and True Suchness Realisation" and to testify to the value associated with both its theory and its true realization. He also established the contemplation of the Middle Way from the perspective of the "Undeniable Consciousness-Only" and the "False Imagination Mere Consciousness System" synergistic functioning, thereby verifying the principle of "the three realms are mind only and all phenomena are mere consciousness" (Zong Jing Lu, 2014; Xiao, 2010; Yu and Chou, 2017). Ultimately, Xuanzang held the eighth consciousness as the core principle of his work all through his life, and promoted the Three-Vehicle Bodhi encompassing the entirety of the Buddha Dharma.

REFERENCES

A Biography of the Tripiṭaka Master of the Great Ci'en Monastery of the Great Tang Dynasty, 2014. *CBETA*, T50, no. 2053, p. 221, b17–p. 280, a5.

Cheng weishi lun, 2014. Vol. 3, *CBETA*, T31, no. 1585, p. 13, c7–8.

Ekottara Āgama, 2014. Vol. 30, *CBETA*, T02, no. 125, p. 713, c15–p. 714, a10.

G.M. Pan, 1999. *Encyclopedia of Chinese Buddhist School Book Publishing*. Buddhist Light Cultural Enterprise Company Ltd. 246–247.

J. Yu and H. Chou, 2017. *Proceedings of the 2017 International Conference on Applied System Innovation*, Paper ID: 0067, Sapporo, Japan.

Madhyama Āgama, 2014. Vol. 47, CBETA, T01, no. 26, p. 723, b16–c2.

Mahāprajñāpāramitā Sūtra, 2014. Vol. 306, *CBETA*, T06, no. 220, p. 558, b18–20.

P.S. Xiao, 2010. *The Undeniable Existence of the Tathāgatagarbha*. True Wisdom Publishing Co., Taipei.

Saṃyukta Āgama, 2014a. Vol. 13, *CBETA*, T02, no. 99, p. 92, c16–25.

Saṃyukta Āgama, 2014b. Vol. 47, *CBETA*, T02, no. 99, p. 317, c17–19.

Zong Jing Lu, 2014. Vol. 51, *CBETA*, T48, no. 2016, p. 717, c28–p. 718, a3.

Advanced material science & engineering

Smart Science, Design & Technology — Lam et al. (eds)
© 2020 Taylor & Francis Group, London, ISBN 978-0-367-17867-3

Study on the application of bodiless lacquerware in Buddha statues

Yuren Cai*
Hongyuan Bodiless Statue Art Factory of Xianyou County, Putian City, Fujian Province, PR China

ABSTRACT: The bodiless lacquerware of ancient Chinese Buddha statues has a long history. The famous sculptor Dai Kui in the Eastern Jin dynasty was the first in the written record to create a bodiless lacquerware of Buddha statue. The art was almost lost after the Song, Yuan, and Ming dynasties. It was not until the reign of Jiajing and Daoguang in the Qing dynasty that this craftsmanship was restored by a Fujian lacquer craftsman. This paper focuses, from the perspective of the technique and material, on the application of bodiless lacquerware in the Buddha statues, the durability and gorgeous artistic effect of the materials and craft in the making process of Buddha statues of bodiless lacquerware, as well as the special expressive force different from other materials, which has far-reaching significance.

Keywords: bodiless lacquerware, lacquerware materials, lacquer craftsmanship

1 INTRODUCTION

Lacquer, a natural white juice from lacquer trees, is a treasure given to human beings by nature. And lacquer art is created by using lacquer as material, which is the earliest understanding of beauty in China. It gives rich colors to utensils. Starting from a vermilion lacquer wooden bowl of Hemudu Clan in the Neolithic age, it has evolved in its long history of seven thousand years. Together with tea, silk, and porcelain, it depicts the most beautiful symbols in China.

2 DEVELOPMENT OF LACQUER CRAFTSMANSHIP IN ANCIENT CHINA

Bodiless lacquerware has a long history in China. Some of them are common, such as raw lacquer, bamboo, wood, rattan, paper, and other kinds of bodiless wares. Next we will mainly discuss raw lacquer bodiless lacquerware. People who can be traced according to ancient allusions such as Dai Kui, a famous sculptor in the Eastern Jin Dynasty, initiated the Ramee-bodiless volume was huge yet its weight was light. Unfortunately, there was no physical lacquerware passed down. The early dry lacquer Ramee-bodiless statues are one of the few remaining existing visible lacquer ones, which are relatively rare in the world and extremely precious. Famous items include the Tang Dynasty's dry lacquer Ramee-bodiless

lacquerware of a Buddha statue in the early 7th century, which has now become the treasure of this museum. In recent years, early bodiless lacquerware statues have also been appearing in international auctions. Even the broken statues have been sold at a sky-high price, which shows the importance of the artistic style of sculpture created by China and the high recognition of it by the whole world. What remains a shame is that after the Song, Yuan, and Ming dynasties, the craftsmanship was almost lost. It was not until the reign of Jiajing and Daoguang in the Qing dynasty that Shen Shao'an, a Fujian lacquer craftsman, was able to restore this craft. However, it was limited to small-scale works. In addition, the national situation in the late Qing dynasty was rather depressed, the national goods were of bad quality, and the local style of modeling was obvious, which could not reflect the extraordinary atmosphere of the flourishing times of the ancient China. First of all, with the learning -of traditional lacquer art, expensive materials, and complex processing procedures, the cost will be inevitably raised, although the product is beautiful. But it's never easy to achieve low prices, which is difficult for consumers to accept. Therefore, innovation is particularly important, which makes it a novel way to explore lacquer technology, new forms of wood, bamboo, metal, ceramics, sculpture, and the perfect integration of comprehensive technology and lacquer technology. The partial coating, on one hand, allows the lacquerware to get a new form of development, and on the other hand, helps it retain the

---

*Corresponding author. E-mail: 214637389@qq.com

properties of the raw materials so as to emphasize the contrast effect, including the contrasts between coarse and fine, black and white, and the complex and the simple, reaching the perfect fit of the overall effect that highlights the emotions expressed by the its maker, which is the new form of integrated process we want (Zhang, 2011; Zhang, 2015; Yang, 2011).

3 MODERN DEVELOPMENT OF LACQUER ART

Lacquer art includes lacquer painting, lacquer ware, and lacquer thread sculpture. Fuzhou's bodiless lacquerware has a strong local character and unique ethnic style. It is also known as the "three treasures of Chinese traditional crafts" along with Beijing's Cloisonne and Jiangxi Jingdezhen Town porcelain. With the change and development of history, people are constantly pursuing the spiritual function above the aesthetic sense, which leads to the weakening and sublimation of the practical functions. In modern society, new materials of convenience and low cost have been invented and applied, which has greatly changed the existential value and usage of lacquer art. Like other handicrafts, most contemporary lacquer art has been separated from the category of general household necessities, turning to the category of handicraft products that are more aesthetically pleasing. The decorative patterns on the lacquerware turned from the body to the planet, and the elements of fashion gradually replaced the traditional ornamentation. In terms of styling, it mostly moved from traditional styles to pure abstract ones, and from traditional styles to contemporary ones.

In the cultural development of Chinese lacquerware, Xianyou lacquerware is a product of folk traditional wisdom with the ancient people's cultural lacquer art used in daily life. The artistic charm of lacquer art has been passed down, which now has become modern art. Apart from being ornamentation, it is also a manifestation of folk artists' own family lifestyle, thinking and creation, emotions, aesthetic styling, and multi-faceted artistic hobbies.

4 LACQUER ART HISTORY IN XIANYOU COUNTY

Fujian lacquerware is divided into Fuzhou bodiless lacquerware and Xianyou practical lacquerware. It has a long history and can be traced back to folk lacquer wares for household usage in the Song Dynasty. Wood, bamboo, and metal are the main materials for making furniture. There is also a traditional history of mutual integration of lacquer art, wood, bamboo, and metal. In wood, bamboo, and metal furniture, the lacquer art and the decorative pattern can be partially made, and the carved black lacquer, carved red lacquer, and other techniques can be used to add to the presentation; or part of the wood grain effect can be retained to obtain artistic innovation so as to become Chinese lacquer

Figure 1. A statue of lacquerware (Maitreya Buddha).

Figure 2. Bodiless raw lacquerware of Maitreya Buddha.

furniture. This puts the emphasis on aesthetics and practicality so that the lacquer furniture has a good effect of combination, which not only opens up the market of furniture, but also improves the cultural value of wood, bamboo, and metal furniture. Classical furniture and lacquer art, lacquerware modeled after antiques, and lacquerware from other places all have their own unique styles and features, and when they are combined they are the "national treasures" integrated with sensuous art, visual art, practical art, hand-feeling art, and characteristics of durability.

Raw lacquer, commonly known as Chinese lacquer, is the main raw material for making bodiless lacquerware. It is extracted from the secretion of the lacquer tree.

Bodiless tire is a kind of tire made of clay, gypsum, wood mould, and so on. Then it is wrapped by layers of grass cloth (linen) or silk cloth and raw lacquer. After drying in the shade, the original tire is broken or taken off, leaving the shape of the ware and lacquer cloth. Then it would be dusted, polished with ground lacquer, and at last decorated with various patterns, hence becoming a bright and gorgeous bodiless lacquerware.

5 THE MAKING OF LACQUERWARE

The main materials are raw lacquer, linen, tile ash, gauze. The production process is as follows:

Step 1. For products less than 2 meters, raw lacquer and linen overlap with each other repeatedly seven times, that is, seven layers. For products more than 2 meters, raw lacquer and linen overlap with each other 14 times, that is, 14 layers. For products over 10 meters, increase layers according to the size.

Step 2. In the fabrication process of bodiless lacquerware, the internal material is the "bone," that is, the "bone structure," using ordinary aplotaxis auriculata as the structure to support, which is not easy to deform. This operation is the most critical part of the bodiless making process.

Step 3. To scrape lacquer ash, first, coarse lacquer ash should be thin and dense. Second, medium lacquer ash should be thick and even. Third, make the edges and corners in such a way as to smooth out the defect (medium lacquer ash is preferred). Fourth, fine lacquer ash, neutral, not too thick nor too thin, should be well-proportioned. And fifth, use fine lacquer ash for edge boundaries, following the order of coarse, medium, and fine. It is a good idea to grind a little once to several times after each step when the lacquer is dried, and then cover it again.

Step 4. After polishing, the surface is then made by chemical synthetic lacquer, and gold lacquer and gold foils are applied after there's particular matter.

Step 5. After the gold pasting is finished, make a protective lacquer to keep off the ash. I take Maitreya Buddha as the main creative inspiration and artistic effect, highlighting with the thickness of the lines the beauty of the lines of Maitreya Buddha statues. The unique features of raw lacquer and the relief-like shape give a more natural texture and aesthetic feeling and present a unique sense of texture and weight in aesthetics, together with extra elegance – nobility on account of the gold foils – presenting an aesthetic value. And in the process of creating this work, I felt deeply the artistic charm of bodiless raw lacquer, of applying lacquer thread to Buddha statues for innovation. Every day I experienced a new feeling, and everything was so interesting! After careful carving and the gold-pasting process, the final gilded body could be fully presented, which allowed me to feel every detail of the work that could withstand deliberation. The deepest sensation that this work presents to people is "exquisite" and "delicate." As the creator, one must have patience and endurance, which also enabled me to deeply realize the importance of patience and carefulness because good work can give people deep sensation and inspiration by relying not simply on materials and techniques, but more on the carefulness and dedication of the creator. Looking at these works, I am filled with a sense of achievement and pride, because every work is put together with painstaking efforts! I feel extremely happy to be able to do what I like in this life! Bodiless Buddha statues have been developing in innovation and creation, and there are countless possibilities in the future.

6 ADVANTAGES OF RAW LACQUER BUDDHA STATUES

Bodiless Buddha statues have raw lacquer, wooden, and stone wares. Compared with the easy-to-crack feature of wooden wares, and the too-heavy-to-move feature of stone wares, the greatest advantages of raw lacquer are: bright and beautiful appearance, not easily affected by water immersion, no deformation, no color fading, hard, and high temperature, acid, alkali, and corrosion resistance. And the greatest feature of the raw lacquer bodiless Buddha statues is their light weight.

7 CONCLUSION

Better development of the Buddha statue bodiless lacquerware lies in innovation on the basis of inheriting the tradition. For those who make lacquerware, days pass by very quickly, and they are as calm as a rock, watching the birth of a good work by their own hands. It's like the feeling of nirvana and very fulfilling. Perhaps this is the happiest moment of creation, leaving a bright mark and beautiful youth in this vibrant time. Perhaps this is the unique charm of bodiless lacquerware! It will write a new chapter in the hands of lacquer sculpture artists, and proceed with better development in innovation.

REFERENCES

L. Zhang, 2011. *Journal of Gugong Studies* 7(1), 156–168.
L. Zhang, 2015. *Journal of Gugong Studies* 14(2), 83–96.
Y. Yang, 2011. *Palace Museum Journal 156*(4), 107–125.

A feasibility study for titanium alloy machining adopted by electrochemical abrasive jet technology

Feng-Che Tsai*, Hsi-Chuan Huang, Ju-Chun Yeh & Wei-Feng Tsai
Department of Creative Product Design, Ling Tung University, Taiwan

Hsi-Chuan Huang
Department of Automation Engineering & Institute of Mechatronoptic Systems, ChienKuo Technology University, Taiwan

Ju-Chun Yeh & Wei-Feng Tsai
Department of Industrial Education and Technology, National Changhua University of Education, Taiwan

ABSTRACT: This study mainly proposes a set of electrochemical abrasive jet machining, ECAJM, to explore the feasibility and mechanism of Ti-6Al-4V workpiece processing. Through the experimental results of the electro-chemical abrasive jet machining system, it was found that the higher the machining current of the ECAJM, without adding SiC abrasive particles, the machining effect of the workpiece is still obvious, but the electrolysis reaction is affected by the gap between the two poles and the TiO_2 film. The effect is that the machining current is prone to instability, which in turn reduces the machining effect and the machining profile. When 0.05 wt% #1000 SiC abrasive particles are added, they can effectively assist the TiO_2 generated by the Ti6Al-4V workpiece's surface. The film can assist the electrolyte to effectively remove the separated metal ions, and at the same time, the kinetic energy of the SiC abrasive particles can be cut by different jet pressures, the material removal of the titanium alloy workpiece can be effectively continued, and a higher jet pressure can be obtained. When the machining gap is 0.4 mm, better machining efficiency can be obtained. It is proved that the high-pressure gas drives the abrasive particles, which can make the abrasive particles generate kinetic energy and then hit the characteristics of the work-piece's surface. Oxide film removal promoter generated the anode surface, effectively facilitate electrolytic reaction is continued, and thus enhance the electrochemical machining efficiency.

1 INTRODUCTION

Poly (3,4-thylenedioxythiophene): poly(styrenesulfo-nate) (PEDOT: PSS) is a conductive polymer blend, which has been widely used in the integration of electronic devices such as organic solar cells, actuators, capacitors, organic light-emitting devices and sensors (Y. Wang et al. 2009). One of the fundamental requirements for operation of all organic optoelectronic devices is a stable anode interface. Indium tin oxide (ITO) has been widely used for organic optoelectronic devices due to its high optical transparency and high conductivity. However, the work function of ITO is low. The most common way to use it is to put a buffer layer on top of the ITO surface. The buffer layer can export the carriers more efficiently.

Among buffer layers, PEDOT: PSS is superior to other materials. It is well known that the conductivity of PEDOT: PSS film can be increased by the addition of polyols, such as sorbitol, DMSO or glycerol (S. K. M. Jönsson et al. 2003, Hu Yan et al. 2009, J. Y. Kim et al. 2002, Sungeun Park et al. 2011). In this work, the mechanism of enhancement in conductivity for the PEDOT: PSS films by adding different molar dilute concentrations of H_2SO_4 was further studied. These experimental results provided further evidence for our proposed mechanism.

2 EXPERIMENTAL

The SIGMA D-Sorbitol (98%) doped Clevios PH 500 PEDOT: PSS was used as solution for preparation of thin films by a spin-coating method. The Sorbitol was added to the PEDOT: PSS directly, and then the doped PEDOT: PSS was stirred for 30 min at room temperature. The mixed solution was doped again by adding different molar concentrations of H_2SO_4. The mixed PEDOT: PSS solution is the so-called double-doped PEDOT: PSS solution. The double-doped PEDOT: PSS solution was coated by spinner on $2 \times 2 \ cm^2$ glass substrates and formed the double-doped PEDOT: PSS film. The glass substrates were precleaned with acetone, methanol, and deionized (DI) water in an ultrasonic

* Corresponding author: tfc@teamail.ltu.edu.tw

bath, sequentially. The spin-coating was performed at a rotation rate of 3500 rpm for 20 sec. The double-doped PEDOT: PSS film was heated at 150°C for 20 min on a hotplate in ambient lab conditions.

The sheet resistance was measured with a four-point sheet resistivity meter (SRM103). The transmittance and surface roughness of the thin films were characterized using the spectrophotometer (HITACHI U-3900) and atomic force microscope (AFM) (XE-70).

3 RESULTS AND DISCUSSION

Figure 1 shows that the sheet resistance and surface roughness of the doped PEDOT: PSS film created through adding 1.5M H_2SO_4 were measured by a four-point probe and AFM, respectively. It is clear that the sheet resistance of the doped PEDOT: PSS film is affected by adding the various ratios of H_2SO_4. The sheet resistance of the doped PEDOT: PSS film is enhanced with increase in the ratio of H_2SO_4, but it drops after the maximum sheet resistance. The sheet resistance value is reduced from 604 to 228 $\Omega/\square$. The surface roughness of the double-doped PEDOT: PSS film is almost unchanged.

Figure 2 shows the sheet resistance and surface roughness of the doped PEDOT: PSS film through adding 1M H_2SO_4. The sheet resistance value is reduced from 604 to 255 $\Omega/\square$. The surface roughness of the 1M H_2SO_4-doped PEDOT: PSS film was smoother as compared with the 1.5M H_2SO_4-doped PEDOT: PSS film. Although the sheet resistance and surface roughness were improved slightly, we can't clearly find the tendency from the 1M and 1.5M H_2SO_4-doped PEDOT: PSS films. This is due to the fact that the reaction rate of high concentrations of H_2SO_4 is too fast. Thus, we reduce the molar concentration of H_2SO_4 again.

The reaction rate of H_2SO_4 was slowed down when the molar concentration of H_2SO_4 was 0.5M. The results are shown in Figure 3. The sheet

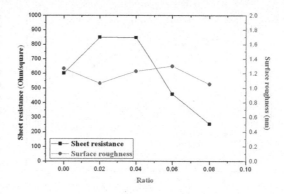

Figure 2. The sheet resistance and surface roughness of doped PEDOT: PSS with different weight ratios of 1M H_2SO_4 to PEDOT: PSS.

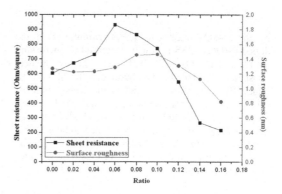

Figure 3. The sheet resistance and surface roughness of doped PEDOT: PSS with different weight ratios of 0.5M H_2SO_4 to PEDOT: PSS.

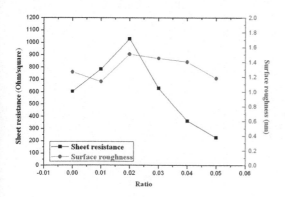

Figure 1. The sheet resistance and surface roughness of doped PEDOT: PSS with different weight ratios of 1.5M H_2SO_4 to PEDOT: PSS.

resistance and surface roughness were reduced significantly. The sheet resistance is reduced from 604 to 216 $\Omega/\square$ and the surface roughness is also reduced from 1.268 to 0.822 nm.

From Figures 1, 2, and 3, it can be seen that there is a common phenomenon. The sheet resistance was slightly increased by adding a small amount of H_2SO_4. The reason for this phenomenon is that the sulfuric acid was reacted with sorbitol preferentially. Thus, the aggregate effect of PEDOT grains combined with sorbitol (S. Timpanaro et al. 2004) was destroyed. After the reaction, the residual H_2SO_4 reacted with PSS. The chemical reaction can be written as $H_2SO_4 + PSS^- \rightarrow HSO_4^- + PSSH$. The non-conductive anions of some PSS^- were substituted by the conductive anions of HSO_4^-. However, the substitution reactions will be favorable for the conductivity enhancement (Yijie Xia et al. 2012).

Figure 4 shows the doped PEDOT: PSS films with different molar H_2SO_4 concentrations as a function of weight ratio of H_2SO_4 to PEDOT: PSS. The transparency of the doped PEDOT: PSS film can be affected

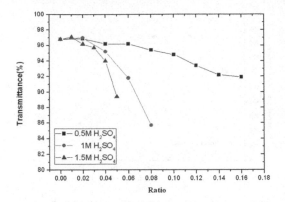

Figure 4. The transmittance of the doped PEDOT: PSS with different molar concentrations of H_2SO_4 as a function of weight ratio of H_2SO_4 to PEDOT: PSS.

by the H_2SO_4 treatment. The transmittance of the doped PEDOT: PSS film was decreased by increasing the ratio of H_2SO_4 to PEDOT: PSS. The lowest sheet resistance of 216 $\Omega/\square$ is obtained at 0.5M H_2SO_4-doped PEDOT: PSS film and at the weight ratio of 0.16, which has transmittance in the visible wavelength range from 400 to 700 nm of 91.9%. The high transparency and low sheet resistance indicated that the 0.5M H_2SO_4-doped PEDOT: PSS films can be used as the transparent conductive electrode of optoelectronic devices.

4 CONCLUSION

This study employed the noncontact AFM, four-point sheet resistivity meter and U-3900 spectroscopy to investigate the origin of the sheet resistance decrease of H_2SO_4-doped PEDOT: PSS films. The

PEDOT: PSS solution doped by adding the different molar concentrations of H_2SO_4 strongly affects surface roughness, sheet resistance, and transmittance. After H_2SO_4 doping, the surface roughness is reduced from 1.268 nm to 0.822 nm. It's indicated that the surface of H_2SO_4-doped PEDOT: PSS film was smoother compared with the doped PEDOT: PSS film. The sheet resistance is improved from 604 to 216 $\Omega/\square$ by adding dilute sulfuric acid. The decrease in the sheet resistance is due to the fact that the nonconductive anions of some PSS^- were substituted by the conductive anions of HSO_4^-, namely that the substitution reactions will be favorable for the conductivity enhancement. The transmittance of the doped PEDOT: PSS film is decreased by increasing the weight ratio of H_2SO_4 to PEDOT: PSS. The transmittance of the 0.5M H_2SO_4-doped PEDOT: PSS film is above 91% in the visible wavelength range from 400 to 700 nm. However, the high transparency and low sheet resistance reveal that the H_2SO_4-doped PEDOT: PSS films can be used as the transparent conductive electrode of optoelectronic devices.

REFERENCES

Y. Wang, J. Phys, 2009. *Conf. Ser.* 152 012023.
S.K.M. Jönsson, J. Birgerson, X. Crispin, G. Greczynski, W. Osikowicz, A.W. Denier van der Gon, W.R. Salaneck, M. Fahlman, 2003. *Synyh. Met.* 139, 1–10.
Hu Yan, Hidenori Okuzaki, 2009. *Synyh. Met.* 159, 2225–2228.
J.Y. Kim, J.H. Jung, D.E. Lee, J. Joo, 2002. *Synyh. Met.* 126, 311–316.
Sungeun Park, Sung Ju Tark, Donghwan Kim, Curr., 2011. *Appl. Phys.* 11, 1299–1301.
S. Timpanaro, M. Kemerink, F.J. Touwslager, M.M. De Kok, S. Schrader, Chem, 2004. *Phys. Lett.* 394, 339–343.
Yijie Xia, Kuan Sun, and Jianyong Ouyang, 2012. *Adv. Mater.* 24, 2436–2440.

Smart Science, Design & Technology — Lam et al. (eds)
© 2020 Taylor & Francis Group, London, ISBN 978-0-367-17867-3

Integration of Chinese and Western medicine in fainting during acupuncture treatment

Jung-Jung Wen*
Ph.D Program of Mechanical and Energy Engineering, Kun Shan University, Tainan, Taiwan

Huannming Chou
Department of Mechanical Engineering, Kun Shan University, Tainan, Taiwan

ABSTRACT: Acupuncture therapy is already a widely accepted practice as a comprehensive alternative therapy internationally through the integration of Chinese and Western medicine. It is used to treat various illnesses such as allergic rhinitis, sudden deafness, facial palsy, brachial plexus injury, and cerebral vascular accident and so forth. All reports show the effective results that may be related to neurological rehabilitation. However, the strong stimulation of De-Qi during acupuncture may cause different types of discomfort in patients and lead to serious side effects such as fainting during acupuncture treatment (FDAT) inducing a possible shock and death. In some cases, the treatment might cause medical disputes and engage in a lawsuit. Additionally, due to the underlying diseases and habit of drugs intake of a patient, it may also increase the incidence rate and severity of FDAT. This paper focuses on reinforcing the safety and completeness to treat fainting during acupuncture through the concept of shock and procedures already established in Western medicine. These include an assessment of medical history prior to the application of acupuncture, the physiological changes during fainting, process management as well as the necessity of physiological monitoring.

Keywords: Acupuncture, Fainting during acupuncture treatment (FDAT), Shock

1 PREFACE

1.1 *Research background and motive*

Acupuncture therapy is an important subject in traditional Chinese medicine. The recent 60 years research about the mechanism of acupuncture effects has been well established in the field of basic physiological medicine. The results showed many similarities with the Western medical theories. Such research focused on the fields of basic electrophysiology and various physiologies of the human body systems. As a form of invasive medical behavior, acupuncture is internationally recognized as a means of auxiliary medical treatment, and it can actually bring about remarkable progress in terms of functional rehabilitation after chronic pain and nerve injury, thus it is covered by the medical insurance system. Although the technique of acupuncture therapy is convenient, safe and has minimal side effects, cases of fainting during acupuncture treatment (FDAT) are not uncommon during the process of acupuncture in clinical practices. FDAT is also among the top two adverse reactions of acupuncture as discovered by various fields of research (Huang, Hsu

& Chang, 2010), and it is often involved in medical disputes (Yang & Yang, 2004). Therefore, if there is any mishandling during the process, the patient may experience severe shock or even face life-threatening danger. Generally, practitioners of Western medicine will set up various physiological monitors in places such as clinics, wards, emergency rooms or intensive care units, so as to evaluate the healthy conditions of patient. The integration of Chinese and Western medicine is currently applied on many types of diseases including cerebral vascular accident (Lee et al., 2009), sudden deafness (Wen, et al., 2014; Wen & Chou, 2016), brachial plexus palsy (Huang, 1993), facial palsy and allergic rhinitis. The Chinese medical education in Taiwan has already incorporated knowledge of Western medicine, and venues for acupuncture therapy are also widely established in Chinese medical clinics, Western medical rehabilitation clinics and medical centers. As medical practitioners are fully responsible for the development of medical quality and the guarantee of patients' safety, this paper aims to study the relevant shock management process as established by Western medicine, so as to explore means of improving the

*Corresponding author: E-mail: rrwen@ms45.hinet.net

safety and integrity of clinical FDAT management process in Chinese medicine.

1.2 Research purpose

Through the mechanism of neurophysiology and cardiovascular circulation system established by modern Western medicine, we hope to study the relationship between these two fields and FDAT. Also, by making immediate judgment through continuous monitoring on a physiological monitor, we can provide adequate drug therapy to avoid incidence of emergency shock. As such, a safe mechanism that encompasses a simple and clear preventive and therapeutic process can be set up.

2 THE RELATIONSHIP BETWEEN FDAT AND PHYSIOLOGICAL MECHANISMS OF NERVES AND BLOOD CIRCULATION

2.1 The relationship between FDAT and shock

According to the definition on quantitative operation by Huang Weisan and Lin Shaogeng, when the needle is applied, the patient may experience excessive stimulation that leads to fainting, and this phenomenon is known as "fainting during acupuncture treatment". Clinical symptoms of FDAT include paled face, irregular heart rate, shortness of breath, cold sweating, chest tightness, dizziness or vertigo and weak pulse. Patients with more serious conditions may even display symptoms like cold limbs, delirium, and barely palpable pulses (Lin, 2009; WHO, 2007). Such conditions are commonly induced by emotional agitation, panic, weakness and fatigue. Hence, practitioners of Chinese medicine have established seven contraindications that are deemed unsuitable for acupuncture therapy.

On the other hand, according to the World Health Organization (WHO), fainting during acupuncture treatment is defined as "an adverse reaction to acupuncture; a feeling of faintness, dizziness, nausea and cold sweating during and/or after needling, also called needle sickness" (Zhang, 1985). From the perspective of Western medicine, FDAT is a condition of functional disorder caused by strong stimulation of the peripheral sensory nervous system and influences the function of central nervous system and cardiac vascular system after needling. By assessing the symptoms of shock and reviewing the research on paled face, weak pulsation with tachycardia and hypotension, Western medicine regards shock as a form of physiological condition that exhibits the symptom of a suppressed circulatory system and subsequently turns into a pathological process that indicates insufficient body tissue perfusion. It can be divided into four types, namely hematogenic, cardiogenic, neurogenic and vasogenic. The compensatory mechanism triggered by the incidence of shock is mainly presented as increased cardiac output of the left ventricle to maintain the stability of arterial blood pressure, and increased heart rate as well as contracted peripheral blood vessels. Concurrently, the sympathetic nervous system is activated by the secretion of epinephrine and norepinephrine. Even though such compensatory mechanism might be beneficial during the early stage, when the heart rate increases excessively, the efficiency of cardiac pumping will ultimately decrease, and the function of vascular constriction might also lead to aggravation of the shock syndrome, such that the benign compensatory baro-receptor reflex gradually deteriorates into the acute symptom of severe vasovagal reflex.

2.2 The physiological changes that leads to FDAT

This process first explores the morphological space of needling and its relevant acupuncture points, and further employs the knowledge of physics to assess the relationship between needling and the introduction of senses into nerve fibers. Upon anatomical observations on the 16 acupuncture points such as Tsusan Li (S36), the components of this space are known to be 35.2% of nerve bundles, 14.8% of free nerve endings, 4.5% of muscle spindles, 45.5% of blood vessels, and many tiny lamellar corpuscles (Bowsher, 1998). Needling sensation is an expression of the frequency messages, which include soreness, distension, heaviness, numbness and aching. It is the physical conduction between the muscle potential produced by needling and the different frequency messages. This is seen in Thompson's neurophysiological study in 1994, and as listed in table 1, the Pacinian corpuscle differentiated from nerve terminal can convey the vibration signals of frequencies between 60~500 Hz, where the Meissner's corpuscle is responsible for conveying vibration signals of extremely low frequencies and those of frequencies lower than 80 Hz (Yang et al., 1994). The muscle potential produced by needling falls into the approximate range of 30 to 350 mV, and it can trigger the exchange of sodium and potassium ions in the intracellular membrane, thereby initiating the action potential produced by depolarization. Therefore, it can convey the message through nerve fibers from the afferent pathway type II (A_β, A_γ fiber), type III (A_δ fiber), and type IV (C fiber) into the spinal cord, and can also enter the brainstem along the ventral lateral cord to convey the sensations of pain and temperature. The De-Qi points mostly distribute at various layers of tissues between the subcutaneous and periosteam of the acupoint, but are mainly located at deep tissues (Macdonald, 1998). Upon in-depth level observations on the periosteal stimulation, the conveyance of messages includes four types of segments such as dermatome, myotome, sclerotome, and viscerotome (Macdonald, 1998). The effects of needling depend on whether De-Qi is achieved as well as the degree of De-Qi. Many studies have revealed that needling can improve the physiological balance of providing feedback to relevant areas of the cerebral cortex, and can affect the functional adjustment of the visceral organ's autonomic nervous system. From the perspective of anatomy, almost every visceral organ is subjected to functional adjustment by the sympathetic nerve and parasympathetic nerve. Hence, FDAT shares a significant relationship with the adjustment of functional differences by the autonomic nervous system.

3 THE ESTABLISHMENT OF A SAFE CHINESE AND WESTERN MEDICAL TREATMENT

The prevention of FDAT may be achieved from various areas such as the patient's medical history, the handling process as well as the medical facilities. In terms of medical history, things to take notice of include previous diagnosis of FDAT, heart diseases, habit of sedatives use and emotional instability. The handling of FDAT should combine both Chinese and Western medical approaches. The medical monitor should include four items of physiological monitoring: electrocardiogram (EKG), non-invasive blood pressure (NIBP), respiratory rate (RR) and saturation of pulse oxygenation (SpO_2). Also, there must be oxygen supply, emergency medications (atropine, ephedrine, epinephrine) and establishment of venous infusion sets.

4 DISCUSSION

Chinese medicine employs the diagnostic methods of observation, listening, interrogation and pulse-taking as well as treatment based on various symptoms. Similarly, Western medicine observation also serves as the initial step, but subsequent diagnosis is based on functions of the physiological system and various physiological monitoring data. For instance, electrocardiogram (EKG) can be used to diagnose abnormal heart rate, arrhythmia and myocardial injury; sphygmomanometer can measure arterial blood pressure to analyze arteriolar resistance, circulating blood volume, and compliance of vessel wall; digital arterial blood oxygen saturation (SpO_2) and respiratory rate can be used as preliminary indications of the aerobic and anoxic states of systemic tissues. When the patient is experiencing arrhythmia, heart valve disease (HVD) or coronary artery disease (CAD), the baroreceptor reflex may not be triggered during FDAT, thus leading to enhanced activity of the parasympathetic nerve caused by the vasovagal reflex. As such, rapidly decreasing heart rate, myocardial ischemia and lowering blood pressure can result in cardiovascular shock. Patients will experience fast and shallow breathing during FDAT and shock, so if there is no oxygen supply, the arterial blood oxygen saturation will fail to the recovery of blood pressure (BP), heart rate (HR) and cardiac rhythm.

Judging from the various symptoms of FDAT defined by the Chinese medicine as mentioned above, they are all relevant to the autonomic nervous system emphasized by Western medicine. The function of the autonomic nervous system is mainly to adjust the differences between blood pressure and heart rate. Timely treatment by practitioners of Chinese medicine allows immediate changes in posture, and strong stimulations can be imposed on various acupuncture points such as Yung Chuan K1, Ho Ku Li4, Nei Kuan Cx6 and Pai Hui VG19 based on traditional experience. Yet, even though such stimulations are known to be able to increase sympathetic nerve activity, there is a lack of credible data record and analysis of the changing process throughout the course of disease, thus, through dynamic monitoring on a physiological monitor, changes in blood pressure (BP), cardiac rhythm, heart rate (HR), respiratory rate and digital blood oxygen saturation (SpO_2) can be simultaneously analyzed. Also, the posture of 30 degrees' head elevation or full-flat can be aptly adopted, mask oxygen supply with a flow rate of 6 L/min can be immediately provided, and drug therapy can be applied by establishing an infusion route. For example, to increase heart rate, Atropine 1 mg IM or 0.5 mg IV can be administered; to counter arrhythmia, Xylocaine 40–60 mg IV can be provided; to balance increased blood pressure, Ephedrine 10 mg~20 mg IV can be provided; and if the condition of shock persists, Epinephrine 0.05 mg~0.2 mg IV can be provided as an emergency medication.

5 CONCLUSION

Acupuncture is an effective type of invasive therapeutic method, but cases of FDAT are not uncommon in clinical practices. Such phenomenon is, to a certain extent, associated with the patient's underlying diseases and medication history, thus we believe that the explanations to various physiological conditions during FDAT based on Western medicine's basic neurophysiology and circulatory physiology are reasonable, and we recommend the establishment of physiological monitors at therapeutic venues to obtain firsthand records of patients' physiological conditions, as we hope to offer immediate judgements and appropriate treatment based on synchronized monitoring. We believe such measures can guarantee the patients' safety and prevent cases of medical disputes.

REFERENCES

C.K. Lee, H.C. Lee, L.P. Chang, et al., 2009. Journal of Integrated Chinese and Western Medicine 11(1) 17–25.
C.Y. Yang & H.I. Yang, 2004. *J Chin Med* 15(1):1–15.
D. Bowsher, 1998. Mechanisms of Acupuncture, in J. Filshie & A. White (Eds.), *Medical Acupuncture: A Western Scientific Approach*, Churchill Livingstone, Edinburgh, Scotland, 69–79.
J.J. Wen & H.M. Chou, 2016. *Proceedings of the IEEE International Conference on Advanced Materials for Science and Engineering IEEE-ICAMSE* 309–312.
J.J. Wen, C.K. Lee, H.C. Lee, et al., 2014. *Innovation, Communication and Engineering* 723–728.
J.R. Macdonald, 1998. Acupuncture's Non-segmental and Segmental Analgesic Effects: The Point of Meridians, in

J. Filshie & A. White (Eds.), *Medical Acupuncture: A Western Scientific Approach*. Churchill Livingstone, Edinburgh, Scotland, 83–96.

J.S. Yang et al., 1994. *Acupuncture*. Taipei: Jyin Publishing Company.

P. Wu, E. Mills, D. Moher, et al., 2010. *Stroke 41*(4) e171–e179.

S.G. Lin, 2009. *A New Edition of Acupuncture with Coloured Pictures*. Taipei: Jyin Publishing Company.

T.H. Huang, K.H. Hsu & H.H. Chang, 2010. *J Chin Med 21*(3,4) 133–142.

W.S. Huang, 1993. *The Science of Acupuncture*. Taipei: Zhengzhong Bookstore.

WHO, 2007. *International Standard Terminologies on Traditional Medicine in the Western Pacific Region*. World Health Organization, Geneva, Switzerland.

X.P. Zhang, 1985. *Mechanism Research on the Functions of Acupuncture*. Anhui Science and Technology Publishing House, Qiye Bookstore Pte Ltd.

Smart Science, Design & Technology — Lam et al. (eds)
© 2020 Taylor & Francis Group, London, ISBN 978-0-367-17867-3

Applying the ASSURE model in designing digital teaching materials - using the prevention of Sexual Assault and Sexual Harassment in junior high schools as an example

Chien-Yuan Chen*
Ph.D Program of Mechanical and Energy Engineering, Kun Shan University, Tainan, Taiwan

Huannming Chou
Department of Mechanical Engineering, Kun Shan University, Tainan, Taiwan

ABSTRACT: This paper applies the ASSURE model teaching approach in the research on the creation of digital teaching materials for the prevention of sexual assault and sexual harassment (SAASH). The design of the course will focus on the learners to improve their learning effectiveness and to achieve the objective of improving their self-protection capabilities pertaining to SAASH. Based on the need to strengthen the education of learners on SAASH, and the statistics on the current situation regarding SVASH as presented in the incident reporting in schools, the ASSURE model is used to design the course materials for prevention of SAASH. This involves six stages, namely: analyzing learners, stating objectives, selecting methods, media and materials, utilising media and materials, requiring learners' participation, and evaluating and revising the course. These are used to understand the learning problems of the learners, and then adjust the teaching content and method accordingly so as to achieve the best learning effectiveness. The digital teaching materials to be developed as discussed in this paper shall be adopted in junior high Schools, and the research results can be offered to the respective junior high school teachers for use as reference materials in conducting their course on the prevention of SAASH. This is to improve the quality of such courses and to achieve the objective of reducing the number of cases of SAASH in schools.

Keywords: ASSURE Model, Digital Teaching Materials, Sexual Assault, Sexual Harassment

1 PREFACE

From 1993 onwards, women's groups in Taiwanese areas have started worrying about the major frequently occurring gender violence events in society, triggering women's groups to fight for legislations preventing and controlling SAASH. Through the long-term efforts of women's groups and female academics on this issue, as well as the efforts of the society and the government, the public quickly saw actual results. Within a few years, in terms of the legislations concerning SAASH prevention and control, legislations such as the "Sexual Assault Prevention Act," "Offence against Sexual Autonomy," "Act of Gender Equality in Employment," "Gender Equity Education Act," and "Sexual Harassment Prevention Act" have all been established. However, according to the sexual assault crimes' statistical information, from 2005 to 2016 in Taiwan, the number of victims that have reported being involved in sexual assault crimes amounted to 96,201 people, with most victims aged in the range of 6 years old to less than 24 years old accounting for 75.86%. Among them, the number of victims aged 12 to less than 18 (middle and high school students) was highest, occupying a total ratio of 55.3% of the victims, followed by those in the age range of 18 to less than 24 (university students), occupying a total victim ratio of 13.65%, while the remaining victims aged 6 to less than 12 (primary school students) occupied a total victim ratio of 8.91%, as shown in Table 1.

From 2007 to 2016 in Taiwan, the number of sexual harassment cases amounted to 3,251 victims [2], with the age concentration of the victims below the age of 30 accounting for 65.45%. Among them, the highest concentration of victims were aged 18 to below 30, accounting for 45.86%, followed by

---

*Corresponding author: E-mail: chien00053@gmail.com

Table 1. The number of reported sexual assault cases in Taiwan categorised based on the age of victims.

Victim's Age Range	%	Victim's Age Range	%	Victim's Age Range	%
Under 6 Years	3.02	6-Under 12 Years	8.91	12-Under 18 Years	55.3
18-Under 24Years	13.65	24-Under 30 Years	6.79	30-Under 40 Years	7.11
40-Under 50 Years	3.36	50-Under 65 Years	1.43	Above 65 Years	0.44

Table 2. The number of established sexual harassment cases in Taiwan categorised based on the age of victims.

Victim's Age Range	%	Victim's Age Range	%	Victim's Age Range	%
Under 18 Years	19.59	18~Under 30 Years	45.86	30~Under 40 Years	21.13
40~Under 50 Years	7.97	50~Under 65	3.17	Over 65 Years	0.74

victims under the age of 18, accounting for 19.59%, as shown in Table 2.

From Tables 1 and 2, in the past 12 years, most sexual assault victims were in the age range concentration of primary, middle and high school as well as university students, accounting for 75.86%. Among them, middle and high school students were the greatest group of victims, accounting for 55.3%; in addition, sexual harassment victims under the age of 30 accounted for 65.45%, and among them those below the age of 18 accounted for 19.59%, indicating that SAASH cases in the country were gradually moving into the school environment. Therefore, to implement SAASH prevention and control in school environments in order to lessen the occurrences of SAASH cases occurring in schools is essential to prevent and control SAASH. If it is possible to implement SAASH prevention's advocacy and education on campus, strengthen teachers' professional knowledge and improve students' cognition of sexual autonomy, sexual harassment awareness, sexual assault awareness and self-protection ability in an environment sensitive to the crisis (Chen and Lin, 2007), this should be the most immediate approach to stopping gender violence behavior completely.

The implementation of on-campus SAASH prevention's advocacy efforts along with increasing the learners' learning efficacy requires good systematic teaching material to achieve the best campus SAASH prevention. Currently, the systematic teaching design theories used in the education field can be divided into two categories - "Systematic Teaching Design" and "Integrated Teaching Design," (Gagne and Merrill, 1990) which consists of the four models that are the ADDIE model, the ASSURE model, the

DICK & CAREY model and the KEMP looping teaching design model. To design digital teaching material on SAASH that are appropriate to the students and that they would like, this paper uses the ASSURE model teaching method to create digital teaching material that targets SAASH, using learners' curriculum design as the core and setting appropriate learning targets and teaching methods to increase the learners' learning efficacy in order to achieve the goal of learners' enhanced ability to protect themselves against the SAASH crisis. Firstly, this paper looks at the educational requirements that need to be strengthened for learners in regards to SAASH, and the situation of on-campus SAASH reports' statistics. The ASSURE model is then used as a method to design a teaching curriculum on SAASH - from analyzing learners' learning background and learning requirements, writing the appropriate learning targets, selecting videos, animations, news reports and multimedia materials such as self-produced PPT teaching materials as the teaching method, using multimedia computer and self-produced PPT teaching materials - all in order to guide reflection, oral queries and awarded quizzes to motivate learners' learning and participation. At the end of the teaching process, it can be determined through various questionnaires if learners' learning efficacy meets the learning targets. Using a statistical method to analyze the questionnaire's results and understand learners' learning issues to further adjust the educational content and methods will also help achieve the best on-campus SAASH prevention and control advocacy.

2 THE SITUATION OF ON-CAMPUS SAASH AND THE EDUCATIONAL REQUIREMENTS

2.1 The current situation of on-campus SAASH

According to Taiwan's Ministry of Education's sexual assault demographic statistical information (Ministry of Education, Department of Statistics, 2016) (the statistical information used prior to 2012 (inclusive) was reports that were investigated and handled, the statistical information after 2013 (inclusive) was cases that had been investigated and confirmed) from 2006 to 2012, middle school victims of sexual assault's classification in female and male victims are as shown in Table 3, where there was an average of 301 victims every year - female victims accounted for 82.23% and male victims accounted for 17.77%. Among them, sexual assault victims went up to 720 in 2012, with male victims gradually increasing from 11 people in 2006 to 139 in 2012. From here, it can be seen that sexual assault prevention and control advocacy works in the middle school phase need to increase in terms of male students' sexual assault chances, and the information contained in the teaching content also needs to be increased. In addition, from 2013 to 2015, sexual

Table 3. The number of male and female middle school sexual assault victims.

The Number of Male and Female Middle School Sexual Assault Victims from 2006 to 2015

Year	2006	2007	2008	2009	2010	2011	2012	2013	2014	2015	Total Number
Female	140	193	201	120	211	354	581	145	177	115	1800
Male	11	45	16	83	54	41	139	43	64	55	389
Number	151	238	217	203	265	395	720	188	241	170	2189

assault victims averaged 200 victims during middle school, with female victims accounting for 73% and male victims accounting for 27%.

From 2017 to 2012, from the number of male and female middle school sexual harassment cases (Ministry of Education, Department of Statistics, 2016) being investigated and handled, as shown in Table 4, there were 487 victims on average every year - female victims accounted for 82%, male victims accounted for 18%, and among them in 2012 the number of sexual harassment victims hit a peak at 1138, with male victims gradually increasing from 14 in 2007 to 237 in 2012. From here, it can be seen that sexual harassment prevention and control advocacy works in the middle school phase need to increase in terms of male students' sexual assault chances, and the information contained in the teaching content also needs to be increased. In addition, from 2013 to 2015, sexual harassment victims averaged 1,006 victims per year during middle school, with female victims accounting for 77.5% and male victims accounting for 22.5%.

From the 2014 to 2016 statistical information of the relationship between sexual assault victims and perpetrators from the age of 12 and below 18, it was shown that the reported cases across the three years were 5933 cases, 5653 cases and 4437 cases respectively. The relationship between the victims and the perpetrators were on average as follows: (1) the highest being boy/girlfriends, about 30.42%, (2) followed by schoolmates, about 12.87%, (3) ex-boy /girlfriends, about 8.13%, (4) cyber acquaintances, about 6.99%, (5) ordinary friends, about 6.38%, (6) lineal relative by blood, about 3.76%, (7) collateral relatives, about 3.39%, (8) strangers, about 2.56%, (9) friends of family members, about 1.44%, (10) teacher-student relationships, about 1.43%, (11) neighbors, about 0.99% and so on. The above shows that the chances of the perpetrator knowing the victim is about 76.8%, which means most perpetrators are people close to the victims, and therefore

it is essential to establish the mindset to "Beware of Sexual Assault Cases Perpetrated by People Close to You" - the concept that danger is just around you.

2.2 On-campus sexual assault and harassment's educational requirements

Research on the educational requirements for SAASH prevention and control in middle school shows (Yen, et al., 2001) that middle students in general lack the knowledge and understanding of the concept of SAASH as well as the resources for the incidents' prevention and control. It is necessary to strengthen sexual assault prevention, sexual harassment prevention, the concept of sexual assault and relevant legislations, with more than 80% of the efforts surrounding the identification of a risky situation, followed by the crisis management of a sexual assault incident, and finally self-protection methods. In addition, according to the Ministry of Education's primary and high school students' SAASH prevention and control educational curriculum, it has been shown that (Ministry of Education, 2012) sexual assault incidents on campus in middle schools usually take the form of forced sexual intercourse or consensual sex; sexual harassment incidents, on the other hand, usually take the form of physical contact, indecent exposure of the lower part of the body, voyeurism and secret photography, disseminating texts or photos with the intention of sexual harassment, excessive pursuit and etc. Students' favorite teaching methods and teaching medium is "Designing a Course to Teach" and "Video Tape, Multimedia and Slides Teaching Materials."

This paper refers to the Ministry of Education's sexual assault prevention education outline and schools' sexual assault prevention or sexual harassment prevention education requirements to establish a teaching curriculum that targets SAASH prevention and control in middle schools, including the four major units of "Understanding Your Body's

Table 4. The number of male and female middle school sexual harassment victims.

The Number of Male and Female Middle School Sexual Harassment Victims from 2007 to 2015

Year	2007	2008	2009	2010	2011	2012	2013	2014	2015	Total Number
Female	93	115	115	411	761	901	798	805	737	2396
Male	14	26	26	97	125	237	228	260	191	525
Number	107	141	141	508	886	1138	1026	1065	928	2921

Table 5. The course outline of the teaching curriculum for SAASH prevention and control advocacy.

Unit Goal	Unit Name	Unit Content
Understanding Your Body's Self-Autonomy and Limits	**Be the Master of Your Own Body**	1. Do not touch me randomly — learn to differentiate good and bad contact 2. Physical Limits - (1) Understand your body's limits, (2) Outline the prohibited areas of your body, (3) Determine who can touch your private parts 3. How to protect yourself — your body is your own to protect, respect your own privacy 4. Respect sexual self-autonomy
Understanding Sexual Harassment Incidents	**Act with Decency**	1. Understand the definition of sexual harassment 2. Understand the types of sexual harassment 3. Understand the current situation of sexual harassment 4. Understand the handling methods of sexual harassment incidents — complaints, mediation and litigation 5. Understand the relevant legislations related to sexual harassment — "Act of Gender Equality in Employment," "Gender Equity Education Act," and "Sexual Harassment Prevention Act"
Understanding Sexual Assault Incidents	**Keep Away from Sexual Assault**	1. Understand the definition of sexual assault — forced sexual intercourse and obscene activities. 2. Understand the current situation of sexual assault — the relationship between the victim and the perpetrator, the different relationships between males and females as well as regular sexual assault incident locations 3. Understand the relevant legislations related to sexual assault — "Category of Criminal Offense," "Nature of Sexual Assault" — "Indictment with/without Complaint" 4. Understand sexual assault incidents — the responsibility of the perpetrator 5. Prevent rapes during dates — how to ensure your safety when going on dates 6. The facts and myths of sexual assault.
Enhancing Environment Sensitivity and Crisis Management	**Friendly Reminders**	1. Identify dangerous scenarios: beware of threats surrounding you, people close to you that might threaten your safety; there might also be traps at home and with cyber acquaintances. 2. The ultimate tips to protecting yourself — the four steps of self-protection 3. Learn to adapt when you are a victim and ask for help — how to handle a sexual assault incident 4. Understand social help organisation's resources and reporting system

Self-Autonomy and Limits," "Understanding Sexual Harassment Incidents," "Understanding Sexual Assault Incidents," and "Enhancing Environment Sensitivity and Crisis Management" - these unit names and contents are all shown in Table 5.

3 DESIGN OF DIGITAL TEACHING MATERIALS

To help learners quickly absorb the educational curriculum, this paper uses the ASSURE model to design SAASH prevention and control educational curriculum's digital teaching materials; the ASSURE model was developed by Heinich, Molenda, Russell & Smaldino. The ASSURE model uses six steps to plan and design learning materials, including teachers' analyses of their learners' characteristics, writing with students as the center of the learning target, selecting the appropriate media and method to make the course more interactive, using media and teaching materials to increase learning interaction, stimulating learners' active participation and chances, and implementing assessments and improvements after each class. This allows the teaching curriculum to be reviewed and improved, with each step being integrated into the

Table 6. Descriptions of each step of the ASSURE model's design.

The Six Phases of the ASSURE Model	Learning Targets
A Analyze learners	1. Common Characteristics: Learners' backgrounds are middle school students. 2. Special Characteristics: Increase sexual assault prevention, sexual harassment prevention, sexual assault and myth's concept and knowledge about relevant legislations. 3. Learning Style: Prefers video tape, multimedia and slides teaching materials.
S State objectives	1. Learners: Middle school students 2. Behavior: Learn to protect themselves and pick up crisis management abilities and skills. 3. Scenario: Understand more about the concept of the body's self-autonomy; understand the reason SAASH occurs, as well as its types and legislative knowledge. 4. Level: Has the awareness to put related prevention knowledge to practice in dangerous situations.
S Select instructional methods, media, and materials	1. Briefing. 2. Picture books. 3. Puppet educational tools, learning lists. 4. News reporting clips and videos. 5. Awarded quizzes.
U Utilize media and materials	1. Arrange multimedia classrooms. 2. Self-produce multimedia presentational teaching materials to use and try teaching.
R Require learner participation	1. Watch videos and news reporting clips to let students reflect and discuss. 2. Use puppets as interactive teaching materials to deepen the impression on students. 3. Use awarded quizzes to emphasize important points. 4. Use oral queries to conduct important discussions.
E Evaluate and revise	1. Evaluate students' learning efficacy. 2. Assess course satisfaction level.

design of the whole program. Descriptions of the steps of this course's ASSURE model design are as shown in Table 6.

4 RESULT AND DISCUSSION

This paper uses the ASSURE model to design the digital teaching materials for a course on SAASH prevention and control, and the learning activities are implemented in real life at two middle schools. The teachers and students that participated in the course totaled 235 people. An analysis of the learners was conducted and learners were stimulated to participate in the course; post-course questionnaires were also conducted. The questionnaires used in this paper are divided into two categories, which are "Post-course Questionnaire 1: Questionnaire on SAASH Prevention and Control's Related Knowledge" and "Post-course Questionnaire 2: Questionnaire on Course Learning Satisfaction Levels," below are the descriptions respectively: (1) Post-course Questionnaire 1: According to the design of this paper' course unit target and unit content, four aspects such as "Understanding Your Body's Self-Autonomy and Limits", "Understanding Sexual Assault Incidents", "Understanding Sexual Harassment Incidents" and "Enhancing Environment

Sensitivity and Crisis Management" were designed as part of the questionnaire's content; (2) Post-course Questionnaire 2: Three aspects such as "Course Content", "Activities Approach" and "Students' Acceptance and Learning Efficacy" were designed as part of the questionnaire's content.

4.1 CASE 1: Analysis of the questionnaire on SAASH prevention and control's related knowledge

This questionnaire consists of four aspects with 18 questions in total, and is used to assess the learners' knowledge of SAASH's prevention and control; among them, other than questions B1, C1, C2, C4 and D3, which are negative questions, the rest are all positive questions. The measurement is based on Likert's (1932) 5-point scale (5-point Likert-type scale) measurement; from strongly agree (5 points) to strongly disagree (1 point). This paper uses the Excel statistical software to conduct questionnaire data analysis, and all of the results of the analysis' variation is 16.19, covariance is 17.60, Cronbach's α value is 0.73, the latter of which is an indicator that the questionnaire's reliability and validity (content validity) are all within good margins. From the lower scores in each question of the questionnaires, it can be said that middle school students in general do not have a clear understanding of the relationship between the SAASH victim and the perpetrator. They also lack legal knowledge and awareness of social welfare resources.

4.2 CASE 2: Analysis of the questionnaire on course learning satisfaction levels

This questionnaire is a survey of the teachers and students involved in this course's satisfaction level towards this course, and is divided into three aspects with 9 questions in total. Using the Excel statistical software to conduct questionnaire information analysis, all of the results of the analysis' variation is 1.97, covariance is 4.81, Cronbach's α value is 0.93, the latter of which is an indicator that the questionnaire's reliability and validity are within a really high range. The average score of this questionnaire's 9 questions' satisfaction levels is 4.68.

5 CONCLUSION

This paper designed digital teaching materials for a SAASH prevention and control advocacy course, and implemented actual learning activities at the campus of a middle school. From analyzing the learners' and stimulating their participation, and conducting post-course questionnaires, the following was found through an analysis of the questionnaires results: (1) The satisfaction levels for the three aspects of this course -

"Course Content," "Activities Approach" and "Students' Acceptance and Learning Efficacy" - were extremely high. (2) Middle school students generally do not have a clear understanding of SAASH as well as the relationships between the victim and the perpetrator; in addition, they also lack a level of understanding and knowledge regarding the relevant legislations and laws. This paper uses results from the analysis as a reference for the next course's planning, revision and improvement, allowing learners to quickly understand and absorb the education content, thereby increasing learners' learning efficacy in order to achieve the goal of learners being able to prevent and protect themselves better against the SAASH crisis.

REFERENCES

H.N. Chen and M.C. Lin, 2007. *NTU Social Work Review 14* 211–260.

R.M. Gagne and M.D. Merrill, 1990. *Educational Technology Research and Development 38* (1), 23–30.

Ministry of Education, Department of Statistics, 2016. *On-Campus Sexual Assault and Sexual Harassment Statistics by Gender/On-Campus Sexual Harassment Incident Statistics - Based on the Relationship of the Subject in Question*. Ministry of Education.

H.W. Yen, et al., 2001. *Formosan Journal of Sexology 7*(1) 40–55.

Ministry of Education, 2012. *Middle and Primary Schools' Sexual Assault and Sexual Harassment Prevention Educational Course Teacher's Outline*. Ministry of Education.

Smart Science, Design & Technology — Lam et al. (eds)
© 2020 Taylor & Francis Group, London, ISBN 978-0-367-17867-3

Research analysis of the effect of digital convergence on competition in the cable TV industry and its competitive strategy

Kuan-Ling Lai*
Ph.D Program of Mechanical and Energy Engineering, Kun Shan University, Tainan, Taiwan

Huannming Chou
Department of Mechanical Engineering, Kun Shan University, Tainan, Taiwan

ABSTRACT: In a landscape of digital convergence, not only do industry players in the cable TV business have to accept the transformation brought about by digital technology, they also have to face the challenges that arose from their competitors on the internet directly. In the face of high speed information network development trends, consumers have access to diversified sources of entertainment media. They are only interested in the continuity of streaming, as well as immediacy and richness of content. As long as industry players are able to provide timely media content that is rich and of high quality coupled with fast comprehensive application services, they will be able to gain the acceptance of consumers. This research redefines industry supply chain post-digitization of cable TV and affirms the relative strategic position of industry players in the cable TV business. The research paper also utilizes the Porter five forces analysis model to shed light on the corresponding situation caused by competition that resulted from the digital convergence of cable TV, projecting a re-aligned framework of cable TV industry competition. The greatest threat to cable TV brought about by digital convergence stems from the internet TV service providers and alternative media products in the form of OTT TV, as well as the customization and mobility of media content all working together to gradually decrease the number of existing and potential consumers of cable TV. This research employs case study analysis methodology and expert interviews to study the impact of digital convergence on competition in the cable TV industry and its competitive strategy, thereby proposing three competitive strategies which cable TV owners can adopt to improve their competitive advantage.

Keywords: Digital convergence, cable TV, digitization, industrial supply chain, Porter five forces analysis

1 TAIWAN'S CABLE TV INDUSTRY ENVIRONMENT

The digitization of cable TV's broadcast technology from analogue to digital represents an important developmental milestone in Taiwan's cable TV industry. Since the comprehensive digitization of cable TV, business operators within the industry have started developing a variety of applications for digital set-top boxes, such as PVR recording, Time Shift playbacks, SMOD on-demand videos, digital photo frames, emails and multi-screen streaming.

Having accepted the revolutionary changes brought by digital technology, cable operators now also need to face the competition from online streaming media providers directly. Under the trend of digital convergence and the development of high-speed information networks, sources of audio and video media are now mobile and diversified. The concerns of consumers are the smooth streaming of audio and video, real-time capabilities and rich content. It does not matter to them if the providers of the media content are cable TV telecommunications network operators or over-the-top content (OTT) platforms. So long as the media provider can provide real-time, high-quality and rich media content, and have comprehensive high-speed application services, consumers are likely to accept it.

With the digitization of cable TV, cable operators have launched many value-added applications services. Except for high-definition quality channels, high-level substitute applications can be found online for all kinds of these application services. Even though cable operators continue to invest in the development of application services, they might not be able to get a return on their

---
*Corresponding author. E-mail: ellen@careearth.com

investment due to the influence of online high-level substitutes. Ultimately, the variety of application services on set-top boxes might just become one of the incentives to attract some consumers to continue watching cable TV.

While cable operators anticipated the variety of application services launched post-digitization to bring them additional profit aside from monthly broadcast fees, a value-added application service that can bring cable TV operators major returns has not appeared since the completion of all-round digitization promotion. Viewers subscribe to cable TV services largely to watch TV programs and for broadband Internet access; additional subscriptions, on the other hand, are mostly for value-added channels and programs. Whether or not cable operators are able to identify an application service that consumers are willing to pay for is still subject to test in the consumer market (National Communications Commission, 2016).

During the process of cable TV's digitization, cable operators invested large amounts of human resources and funds to build digital headed machine rooms, expand cable Internet to contain digital channels, as well as develop digital set-top boxes and send professionals to install the boxes at customers' homes. Upon all-round digitization, they then had to face the annual set-top box trade-ins and capital investment in application services development. The operating cost of cable operators' post-digitization far exceeds their cost pre-digitization. At the same time, cable TV operators have to face increasing internal operating costs, as well as strong competition from cable TV industry peers, telecommunications operators and online media providers. In this tough environment to operate a cable TV business, it is essential to figure out how to respond to the changes in the industry as well as the operational challenges - and that is the main research motivation for this thesis.

2 THE SUPPLY CHAIN OF THE CABLE TV INDUSTRY

Taiwan's cable TV industry chain is as shown in Figure 1, where it can be seen that program content production and media transmission is operated separately. From the top to bottom, the industry chain can be categorized into the upstream, i.e., program content production and provision; the midstream, i.e., channel distributors and suppliers, Internet equipment manufacturing, and ISP broadband Internet services; and finally the downstream, i.e., cable TV system operators. Cable operators integrate the each level of the industry chain's media services content, local community-produced programs and self-produced programs, and finally transmit the media services to the homes of end consumers.

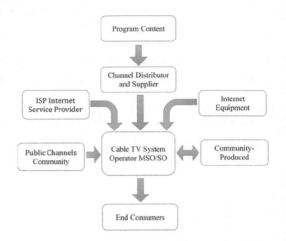

Figure 1. Taiwan's cable TV industry supply chain diagram.

The industry chain of Taiwan's cable TV industry coexists in a close symbiotic relationship, with cable TV system operators being the industry's end channel - building a transmission route into the homes of each customer, having the most face-to-face contact opportunities with consumers, and also equipped with valuable customer information and management abilities. Therefore, cable TV system operators have the potential to provide multi-element Internet services and customer value, integrating the upstream, midstream and downstream of the cable TV system ecosystem, creating unlimited development opportunities for the industries digitized future.

Digitization is a crucial skill and technology revolution and development in the cable TV industry's operating environment. Converting from analogue to digital and high-definition, cable TV's media transmission is going to further advance towards 4K resolution and supply chain in the future. In line with the changes brought by the digital era, the situation has gradually transformed. Other than the original cable TV network and telecommunications network in households these days, with 3G and 4G mobile networks becoming more and more commonplace, wireless networks have solidified its presence in households - not only as a family network, but a personal network as well.

3 ANALYSIS OF CABLE TV'S COMPETITIVE ABILITIES

Michael Porter (1979) stated in "Competitive Strategy" that the aim of enterprises adopting competitive strategies is to find their own position within the industry's competition, or to identify the competitive positions that affect their ability to continue making profit or disable their advantages. This allows enterprises to have the best abilities to resist these forces as well as analyze the industry's structure, competitors and industrial evolution. After establishing a complete industrial competition analysis model, Porter categorized the competitive factors in the industry into Five Forces, which are Industry, Buyers, Suppliers, Substitutes and Potential Entrants (Porter, 1979; Porter, 1980; Li and Chiu, 2010).

Using Porter's Five Forces analysis model, this research restructures the five forces analysis of cable TV industry's competition in line with the industry's competitive scenarios that have arisen in response to digital convergence, as shown in Figure 2.

3.1 Competition within the industry

Taiwan's cable TV industry is an open market for competition. Other than Chunghwa Telecom's MOD, cable TV's operating region allows multiple operators to submit an operating application. Operators need only submit a cable TV operation application to the National Communications Commission (NCC) according to law, with the county as the unit, and start operating upon approval. The competition within Taiwan's cable TV market and industry is very fierce. Even though most regions now only have one cable TV operator competing with Chunghwa Telecom's MOD, this market competition situation is unlikely to last into the future under the open market conditions.

3.2 Potential entrants' market entry pressure

After digital convergence, media has become digitized and online media blossomed like spring flowers. Other than the broadcasting of audio and video through set-top boxes on TV, the trend is also advancing towards mobile and personal viewing on Smartphone and tablets. Some channel providers have even launched channel content apps in line with the trend, seizing the mobile and personal market. There are no OTT TV laws and regulations in Taiwan currently, and due to the convenience by which consumers can access digital media, most of the media OTT TV or apps in the market now obtain digital media through illegal methods and convert them for broadcasting. No fees are paid for authorized content under these illegal actions, thus requiring little amount of equipment funds. And this has had a huge impact and influence on legal operators.

3.3 Cable TV suppliers

Cable TV suppliers include program/channel suppliers, public channels and community programs, community-produced programs, Internet hardware equipment manufacturers, and ISP Internet service providers. Among them, there are more options for hardware manufacturers and ISP Internet service providers, which lower the suppliers' bargaining power. During the process of digitization, cable TV operators have had to select a supplier for condition access (CA), middleware (MW) and set-top boxes (STB), and after confirming the supplier and completing the STB installation and digital conversion, start making yearly maintenance fee payments to maintain the cable digital system's stability and service developments. Under the circumstances of each supplier's system incompatibility, cable operators have to either continue working with the same supplier or replace the supplier, which decreases operators' bargaining power.

In terms of channel distributors and suppliers, there are three main categories: (1) free-to-air channels, e.g. Taiwan Television (TTV), China Television (CTV), Chinese Television System (CTS), Formosa Television (FTV), Hakka TV and etc, which are must-have channels under NCC regulations; (2) channels that operators have to pay for to purchase, e.g. HBO, SET, EBC, Videoland and etc, which operators can broadcast legally only after they have settled the airing authorization fees with the channel distributors or suppliers; (3) channels that can only air if the channel providers pay, e.g. Eastern Home, ViVa TV and other shopping channels.

Each year, cable operators face renewed airing fees negotiations with channel providers, authorization fees adjustments, cuts in fees for paid channels... these have resulted in airing fees disputes for two consecutive years that had to be mediated by the NCC.

3.4 Consumers' bargaining power

Cable TV's monthly fees are regulated by regional NCC or the local government's administration. Each year, cable operators also have to also face the fees upper limit for the next year set by the Rates Review Committee set up by the NCC or the local government - the rates are therefore publicly known in the market. The operators of some buildings, dormitories and hotels are considered a category of customers that purchase their subscriptions for multiple users at

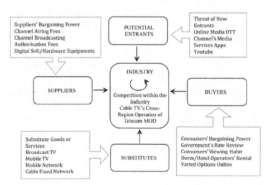

Figure 2. Five Forces analysis of cable TV industry's competition.

one time, and have the power to negotiate a lower-cost deal; regular consumers do not have such bargaining power. Even though consumers' cable TV habits are unlikely to be easily changed currently, cable operators should pay attention to future consumers, especially the younger demographics' habits of watching free TV on the Internet. This will influence consumers' bargaining power in the long-run.

3.5 Substitute goods or services' substitution pressure

After digital convergence, competition appears in the form of online competitors with media sources from all over the world. Users are only concerned with fast, real-time services and information - they do not care if the provider is cable TV or telecommunications operators. Especially with 4G mobile network becoming commonplace, mobile users spend more time browsing their Smartphone than watching TV. In addition, most media websites provide all kinds of programs with free streaming services - this constitutes an even greater threat and impact on cable TV.

4 CASE STUDY ANALYSIS AND EXPERT INTERVIEWS

This thesis' subject, Company H, is the first cable TV company in Taiwan to complete the transformation from analogue to digital, and was recognized by the Ministry of Culture in Taiwan, winning the Golden Visual Awards' Best Annual System Performance award six times. According to Company H's website, it invested in the construction of the modernized, multi-function digital machine room in 2009 to provide high-definition channels, digital program timetables and other digital TV services. In June 2014, it became the first cable TV company in the country to achieve all-round digitization - from traditional media transmission, it expanded to become interactive multimedia containing information, communication and digital video services, satisfying the demands for different-focus, multi-element programs.

According to the NCC's statistics, Company H's users grew from 128,700 in 2009 when it first started promoting digitization to 149,600 upon the completion of its all-round digitization, making up 69.06% of the market share. However, under the digital convergence trend, viewers shrank to 148,700 in 2017, making up 67.47% of the market share.

Interviewee at Company H is as shown in Table 1. The interview direction of this thesis is conducted according to an analysis of the industry's abilities, including core competencies and development, competition and competitiveness, viewer demographics and service substitutions, digital convergence's influence, and competitive strategies under the influence of global and personal mobile media availability.

Table 1. Interviewee at Company H.

Interviewee's Code	Job Position	Professional Background
A	Top-Level Management	Viewing and Online Sales Operation
B	Top-Level Management	Customer Service and Sales Operation
C	Top-Level Management	Online Information Sales and Technical Development
D	Top-Level Management	Treasury Management
E	Mid-Level Management	News Production and Program Promotion
F	Mid-Level Management	Engineering System Operation

5 STRATEGIES THAT SHOULD BE IMPLEMENTED BY THE INDUSTRY AND RECOMMENDATIONS

Through case study analysis, expert interviews and competitor analysis, this thesis finds that online media services have had the greatest impact on the industry's competition. With digital convergence, cable operators face the biggest threat from online TV services and OTT TV's substitutes, with online media services' mobile and personal features corroding cable TV's original or potential user demographics. If cable operators hope to turn the tide in their favors, they will have to find ways to conform to changes in the environment and consumer behavior in order to utilize the changes wisely to expand their scope of operations.

Just as the flow and change in the TV viewing demographics, viewers in Taiwan mainly watched broadcast TV prior to 1993; broadcast TV's viewing market, including promotional income and program production, only gradually switched to cable TV when the implementation of cable TV was officially announced in 1993.

In 2010, to accelerate the launch of Taiwan's digital convergence development, the Executive Yuan not only launched a "Digital Convergence Development Plan", but also set up a "Digital Convergence Committee" to be responsible for the supervision, coordination and promotion of digital convergence efforts (Yuan, 2010). And when networks became high-speed enough for the smooth delivery of media services online, OTT TV, YouTube and other online media started to rise up, constituting a real threat to cable TV's viewer growth. And this substitutes' threat will only grow as time passes.

The industry's competitiveness and profit-making abilities depend on the industry's structure. Enterprises provide products or services; whether an industry is new or mature, high-tech or basic technology, or the government's administration extent - none of these are the final deciding factor of competitiveness and profit-making abilities. Many of these factors might influence competitiveness and profit-making abilities in the short-term, but in the middle or long-run, it is the industry's structure that decides its profit-making abilities. Therefore, understanding the structure of an industry is conducive to establishing effective strategic orientation (Porter, 2008).

Below are several recommendations on the strategies that can be adopted by the cable TV industry in response to future competition in order to create a favorable environment for enterprises:

(1) The government should establish laws to regulate online unauthorized content use in order to protect audio and video proprietary rights, and also lay down the relevant decrees for online TV broadcasting so that operators can operate online TV legally.
(2) Leverage the changes in the digital era to one's own benefit by repositioning the company's operations, expanding from cable TV broadcasting and broadband Internet services online media viewing platforms.
(3) Reinforce the industry's online structure, advancing towards 1G high-speed network development to prepare for 4K resolution media services, and putting a greater distance between oneself and competitors. Users should be able to use mobile networks to view media when they are outside, then use cable TV's fixed Internet services to continue viewing when they arrive home. Wifi hotspots can even be set up for consumers to best leverage the features of cable TV's Internet services.
(4) There are no boundaries in the world of network communications - redefine the competition stage, using online social networks and communication apps to launch online and media businesses, broadcasting media real-time from cable TV. This is different from online TV platforms that require conversion, and can attract young users. Another approach is to adopt collaborations with mobile networks or online platforms to expand the scope of business operations, and put increased pressure on new competitors seeking to enter the market.

6 CONCLUSION

After the digitization of the cable TV industry, operators invested large amounts of resources and capital into importing new technology and new services. However, the industry's profit-making potential is limited by the threat of potential and new entrants into the industry in this digital convergence era.

To prevent these threats, enterprises have to either greatly reduce their fees or make large investments to deter new competitors from entering the market, but these are strategies that can be adopted by new entrants at the same time. For example, the ultimate winner in the competition between Taiwan's New Taipei City cable TV industry's new and existing operators were the consumers only.

To conclude, this thesis suggests that future cable operators can apply three competitive strategies to increase their competitiveness as described in the following: (1) conform to the changes in the industry's environment and consumer behavior, and utilize those changes wisely to expand the existing scope of operations; (2) increase entry barriers and the distance with competitors, and adopt industry or cross-industry collaborations to put pressure on potential and new entrants; (3) leverage the role of local services and digital convergence technology, using the close connection between a variety of transmission media and the consumers to increase the adhesion in the industry and also corporate identity recognition.

REFERENCES

Executive Yuan, 2010. *Digital Convergence Development Plan 2010–2015*, Taipei City, 2010.
M.E. Porter, 1979. *Harvard Business Review 57*(2) 137–145.
M.E. Porter, 1980. *Competitive Strategy: Techniques for Analyzing Industries and Competitors*. New York: Free Press.
M.E. Porter, 2008. *Harvard Business Review 86*(1) 57–71.
M.H. Li and R.M. Chiu, 2010. *Competitive Advantage*. Commonwealth Publishing Group, 2010.
National Communications Commission, 2016. *2016 Communications Business Overview*. Taipei City: National Communications Commission.

Communication science & engineering

Smart Science, Design & Technology — Lam et al. (eds)

Performance of fuzzy inference system for resource allocation in underlying device-to-device communication systems

Yung-Fa Huang*
Department of Information and Communication Engineering, Chaoyang University of Technology, Taichung, Taiwan

Tan-Hsu Tan & Geng-Wei Lin
Department of Electrical Engineering, National Taipei University of Technology, Taipei, Taiwan

Hsinying Liang & Sheng-Qiang You
Department of Information and Communication Engineering, Chaoyang University of Technology, Taichung, Taiwan

ABSTRACT: In this study, the fuzzy inference machine (FIM) is applied to the Conventional Heuristic Algorithm for resource allocation in the Device-to-Device (D2D) communication system to improve the system throughput. In the middle and heavy load environment where the number of channels cannot be allocated to all D2D users, the system can be operated in the Reuse Mode. At this mode, cellular systems will suffer cochannel interference between users due to the underlying communication. Therefore, this study applies FIM with conventional method in heavy load. In this paper, the proposed FIM is applied to the conventional method, which can not only improve energy efficiency but also improve system throughput.

1 INTRODUCTION

In recent years, the increasing service demand for smart devices has led to a dramatic growth in the transmission rate requirements of wireless communication systems (S. Koh, et al. 2017). The Device-to-Device (D2D) communication system is an important technology used in 5G communication systems. The D2D communication system does not need to be transferred through the base station. The communication method is a direct link, thereby improving spectral efficiency of wireless communication systems (L. Melki, et al. 2015). D2D communication with mode selection has received extensive attention (S. Hakola, et al. 2010, G. Yu, et al. 2014, M. Jung, et al. 2012). Through D2D proximity communication, the communication device can accurately search and quickly match in a certain area, and the user can conveniently interact with and share data with surrounding friends (C.-P. Chien, et al. 2012, L. Lei, et al. 2012).

In cellular network systems, using mode selection can alleviate the interference caused by D2D communication according to the communication quality of D2D users and cellular users (K. Doppler, et al. 2010, D. Wu, et al. 2017). Therefore, this study investigates the optimization of resource allocation in D2D communication system, in addition to considering the impact of D2D Pair and Cellular User on the performance, and proposes an FIM for power control to effectively improve system throughput and energy efficiency.

2 SYSTEM MODELS

In this study, the impact of mode selection, resource allocation, and power allocation on the performance of the uplink transmission of the D2D communication system is investigated. The system model is shown in Figure 1.

In the D2D communication modes of Figure 1, Figure 1(a) is a Cellular Mode, which communicates with the cellular user through the base station. The resources consumed in this mode are the highest in the three modes. In Figure 1 (b), a dedicated mode is shown where D2D users use a dedicated channel for direct transmission in which the channel is allocated to D2D users for transmission. It has less resource cost than the cell mode because the positions of the transmitting end and the receiving end of the D2D user are in a neighboring relationship. Figure 1 (c) shows the Reuse Mode that the two D2D users share with the

---

*Corresponding author. E-mail: yfahuang@cyut.edu.tw

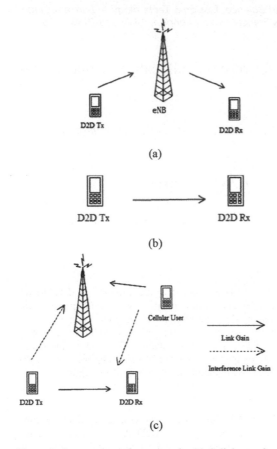

(a)

(b)

(c)

Figure 1. System channel usage mode: (a) Cellular mode, (b) Dedicated mode, (c) Reuse mode.

Table 1. Simulation parameters.

Base station coverage (m^2)	500 × 500
The distance between the transmitter and receiver of the D2D pairs (d_D, m)	20
Number of uplink channels	20
Number of downlink channels	20
Number of device-to-device user pairs (K)	1, 2,..., 15
Number of cell users (M)	20
System bandwidth (B_T, MHz)	3
Maximum transmission power of all users (p^D_{max}, dB m)	24
Minimum transmission power for all users (p^D_{min}, dB m)	13
Standard deviation of shadow fading (dB)	10
Noise power spectral density (dBm/Hz)	-174

cellular user's channels. This mode can improve spectrum usage efficiency because it does not use any dedicated channel. It needs to be carefully managed, however; otherwise, the throughput of D2D users and cellular users will deteriorate due to cochannel interference.

In this paper, it is assumed that the coverage of the base station is a circular area with a radius of 500 m. The base station is the center, and the cellular users and D2D users are uniformly distributed in the coverage area of the base station. The distance be-tween the transmitting end and the receiving end is d_D for a D2D pair. It is assumed that the base station knows the channel status of all links. Other simulation parameters are shown in Table 1.

3 FUZZY INFERENCE SYSTEMS

Figure 2 shows the function block diagram of the proposed FIM (E. H. Mamdani, et al. 1975). After the input parameters are fuzzified by the fuzzifier, the input data are fuzzy inferred with the fuzzy rules base. Finally, the inferred data are defuzzified to output parameters.

Figure 3 shows the membership function of the input fuzzy parameters. Figure 4 shows the membership function of the output fuzzy parameters. The fuzzy parameters of the output and input are all triangle functions. We assume that in the input fuzzy parameters $d_{D,B}$, $d_{D,C}$, $d_{C,B}$, and d_D, the terms of the membership functions are named by N (Near), M (Medium), and F (Far) as shown in Figure 3, respectively. The terms of L (Low) and H (High) membership functions of output parameters $P^c_{k,m}$ and $P^{(3)}_{k,m}$ are shown in Figure 4, respectively.

The fuzzy rule base of this study has four IF parts and one THEN part. In each IF part, the three terms of N, M, and F respectively represent the near (Near), middle (Middle), and far (Far) of the distance. In each THEN part, the two terms L and H represent the high and low powers,

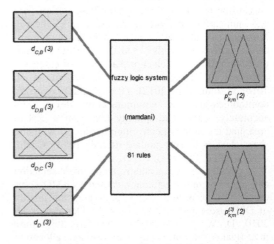

System fuzzy logic system: 4 inputs, 2 outputs, 81 rules

Figure 2. Function block diagram of the fuzzy inference ma-chine.

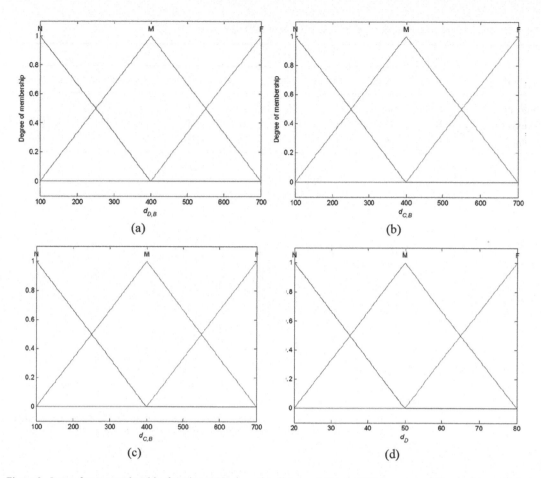

Figure 3. Input fuzzy membership functions: (a) $d_{D,B}$, the distance between D2D user transmitter and base station; (b) $d_{C,D}$, the distance between cellular user and D2D user receiver; (c) $d_{C,B}$, the distance between cellular user and base station; (d) d_D, D2D user transmitter and receiver.

respectively. The inference rule base is shown in Tables 2 and 3.

4 SIMULATION RESULTS

The performance of D2D Reuse Gain (DRG), Cellular Reuse Gain (CRG), and System Reuse Gain (SRG) is compared for conventional method and Fuzzy as shown in Figure 5. The reuse mode of D2D users is working at $K = 1$. The transmit power of all users of conventional method is the highest value (24 dBm) expressed with HP. The DRG is the total throughput obtained by the D2D user operating in the reuse mode. The CRG is the total throughput reduced by the interference in the reuse mode. The SRG is the sum of D2D DRG and CRG. It can be

Table 2. The inference rule base for the transmission power of cellular users.

IF/ THEN		N($d_{C,B}$)			M($d_{C,B}$)			F($d_{C,B}$)		
		N ($d_{D,B}$)	M ($d_{D,B}$)	F ($d_{D,B}$)	N ($d_{D,B}$)	M ($d_{D,B}$)	F ($d_{D,B}$)	N ($d_{D,B}$)	M ($d_{D,B}$)	F ($d_{D,B}$)
N ($d_{D,C}$)		H	H	L	H	H	L	L	L	H
M ($d_{D,C}$)	N (d_D)	L	H	L	L	L	L	L	L	L
F ($d_{D,C}$)		L	H	L	L	H	H	L	L	L

Table 3. The inference rule base for the transmission power of D2D pair users.

IF/ THEN		$N(d_{C,B})$			$M(d_{C,B})$			$F(d_{C,B})$		
		$N(d_{D,B})$	$M(d_{D,B})$	$F(d_{D,B})$	$N(d_{D,B})$	$M(d_{D,B})$	$F(d_{D,B})$	$N(d_{D,B})$	$M(d_{D,B})$	$F(d_{D,B})$
$N(d_{D,C})$		L	L	H	H	L	H	H	H	H
$M(d_{D,C})$	$N(d_D)$	H	H	H	H	H	H	H	H	H
$F(d_{D,C})$		H	H	H	H	H	H	H	H	H

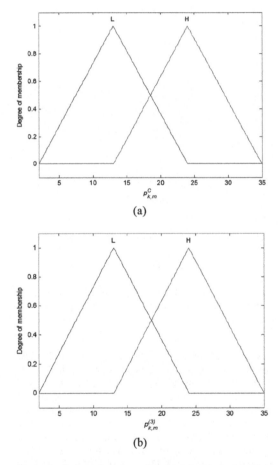

(a)

(b)

Figure 4. Output fuzzy membership function: (a) $P^c_{k,m}$, **dBm** the transmission power of cellular users; (b) $P^{(3)}_{k,m}$, **dBm** the transmission power of D2D users.

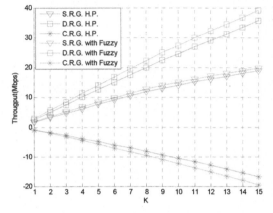

Figure 5. The comparisons of System Reuse Gain (SRG) and D2D Reuse Gain (D2D) for $M = 20$ ($d_D = 20$ m).

seen from Figure 5, at $M = 20$, the D2D DRG performance of proposed fuzzy outperforms the conventional method. The, CRG, however, is inferior to the conventional method, indicating that the interference is high due to the D2D reusing the channels.

5 CONCLUSION

In the study of the Device-to-Device (D2D) communication system, we combine the fuzzy inference machine with the conventional method. With $M = 20$, although the CRG is 15.8% lower than that of the conventional one, the D2D DRG is increased by 9.6%. Then the SRG increases by 2.2%. Therefore, this study combined with the FIM to inspire the conventional heuristics method, to improve system performance.

ACKNOWLEDGMENT

This research is partially sponsored by Chaoyang University of Technology (CYUT) and Higher Education Sprout Project, Ministry of Education, Taiwan, under the project name "The R&D and the cultivation of talent for Health-Enhancement Products." This study was also partially supported by Ministry of Science and Technology of Taiwan under the Contract MOST 106-2221-E-324-020.

REFERENCES

S. Koh, and S. Lee, 2017. Implementation of Open Air Interface control software for 4G network, *Inter. Conf. Ubiquitous and Future Networks*, 747–749.

L. Melki, S. Najeh, and H. Besbes, 2015. System Performance of Cellular Network Underlaying D2D Multi-Hop Communication, *Inter. Conf. Commun. and Networking*, 1–5.

S. Hakola, T. Chen, J. Lehtomaki, and T. Koskela, 2010. Device-to-Device (D2D) Communication in Cellular Network—Performance Analysis of Optimum and Practical Communication Mode Selection, *IEEE Inter. Conf. Wireless Commun. and Networking*.

G. Yu, L. Xu, D. Feng, R. Yin, G.Y. Li, and Y. Jiang, 2014. *IEEE Trans. Commun.*, 62, 11, 3814–3824.

M. Jung, K. Hwang, and S. Choi, 2012. Joint Mode Selection and Power Allocation Scheme for Power-Efficient Device-to-Device (D2D) Communication, *IEEE Inter. Conf. Vehi. Techno.*, 1–5.

C.-P. Chien, Y.-C. Chen, and H.-Y. Hsieh, 2012. Exploiting Spatial Reuse Gain through Joint Mode Selection and Resource Allocation for Underlay Device-to-Device Communications, Inter. Sym. Wireless Per. Multimedia Commun., 80–84.

L. Lei, Z. Zhong, C. Lin, and X. Shen, 2012. *IEEE Wireless Commun.*, 19, 96–104.

K. Doppler, C.H. Yu, C.-B. Ribeiro, and P. Janis, 2010. Mode Selection for Device-to-Device Communication Underlaying an LTE-Advanced Network, *IEEE Inter. Conf. Wireless Commun. and Networking*, 1–6.

D. Wu, L. Zhou, J. Lehtomakin, and L. Zhou, 2017. Social-Aware Distributed Joint Mode Selection and Link Allocation for Mobile D2D Communications, *IEEE Inter. Conf. Commun.*, 851–856.

E.H. Mamdani, and S.Assilian, 1975. *Inter. J. Man-Machine Studies*, 7, 1–13.

Smart Science, Design & Technology — Lam et al. (eds)
© 2020 Taylor & Francis Group, London, ISBN 978-0-367-17867-3

Software-defined network for multipath routing of IoT data

Chuan-Bi Lin* & Yu-Chiang Chang
Department of Information and Communication Engineering, Chaoyang University of Technology, Taichung, Taiwan

ABSTRACT: As the Internet of Things (IoT) rapidly evolves, a large amount of data is generated. However, data loss and network security problems are caused by traffic congestion due to selecting the best or shortest path for transmission in the general network. Therefore, to solve the above problems, a novel network architecture, called Software-Defined Network (SDN), is proposed. The SDN uses the controller and Openflow protocol to divide the routes into a control plane and a data plane. In the SDN, we can adopt multipath to deal with a large amount of IoT data without changing the number of hardware devices. Therefore, in this paper, a SDN architecture consisting of Segment Routing (SR), multipath routing, and weight-based priority mechanisms is proposed to transmit IoT data with. We expect that the performance of IoT data transmission in the future can achieve efficient traffic management, bandwidth usage, and quality of service by mininet simulation.

1 INTRODUCTION

Because of the rapid development of the Internet of Things (IoT) (Hassanein et al. 2017; Wang et al. 2015; Yue Pan et al. 2014), a large amount of data has been generated. However, the general network transmission, adopting the shortest path or the best path to deal with the large amount of IoT data, may cause problems of traffic congestion and network security (Liu et al. 2018). Although multipath transmission is proposed (Chiang et al. 2017), the complex settings of network devices can increase labor and time costs. Therefore, to solve the problems of data loss and network security, a novel network architecture, called Software-Defined Network (SDN) (Kreutz et al. 2015), is proposed. The SDN is divided into control plane and data plane by the controller and openflow protocol (Malishevskiy et al. 2014). It mainly adopts centralized management by the controller to use the bandwidth effectively. Therefore, the SDN can dispatch the multipath to transmit IoT data by using a software program in the controller without setting the number of the hardware devices.

However, because the number of forwarding data flows increases, the flow tables in the SDN switches also grow. As the flow tables grow, more storage resources are needed. To meet this demand, the TCAM (ternary content addressable memory) (Long et al. 2012) is used in the SDN. But because the cost of TCMA is too high, we adopt segment routing (SR) (Dugeon et al. 2017; Davoli et al. 2015) to reduce cost in the SDN. The SR is used mainly to tag packets and then transmit them via the source routing.

In addition, if the sensing data is too much to allow transmitting more important data first, the users may not receive instant information to monitor and prevent something. To let important packets transmit first in the congestion traffic, we adopt priority (Jiawei et al. 2019) values for the data, called weight-based priority. The more important the data, the bigger its weight value, that is, the data is transmitted first according to the biggest value.

It is of interest to know if a network can achieve high throughput and low delay under IoT data traffic without setting the number of hardware devices.

Therefore, in this study, we propose a network architecture based on SDN and consisting of SR, multipath routing, and weight-based priority methods to send sensing data to the destination. We expect that in the future, the performance can achieve high throughput, low delay, and efficient bandwidth management for IoT data transmission by using mininet simulator.

The rest of this paper is organized as follows. We introduce related works in the second section. The third section is our system architecture and weight-based priority method. Finally, we discuss the paper and future works.

2 RELATED WORK

In this section, we introduce the related studies of SR and SDN. The related descriptions are as follows.

A. SDN architecture

The SDN uses the concept of openflow protocol to separate the network architecture, the control

---

*Corresponding author. E-mail: cblin@cyut.edu.tw

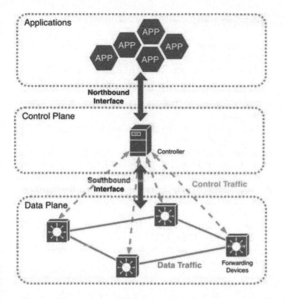

Figure 1. SDN architecture.

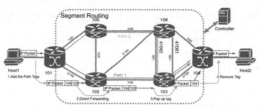

Figure 2. The routing of SR combined with SDN.

plane, and the data plane, as shown in Figure 1. The control plane has a controller to control the entire network to implement a centralized management and exchange information between network devices and applications. The data plane includes some openflow switches to implement the instruction from the controller and forward the packets to destinations. Besides, the SDN is divided into the northbound and southbound interfaces.

In the southbound interface, the openflow protocol is usually used, which is one of the communication standards of SDN. The controller can modify the flow table or routing table through openflow, control the forwarding device's scheduling of traffic, and improve overall network performance and utilization. In the northbound interface, the controller can provide a unified interface for the upper application layer. This is because it lets the applications manage the underlying forwarding devices centrally and reduce the complexity of management and maintenance effectively.

B. Segment Routing (SR)
The SR is based on the SR-MPLS (Multi-Protocol Label Switching) for the forwarding process (Sinha et al. 2017; Maila et al. 2017) in the SDN. It adopts the label, as MPLS label stacking, and SPF (Shortest Path First) algorithm (Jiang et al. 2014; Jasika et al. 2012; Rosyidi et al. 2014) to route the data. The SR is used to convert the paths of passed nodes to the segment list, and the data accords the list for routing. Here, an example is used to explain the SR, as shown in Figure 2. When Host 1 sends an IP

packet to Host 2, the paths are converted to a segment list according to SR, and the list is stored in the controller, e.g., Path 1 = {101, 102, 103, 104} and Path 2 = {101, 105, 106, 103, 104}. Then, the controller gives the SPF information to Host 1 after inquiring. Besides, the packet drops the related label and changes the segment list when passing through the node (switch). For instance, the 101 label is dropped and the list becomes Path 1 = {102, 103, 104} when the data passes through node 101. Therefore, in this study, we will adopt the SPF of SR to route the data with the biggest weighted value first.

3 SYSTEM ARCHITECTURE

To simulate the proposed methods by mininet, we propose a system architecture including three sensors, a SDN-based network architecture SR, multipath routing, and weight-based priority methods, as shown in Figure 3. The functions of three sensors are mainly for temperature, smoke, and carbon dioxide. Next, the executed steps of the proposed system architecture are described as follows.

Step 1: The three sensors collect sensing data and send it to the gateway by wired or wireless networks. Then, the sensing data is stored temporarily in the gateway.

Step 2: Before beginning to transmit, the sensing data in the gateway sends the inquiry to the controller.

Step 3: When the controller receives the inquiry from the gateway, it gives the information of multipath and weighted value to sensing data and selects

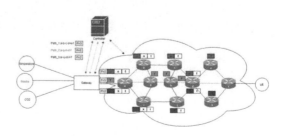

Figure 3. Experimental architecture.

the shortest path to the data with the biggest weighted value by SR.

Step 4: The IoT data is routed according to the rules from the controller under the data plan of a SDN.

4 DISCUSSION

In this paper, we propose a new network architecture based on SDN for IoT data transmission. This is because a large amount of sensing data causes traffic congestion and is easily lost via a single path or an optimal path transmitting in the general networks. The proposed SDN architecture consists mainly of segment routing, multipath routing, and weight-based priority mechanisms to send IoT data to the destination. In future work, we expect that the performance of IoT data can achieve high throughput, low delay, and efficient bandwidth management by using a mininet simulator.

ACKNOWLEDGMENT

This research is partially sponsored by Chaoyang University of Technology (CYUT) and Higher Education Sprout Project, Ministry of Education, Taiwan, under the project "The R&D and the cultivation of talent for Health-Enhancement Products."

REFERENCES

H.S. Hassanein and S.M.A. Oteafy, 2017. *IEEE. DCOSS.* 207–208.

L. Wang and R. Ranjan, 2015. *IEEE. MCC.* 2 76–80.

Y. Pan, Y. Li and J. Zhang, 2014. *IEEE. CCIOT.* 131–134.

Y. Liu, Y. Kuang, Y. Xiao and G. Xu, 2018. *IEEE. JIOT.* 5 257–268.

Y. Chiang, C. Ke, Y. Yu, Y. Chen and C. Pan, 2017. *IEEE. ICASI.* 1247–1250.

D. Kreutz, F.M.V. Ramos, P.E. Veríssimo, C.E. Rothenberg, S. Azodolmolky and S. Uhlig, 2015. *IEEE. JPROC.* 103 14–76.

A. Malishevskiy, D. Gurkan, L. Dane, R. Narisetty, S. Narayan and S. Bailey, 2014. *IEEE. GREE.* 73–74.

F. Long, Z. Sun, Z. Zhang, H. Chen and L. Liao, 2012. *IEEE. ICSAI.* 1218–1221.

O. Dugeon, R. Guedrez, S. Lahoud and G. Texier, 2017. *IEEE ICIN.* 143–145.

L. Davoli, L. Veltri, P.L. Ventre, G. Siracusano and S. Salsano, 2015. *IEEE EWSDN.* 111–112.

W. Jiawei, Q. Xiuquan and J. Chen, 2019. *IEEE. iet-com.* 13 179–185.

Y. Sinha, S. Bhatia, V.S. Shekhawat and G.S.S. Chalapathi, 2017. *IEEE. SDS.* 156–161.

G. Maila, I. Marius and C. Victor, 2017. *IEEE. ATEE.* 34–38.

J. Jiang, H. Huang, J. Liao and S. Chen, 2014. *IEEE. APNOMS.* 1–4.

N. Jasika, N. Alispahic, A. Elma, K. Ilvana, L. Elma and N. Nosovic, 2012. *IEEE. MIPRO.* 1811–1815.

L. Rosyidi, H.P. Pradityo, D. Gunawan and R.F. Sari, 2014. *IEEE. IGBSG.* 1–4.

Computer science & information technology

Smart Science, Design & Technology — Lam et al. (eds)
© 2020 Taylor & Francis Group, London, ISBN 978-0-367-17867-3

A secure ambulance communication protocol in VANET

Chin-Ling Chen
School of Computer and Information Engineering,Xiamen University of Technology, Xiamen, Fujian Province, PR China
School of Information Engineering, Changchun Sci-Tech University, Changchun City, Jilin Province, PR China
Department of Computer Science and Information Engineering, Chaoyang University of Technology, Taichung, Taiwan

Yong-Yuan Deng*, Kai-Wen Zheng, Ting-Lun Yang & Jian-Zhi Huang
Department of Computer Science and Information Engineering, Chaoyang University of Technology, Taichung, Taiwan

ABSTRACT: VANET has been the focus of research in recent years. The main application of VANET is to improve road safety and reduce traffic accidents. In addition, VANET can also reduce the time for the rescue vehicle to arrive at the accident location. However, the transmitted messages are vulnerable to unauthorized access, affect the efficiency of rescue, and even threaten the privacy of the injured. Therefore, it will be important to provide a protected communication protocol for ambulances during the rescue process. In this study, we propose a secure ambulance communication protocol in VANET through a cryptography mechanism. The proposed scheme can not only protect the privacy of the injured, but also provide instant traffic information for ambulances during the rescue process. Our proposed scheme achieves data integrity, user anonymity, user untraceability, forward and backward secrecy, and resistance to replay attack.

Keywords: ambulance, VANET, authentication, security

1 INTRODUCTION

The development of vehicles has been rapid, and over the years vehicles have become the main means of transportation. However, the occurrence of accidents in vehicles has also increased year by year. In order to reduce the incidence of accidents, there have been many studies of VANET (vehicular ad hoc network). VANET evolved from a handheld roaming network, but because handheld roaming network is limited in speed, there is an improvement with VANET (Liu et al. 2015; Huang et al. 2013).

In 2006, Jungels et al. proposed that VANET can be divided into two types, vehicle-to-vehicle (V2V) and vehicle-to-road communication (V2R) (Jungels et al. 2006). The main application in VANET is emphasis on traffic safety warnings, reducing traffic accidents and traffic flow control. For example, ambulances can increase the efficiency of rescue through VANET.

As mentioned above, we can provide a communication protocol through VANET that allows the ambulance to communicate with the hospital during the rescue process. The hospital obtains immediate traffic information through the road condition center, then the ambulance gets this information from the hospital. These transmitted messages, however, are vulnerable to unauthorized access, affect the efficiency of rescue, and even threaten the privacy of the injured (Fu et al. 2014; Liu et al. 2015). Therefore, it will be important to provide a protected communication protocol for ambulances during the rescue process.

In 2005, Raya et al. proposed three security issues for VANET: attacking security-related applications, attacking payment mechanisms, and attacking privacy issues of users (Raya et al. 2005). In 2016, Yang et al. proposed a secure and efficient handover authentication scheme for the mobile network environment (Yang et al. 2016). The mobile device needs to register only once to the authentication server, then other service servers can check the legality of the mobile device.

Yang et al. claimed that the proposed scheme achieves several security requirements, such as access grant, key establishment, data integrity, user anonymity, user untraceability, forward and backward secrecy,

* Corresponding author. allen.nubi@gmail.com

and attack resistance. We find that the proposed scheme satisfies lots of security requirements, but does not fully satisfy two of them: one is user untraceability, and the other is forward and backward secrecy.

In Yang et al.'s scheme, the mobile user picks a different pseudonym identity to register to the new service server every time the mobile device links to a different network server. The authors said that this can avoid user untraceability. However, when the mobile device and the service server authenticate with each other successfully, they negotiate a session key. The session key won't change until the mobile device links to another new service server. Therefore, Yang et al.'s scheme doesn't fully satisfy user untraceability, even forward and backward secrecy.

We made an improvement on their proposed scheme and applied it in the VANET environment. We proposed a secure ambulance communication protocol that allows the ambulance to communicate with the hospital during the rescue process. The ambulance can obtain immediate traffic information from the hospital. Our proposed authentication mechanism achieves security, privacy, and efficiency.

2 THE PROPOSED SCHEME

2.1 System architecture

The system framework of the proposed scheme is shown in Figure1.

1) All ambulances and hospital servers must register with the authentication server through a secure channel. The ambulances and hospital servers send their ID to the authentication server. The authentication server returns information that includes parameters calculated by Elliptic Curve Group technology.

2) Once the ambulance wants to connect to the hospital server, it sends its pseudo ID and parameters calculated by Elliptic Curve Group technology to the hospital server. The hospital server checks the legality of the ambulance and responds with information that includes its ID, encrypted transaction number, and parameters calculated by Elliptic Curve Group technology. The ambulance checks the legality of the hospital server and responds with information to the hospital server.

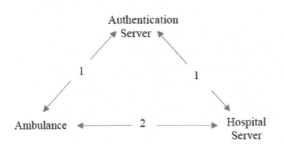

Figure1. System framework of the proposed scheme.

2.2 Notation

q : a k-bit prime
F_q : a prime finite field
E/F_q : an elliptic curve E over F_q
G : a cyclic additive group of composite order q
P : a generator for the group G
S : a secret key of the system
PK : a public key of the system, $PK = sP$
$H_i()$: ith one-way hash function
PID_x : x's pseudonym identity of the ambulance
ID_x : x's identity of the hospital server
r_i, a, b: a random number of elliptic curve group
S_x : x's signature of elliptic curve group
TID : a transaction number, which changes every round
PSK : a session key established by ambulance and hospital server
$E_x()$: use a session key x to encrypt the message
$D_x()$: use a session key x to decrypt the message
CHK_x : the x's verified message
$A \overset{?}{=} B$: determines if A is equal to B

2.3 System initialization

In system initialization stage, the authentication server calculates some parameters and publishes the public parameters for ambulances and hospital servers.

Step 1: The authentication server chooses a k-bit prime p, and determines the tuple of elliptic curve group $(F_p, E/F_p, G, P)$.

Step 2: The authentication server then choosesssas a system secret key and computes

$$PK = sP \qquad (1)$$

as a public key of the system.

Step 3: Finally, the authentication server chooses the hash function $(H_1(), H_2(), H_3(), H_4())$, and then publishes
$(F_p, E/F_p, G, P, PK, H_1(), H_2(), H_3(), H_4())$ to all ambulances and hospital servers.

2.4 Registration phase

2.4.1 Ambulance registration

Step 1: The ambulance chooses a pseudonym identity PID_{AM_i} and sends it to the authentication server.

Step 2: The authentication server chooses a random number r, calculates

$$R_{AM_i} = rP, \qquad (2)$$

$$h_{AM_i} = H_1(PID_{AM_i} \| R_{AM_i}), \qquad (3)$$

$$S_{AM_i} = r + h_{AM_i}s, \qquad (4)$$

and then sends (R_{AM_i}, S_{AM_i}) to the ambulance.
Step 3: The ambulance verifies

$$S_{AM_i}P \overset{?}{=} R_{AM_i} + H_1(PID_{AM}\|R_{AM})PK, \qquad (5)$$

if it passes the verification, then the ambulance stores (R_{AM_i}, S_{AM_i}).

2.4.2 *Hospital server registration*

Step 1: The hospital server chooses an identity ID_{HS}, and sends it to the authentication server.

Step 2: The authentication server chooses a random number r, calculates

$$R_{HS} = rP, \qquad (6)$$

$$h_{HS} = H_1(ID_{HS}\|R_{HS}), \qquad (7)$$

$$S_{HS} = r + h_{HS}s, \qquad (8)$$

and then sends (R_{HS}, S_{HS}) to the hospital server.
Step 3: The hospital server verifies

$$S_{HS}P \overset{?}{\underset{HS}{=}} R + H_1(ID_{HS}\|R_{HS})PK, \qquad (9)$$

if it passes the verification, then the hospital server stores (R_{HS}, S_{HS}).

2.5 *Authentication phase*

When the ambulance wants to connect to the hospital server for traffic information, they need to authenticate each other's legality. In our proposed scheme, the ambulance and the hospital server can authenticate each other directly, without connecting to the authentication server.

Step 1: The ambulance chooses a random number a, calculates

$$T_{AM_i} = aP, \qquad (10)$$

and then sends $(PID_{AM_i}, R_{AM_i}, T_{AM_i})$ to the hospital server.

Step 2: The hospital server chooses a random number b, and calculates

$$T_{HS} = bP, \qquad (11)$$

$$PK_{AM_i} = R_{AM_i} + H_1(PID_{AM_i}\|R_{AM_i})PK, \qquad (12)$$

$$K_{HA1} = S_{HS}T_{AM_i} + bPK_{AM_i}, \qquad (13)$$

$$K_{HA2} = bT_{AM_i}, \qquad (14)$$

and the session key

$$PSK = H_2(K_{HA1}\|K_{HA2}). \qquad (15)$$

Step 3: The hospital server then chooses a transaction number *TID*, calculates

$$c = E_{PSK}(TID), \qquad (16)$$

$$CHK_{AH} = H_3(PSK\|T_{AM_i}), \qquad (17)$$

and sends $(ID_{HS}, R_{HS}, T_{HS}, c, CHK_{AH})$ to the ambulance.

Step 4: The ambulance calculates

$$PK_{HS} = R_{HS} + H_1(ID_{HS}\|R_{HS})PK, \qquad (18)$$

$$K_{AH1} = S_{AM_i}R_{HS} + aPK_{HS}, \qquad (19)$$

$$K_{AH2} = aT_{HS}, \qquad (20)$$

and the session key

$$PSK = H_2(K_{AH1}\|K_{AH1}). \qquad (21)$$

The ambulance verifies

$$CHK_{AH} \overset{?}{=} H_3(PSK\|T_{AM_i}), \qquad (22)$$

if it passes the verification, then the ambulance calculates

$$TID = D_{PSK}(c), \qquad (23)$$

$$CHK_{HA} = H_3(PSK\|T_{HS}\|TID), \qquad (24)$$

$$TID_{new} = H_4(TID), \qquad (25)$$

and sends (PID_{AM_i}, CHK_{HA}) to the hospital server.
Step 5: The hospital server verifies

141

$$CHK_{HA} \stackrel{?}{=} H_3(PSK||T_{HS}||TID), \qquad (26)$$

if it passes the verification, then the session key-PSKbetween the ambulance and the hospital server is established successfully. The hospital server also updates the transmission number TID to TID_{new} by

$$TID_{new} = H_4(TID), \qquad (27)$$

for future communication.

3 SECURITY ANALYSIS

3.1 Data integrity

To achieve the integrity of the transaction data, we use elliptic curve cryptography to calculate the session key PSK and also protect the data integrity. The malicious attacker can't use the signatures (K_{HA1}, K_{HA2}) and (K_{AH1}, K_{AH2}) to calculate the correct session key PSK. Only with the correct session key can parties communicate with each other correctly. Thus, attackers can't modify the transmitted message.

3.2 User anonymity

In the proposed scheme, ambulances do not use their true identities. Pseudonyms are used in the proposed protocol, so there is no anonymity problem. However, the ambulance must register with the authentication server in advance, and the authentication server stores the secret key of the ambulance. Thus, the hospital server can authenticate the legality of the ambulance, but it doesn't know the true identity of the ambulance. The proposed scheme achieves anonymity for the ambulance and the injured.

3.3 User untraceability

In the real world environment, some malicious people try to get the location of the injured by tracing the ambulance. If the ambulance sends the same message continuously, a malicious attacker can trace the location of an ambulance. In the proposed architecture, we use random values to avoid location tracing. The pseudonym identity PID_{AM_i} and the transmission number TID are changed every round. Malicious people can't trace the ambulance. Thus, location privacy is protected, and user untraceability is achieved.

3.4 Resist replay attack

A malicious person may get the transmitted message between the ambulance and the hospital server. If he/she pretends to be a legal ambulance or hospital server, then sends the same message again to the other side, he/she can't succeed in the proposed scheme. All messages between the ambulance and the hospital server are protected with the session key PSK, so the malicious attacker can't calculate the correct session key PSK. Because the transmitted messages are changed every round, the same messages can't be sent twice. The attackers will not succeed in replay attack.

3.5 Forward and backward secrecy

Even if the session key PSK between the ambulance and the hospital server is compromised at any point in time by malicious people, the system still satisfies forward and backward secrecy. The attackers may use the session key PSK to make a future communication or use it to get the previous message. In the proposed scheme, the session key PSK is randomly chosen by the ambulance and the hospital server, and can only be used in the current round. The attacker can't use the same session key PSK to make a future communication, or use it to get the previous message. Thus, a secure ambulance communication protocol achieves forward and backward secrecy.

4 CONCLUSION

In the VANET environment, lots of services can be provided through network servers. For example, the ambulance can choose the best route to reach the accident location, increasing the chance of the injured surviving through VANET. However, lots of users' sensitive data are transmitted through the network. Besides legal users, malicious people also want to get these data. There must be a robust authentication mechanism for ambulances and hospital servers to protect the system's security and users' privacy, while also keeping the efficiency.

Previously, Yang et al. proposed an authentication protocol for mobile cloud computing environment. In their proposed scheme, when a user needs to communicate with a service server, the service server can check the legality of the user without connecting to the authentication server. We made an improvement on their proposed scheme and applied it in the VANET environment. We proposed a secure ambulance communication protocol that allows the ambulance to communicate with the hospital during the rescue process. The ambulance can obtain immediate traffic information from the hospital. The proposed scheme meets lots of security requirements, which is suitable for the VANET environment in the real world.

REFERENCES

J. K. Liu, M. H. Au, W. Susilo, K. Liang, R. Lu, B. Srinivasan, "Secure sharing and searching for real-time video data in mobile cloud," *IEEE Network* 29 (2015), 46–50.

K. L. Huang, K. H. Chi, J. T. Wang, C. C. Tseng, "A fast authentication scheme for wimax–wlan vertical handover," *Wireless Personal Communications* 71 (2013), 555–575.

D. Jungels, M. Raya, P. Papadimitratos, I. Aad, J.-P. Hubaux, "Certificate revocation in vehicular ad hoc networks," Technical LCA-Report-2006-006, *LCA*, 2006.

A. Fu, G. Zhang, Z. Zhu, Y. Zhang, "Fast and secure handover authentication scheme based on ticket for WiMAX and WiFi heterogeneous networks," *Wireless Personal Communications* 79 (2014), 1277–1299.

J. K. Liu, C. Chu, S. S. M. Chow, X. Huang, M. H. Au, J. Zhou, "Time-bound anonymous authentication for roaming networks," *IEEE Transactions on Information Forensics and Security* 10 (2015), 178–189.

M. Raya, J.-P. Hubaux, "The security of VANETs", *Proceedings of the 2nd ACM international workshop on Vehicular ad hoc networks* (2005), 93–94.

X. Yang, X. Huang, J. K. Liu, "Efficient handover authentication with user anonymity and untraceability for Mobile Cloud Computing," *Future Generation Computer Systems* 62 (2016) 190–195.

Smart Science, Design & Technology — Lam et al. (eds)
© 2020 Taylor & Francis Group, London, ISBN 978-0-367-17867-3

Indoor positioning system based on improved Weighted K-nearest Neighbor

Chuanbi Lin* & Yongyu Peng
Department of Information and Communication Engineering, Chaoyang University of Technology, Taichung, Taiwan

ABSTRACT: With the rapid development of modern technology in information and communication technology, Location-Based Services (LBS) have been widely used to make life more intelligent and convenient. Indoor positioning technology plays an important role in LBS by providing more service applications. In wireless indoor positioning systems, however, signals are affected by environmental noise during transmission. Therefore, to avoid the problem of interference and improve the accuracy of indoor positioning, we propose in this paper a beacon-based indoor positioning system consisting of fingerprint positioning method, improved Weighted K-nearest Neighbor (WKNN) algorithm, and cutting fingerprint map.

1 INTRODUCTION

Positioning technology has been widely discussed in recent years because the market of Location-Based Services (LBS) (Basiri et al. 2014) has grown from $8.12 billion in 2014 to $39.87 billion in 2019. Global Positioning System (GPS) (S. Sananmongkhonchai et al. 2009) is one of the positioning technologies that is widely used in various outdoor scenes, such as traffic navigation and logistics systems. In GPS, however, satellite signals can be greatly affected by buildings, and so it becomes impossible to use this technology indoors. Therefore, we must rely on indoor positioning technology. Indoor positioning uses short-range wireless communication technologies, such as WiFi (Wireless Fidelity) (Ohta et al. 2015), Bluetooth Low Energy (BLE) (Memon et al. 2017), and other wireless communication technologies. Among these, WiFi positioning and BLE positioning are used by many people. Although WiFi transmission speed is faster, the power consumption and cost of WiFi positioning are higher than BLE positioning. BLE has low cost and low power consumption. BLE is a standard proposed by the Bluetooth Special Interest Group, and it is applicable to Bluetooth version 4.0 or higher.

A beacon is a device, and it transmits the signal by Bluetooth technology to achieve indoor positioning. The special features of beacons are low cost, small size, and spreading various messages within the signal range.

Because Bluetooth has a noise problem, it needs fingerprint positioning to solve it. The fingerprint location method (Qianqian et al. 2009) collects signal strengths from different locations as signal features for each location and stores them in a fingerprint database. Then, the new signal strength is compared with the fingerprint database to know the current location. In addition, to reduce signal instability, we need to collect the fingerprint of the location area first and establish the database. The signal of fingerprint location method passes a Kalman Filter (KF) and uses K-Nearest Neighbor (KNN) algorithm to locate (Zhu et al. 2015). The above methods, however, may cause inaccurate positioning, because the signal has outliers and the algorithm does not consider weights. Therefore, in this paper, we detect the abnormal value after passing a KF, and use the machine learning WKNN algorithm to increase the accuracy of BLE indoor positioning.

This paper is organized as follows: Section 2 describes the work related to indoor positioning; Section 3 introduces the system architecture; Section 4 explains the positioning method; Section 5 presents the discussion.

2 BACKGROUND

A. Wireless sensor network

Wireless sensing networks (WSN) have continued to evolve due to advances in wireless communication technologies, microcontrollers, and battery technologies. The WSN places sensor nodes in the environment, and the nodes use wireless communication technology to transmit collected data to the server for operation as shown in Figure 1. It mainly includes three parts: induction, transmission, and calculation. WSN application (Zhang et al. 2014) is in the fields of smart home, health detection, and environmental monitoring. The network formed by the beacon device through Bluetooth technology in this article is one of the applications belonging to WSN.

---

*Corresponding author: E-mail:cblin@cyut.edu.tw

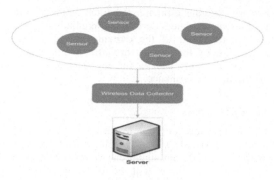

Figure 1. Wireless sensing network system.

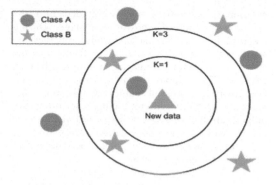

Figure 2. KNN classification diagram.

B. Fingerprint positioning

The fingerprint location method receives signals from different locations as different values and establishes the fingerprint of each location. The image strength obtained by the instant collection is compared with the fingerprint database according to the algorithm. The fingerprint location method is divided into two phases (De Schepper et al. 2017): an offline phase and a positioning phase

In the offline phase, the main purpose is to establish a location fingerprint database before the positioning phase. According to the distribution of indoor environment design nodes, each node collects signal strength and location information, and records them in the database.

In the positioning phase, the beacon signal in the environment is scanned by the smart phone or the smart device. Then, the value of received signal strength indication (RSSI) is uploaded to the server and compared with the fingerprint database stored in the offline phase. After the calculation of the algorithm, the current position information is returned to the user.

C. Machine learning

Machine learning is divided into supervised learning, unsupervised learning, and semisupervised learning. The KNN belongs to supervised learning in machine learning (Dong et al. 2010), and it can judge the input data. The KNN calculates the distance between the unclassified node and the label data, selects the nearest K data according to the distance, and calculates the label data with the largest number of the K data as the predicted classification. Here, the KNN is used to solve the problems of classification and regression. In the classification, the new data is classified at the time of output, and the classification of the new data is determined by the majority of the nearest neighbors, as shown in Figure 2. After inputting new unclassified data, when K = 1, the nearest one node is selected, and thus it is judged as category A. When K = 3, the most recent three nodes are selected, where the number of category B is greater

than category A, so it is judged as category B. In the regression problem, the attribute value of the output data is calculated by the average of the K-nearest neighbors.

D. Weighted K-Nearest Neighbor

The Weighted K-Nearest Neighbor (WKNN) algorithm (Chernoff et al. 2010) based on the KNN algorithm classifies the weights to increase the accuracy of the classification. In the WKNN algorithm, the weights of the nearest neighbors of K are controlled according to the distance. Neighbors with close distances get larger weights, and neighbors with far distances get smaller weights. Then, the weighted average calculation results in the classification of the data.

3 SYSTEM ARCHITECTURE

In this section, we propose an indoor positioning system, which consists of methods presented in section 2. The indoor positioning system architecture is shown in Figure 3. The indoor

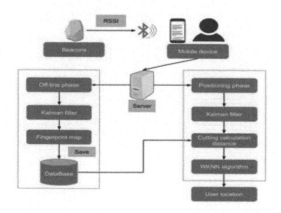

Figure 3. Indoor positioning system architecture.

positioning system is mainly divided into the offline stage and the positioning stage. During the offline phase, the mobile device receives the signal strength values and filters them, and the indoor environment places multiple beacon devices. The signal strength values sent by beacon have different values at different distances. Therefore, different indoor locations are judged on the fingerprint map, and the mobile device will receive the location information of the beacon device and store it in the database to establish a fingerprint map.

In the positioning phase, the fingerprint map of the offline phase is matched with the instantaneous received signal strength value through the WKNN algorithm. The WKNN judges the closest K neighbors, and the weight is determined from these nodes to determine the location. In this paper, the fingerprint map is divided during calculation, which reduces the burden on calculation. Both the offline phase and the positioning phase pass Kalman filtering (KF) to improve signal stability and increase positioning performance.

4 POSITIONING METHOD

A. *Signal filtering*

There is a problem with noise when the mobile device receives the RSSI transmitted by the beacon. This situation causes the RSSI value to be unstable. In this paper, KF is used to improve the stability of the signal. KF (Q. Li et al. 2015) is an optimal recursive data processing algorithm, which is widely used in control systems and image recognition. The KF includes a prediction phase and an update phase as shown in Figure 4. In the prediction phase, the KF uses an estimate of the previous state to make an estimate of the current state. During the update phase, to obtain

a more accurate estimate, the filter optimizes the predictions obtained during the prediction phase with observations of the current state. The pre-processing of the signal through the KF makes the signal more stable, thereby increasing positioning performance.

B. *Offline phase*

The node undergoes KF to turn the signal into a relatively stable state. The fingerprint location method has different signal characteristics in each coordinate, so there will be different periodic laws. After the analysis, the signal characteristics of each node are used to judge the noise, and when the signal is noise, the filtering is performed. The fingerprint map is transmitted to the database for storage. When the fingerprint map is established, cross-validation is performed to check the practicability of the fingerprint map for the positioning phase.

C. *Positioning phase*

The positioning phase is based on the WKNN algorithm for comparison. The WKNN algorithm determines the classification by assigning weights. Multiple beacon devices are used to broadcast signals based on fingerprint location. The RSSI values of each node have different characteristic values. In the indoor environment, the signal is disturbed and the KF is used to improve the signal stability. After filtering, the fingerprint map is used for segmentation calculation. The load is reduced when calculating the distance, then K neighbors are found and weights are assigned. The process of the positioning phase is shown in Figure 5. The following describes the process of cutting fingerprint map, distance calculation, K neighbors, and weight assignment.

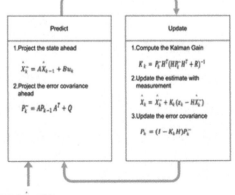

Figure 4. Kalman filter.

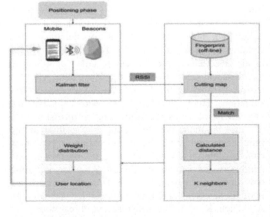

Figure 5. Positioning phase process.

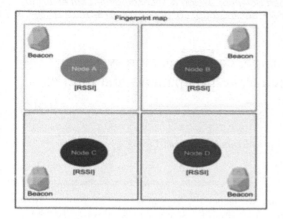

Figure 6. Finger map.

D. Cutting fingerprint map

After the signal is filtered, the instantaneous RSSI signal is compared with the fingerprint map previously established in the offline phase. The instant RSSI signal represents a new node that is not classified. The distance is calculated before the new node finds K neighbors. In the general method, the distance is calculated with each node to find the nearest K nodes. In this paper, the signal features of fingerprint positioning are used, and each node will have different values. The fingerprint map is divided into different blocks as shown in Figure 6. Before segmenting, each node needs to be calculated. After segmenting, only some nodes need to be calculated for the distance. This can reduce the computational burden and improve the performance of positioning.

E. Calculated distance

The new node calculates its distance from the fingerprint map. The closer the distance is, the more likely it is to be similar. After calculating the distance, it is possible to find the K-nearest neighbor nodes. The calculation is performed using the Euclidean distance formula (D. Zou et al. 2013), as shown in (1). The similarity between two nodes is calculated, and then the nearest K nodes are selected and weighted.

Euclidean distance:

$$d(x,y) = \sqrt{(x_1 - y_1)^2 + (x_2 - y_2)^2 + \cdots (x_n - y_n)^2}$$

(1)

F. K neighbors

The new node calculates the distance and selects the nearest K nodes, after a number of positioning measurements to determine the value of K. Generally, the K value is not set to an even number because the value of K will not make a decision if the distance is the same in the situation of the even number. If K = 1, overfitting will occur, and it seriously influences the effect of classification. Because the message characteristics of other nodes are ignored, the decision of K is usually verified from 3.

G. Weight distribution

The distance is calculated according to the Euclidean distance formula. If the distance is farther, the node gets less weight. If the node is closer, the weighted value is larger. The distance represents the same degree of the new node and the fingerprint map, and the weights are from the reciprocal of the distance. Each node weight is weighted and averaged to determine the classification position of the new node.

5 DISCUSSION

This paper proposes a beacon-based indoor positioning system, consisting of fingerprint positioning method, improved Weighted K-Nearest Neighbor (WKNN) machine learning algorithm, and cutting fingerprint map to avoid distance problems and to improve positioning accuracy. In future work, we will evaluate the K values of the improved WKNN algorithm to increase indoor positioning accuracy.

REFERENCES

A. Basiri, T. Moore, C. Hill, P. Bhatia, 2014. *Springer*. 279–282.

S. Sananmongkhonchai, P. Tangamchit and P. Pongpaibool, 2009. *IEEE. TENCON*. 1–6.

M. Ohta, J. Sasaki, S. Takahashi and K. Yamashita, 2015. *IEEE. GCCE*. 483–484.

S. Memon, M.M. Memon, F.K. Shaikh and S. Laghari, 2017. *IEEE. ICETAS*. 1–5.

L. Qianqian, X. Yubin, Z. Mu, D. Zhian and L. Yao, 2009. *IEEE. ITITA*. 40–43.

B. Zhu, L. Yang, X. Wu and T. Guo, 2015. *IEEE. ISCBI*. 74–78.

Y. Zhang, L. Sun, H. Song and X. Cao, 2014. *IEEE. IoT*. 1 311–318.

T. De Schepper, A. Vanhulle and S. Latré, 2017. *IEEE. SCVT*. 1–6.

C. Dong, P.P.K. Chan, W.W.Y. Ng and D.S. Yeung, 2010. *IEEE. ICMLC*. 134–140.

K. Chernoff and M. Nielsen, 2010. *IEEE. ICPR*. 666–669.

Q. Li, R. Li, K. Ji and W. Dai, 2015. *IEEE. ICINIS*. 74–77.

D. Zou, W. Meng and S. Han, 2013. *IEEE. WCNC*. 1564–1568.

Smart Science, Design & Technology — Lam et al. (eds)
© 2020 Taylor & Francis Group, London, ISBN 978-0-367-17867-3

Eye-controlled augmentative and alternative communication system to improve communication quality between dysarthria patients and foreign caregivers in Taiwan

Lung Kuo & Chunching Chen*
Department of Interaction Design, National Taipei University of Technology, Taipei, Taiwan

ABSTRACT: Assistive communication devices are used to enhance the lives of many ALS patients. However, Taiwan has numerous foreign caregivers who cannot read the Chinese text on such devices. To address this language gap problem, we developed an eye-controlled augmentative and alternative communication system, called human eye notification rapid interface (HENRI), with a user-friendly graphical user interface and multilingual pretranslated commands. We conducted experiments by recruiting seven ALS patients with dysarthria who used or planned to use eye-controlled devices. In the experiments, the patients were instructed to execute communication tasks by using their typical communication methods or equipment (called System I) and by using the HENRI system (called System II); subsequently, their performance levels under the two systems were compared. The results showed that the HENRI system improved the communication efficiency, and the HENRI system is suitable as a communication medium between ALS patients and foreign caregivers.

Keywords: eye-controlled interface, amyotrophic lateral sclerosis, multilingual augmentative, alternative communication

1 INTRODUCTION

Amyotrophic lateral sclerosis (ALS), also referred to as motor neuron disease (MND), is a critical neurodegenerative disease. The postonset course of ALS is characterized by gradual difficulty in speaking and swallowing in patients. The disease also compromises limb movement. Eventually, the respiratory muscles are weakened, leading to breathing failure; thus, sustaining life is a difficult task (Murray, 2006). Because patients with ALS have complete mental and sensory functions, they suffer during the course of the disease. In Taiwan, the average annual incidence of ALS was 1.05 per 100,000, with the average age of onset being 52.5 years and the most common age of onset being between 65 and 69 years (Lai and Tseng, 2008). The average survival period after onset is 5.6 years (Lee, et al., 2013). Many studies have indicated that the quality of life of ALS patients is considerably affected by psychological factors (Felgoise, et al., 2016). Approximately 80% of patients with ALS progress to dysarthria, a speech disorder (Tomik and Guiloff, 2010). Most patients experience dysarthria and tetraplegia in the middle and late stages of the diseases. Patients who have lost oral communication ability might feel marginalized and ignored, leading to depression. In addition to experiencing communication problems in the normal course of the disease, many ALS patients receive tracheostomy procedures, which also cause difficulties in oral communication. Accordingly, ALS patients need caregivers' help on all aspects of daily life. They require help in tasks such as drinking water, adjusting their postures, or ameliorating physical discomfort.

2 GUI FOR EYE-CONTROLLED

Since the introduction of eye-controlled systems, many studies have been conducted to improve the corresponding technology. The following regulations govern eye-controlled computer access systems: (a) The average dwell time must be 400–1,000 ms (Kumar, Menges and Staab, 2016). (b) The access method must be accelerated through continual learning (Majaranta, Ahola and Špakov, 2009). (c) During the control process, received and intended messages must be processed simultaneously (MacKenzie, 2014) because the eye is both a sensory receptor and a command applicator, and the interface of an eye-controlled system should not be cluttered with excessive information. (d) The "Midas touch problem," in

---

*Corresponding author. E-mail: cceugene@ntut.edu.tw

which the interface cannot be operated correctly even when it is functionally interactive, must be avoided. (e) The screen of an eye-controlled system must provide visualized feedback, enabling it to act as a "magnet" for attaching to buttons. The screen must also have a zoom-in window or button animation function (Lankford, 2000). (f) Animations or colors can be used at the gaze point to enhance the user experience. Such screen effects help users see their gaze positions, which is vital because an adequate visual feedback mechanism can prevent confusion-based user errors (Majaranta, et al., 2003).

Current research on eye-controlled systems has focused on improving the precision of the corresponding technology. Electrooculography-based human–computer interaction (HCI) and AAC systems have been studied for several years, resulting in a rich body of literature (Kumar, Menges and Staab, 2016; Käthner, Kübler and Halder, 2015; Tsai, et al., 2008; Miniotas, Špakov and MacKenzie, 2004; Bates, et al., 2007). Although ALS-patient-centered experiments have been conducted to promote the development of eye-controlled AAC devices and software (Käthner, Kübler and Halder, 2015; Barresi, et al., 2016; Caligari, et al., 2013; Liu, Huang and Huang, 2017; Spataro, et al., 2014), few of these studies have focused on user communication traits or life care needs. To bridge this research gap, we developed an eye-controlled rapid communication command system—called human eye notification rapid interface (HENRI)—that has a user-friendly graphical user interface (GUI), namely EyeGUI, and multilingual pretranslated commands to overcome communication problems between patients and foreign caregivers and thus facilitate the fulfillment of their needs.

3 EYE-CONTROLLED AAC SYSTEM

3.1 HENRI system

The HENRI system is a multilingual AAC system that allows multiple access methods (e.g., eye gazing, touchscreen, or muscle movement) to control a mouse or an adapted mouse. The primary operating mode of the system is eye control. Moreover, the system was designed with the aim of enabling ALS patients to call for assistance rapidly. It includes four languages, namely Chinese, English, Indonesian, and Vietnamese, but its operating language is Chinese. Speech or voice prompts are generated in Chinese and English, and English, Indonesian, and Vietnamese subtitles are provided. The multilingual function is for patients who employ foreign caregivers in Taiwan. The translated subtitles enable foreign caregivers to understand patients' requests immediately. The user-centered design of the proposed system is aimed at providing patients with a functionality that enables them to rapidly express everyday commands (or routine commands) and provide translations to foreign caregivers to reduce the

time spent on guessing; thus, caregivers can rapidly understand patients' needs, which improves patients' quality of life. To determine whether the proposed system can enable ALS patients rapidly express everyday commands, we conducted live observations and interviews with ALS patients, their family members, and their caregivers with the assistance of the Taiwan Motor Neuron Disease Association.

The everyday commands of ALS patients were collected through interviews and observation. The collected commands were classified in terms of repetitiveness, urgency, accuracy, live entertainment, opinions, and personal needs to produce a set of approximately 200 commonly used items. These items were divided into 12 categories: (1) daily care, (2) special conditions, (3) body sensation, (4) diet, (5) drugs, (6) personal hygiene, (7) sitting position adjustment, (8) talking and chatting, (9) opinion expression, (10) entertainment, (11) movement, and (12) private demands.

The GUI was designed in accordance with the interactive design specifications provided by previous research[19], for eye-controlled systems: (1) designing sufficiently large buttons to avoid inaccurate triggering, (2) providing visualized feedback (animated responses) to enable a user to perceive that a button has been touched, (3) implementing magnetic control to enhance accuracy in the presence of rapid eye jitters, and (4) animating concentric shrinking circles to maintain a user's gaze at a specific position. An animated countdown timer can also be used to determine the amount of time (dwell time) a patient should wait before executing another click. One loop of the timer lasts 0.8 seconds (MacKenzie, 2014).

Buttons are color coded according to their categories to facilitate their identification by users. For example, red represents urgency, green represents diet, blue represents hygiene, and yellow represents entertainment.

3.2 Main interface

After the patient selects the main command button, a follow-up page showing related commands is displayed. For example, if a patient intends to take medicine, the following sequence is followed: (1) the patient selects the "drug" button and then (2) selects the "take medicine" button. Subsequently, the screen displays the text "drug, take medicine" in Chinese, English, and Indonesian or Vietnamese. Concurrently, a voice prompt of the phrase "I want to take medicine" is provided in Mandarin and repeated in English. These voice prompts are repeated three times. The Chinese, English, and Indonesian or Vietnamese texts are displayed in different sizes and colors. Accordingly, this new multilingual design displays three languages on the same page. Each command is quantified and stored in a data file; the commands in the data file can be sorted to expedite the retrieval of repeated user commands.

If the patient has an urgent request that is not part of the menu, he or she can select a service bell (call bell) situated at the top left corner of the main screen to produce a voice prompt of the phrase "I need help" in Chinese followed by English; the prompt is repeated three times. The patient can select the "turn on the alert sound" button located at the lower left corner of the screen; this button activates a loud alert ("ding-ding") that continues until the patient or caregiver deactivates it.

4 COMMUNICATION WITH FOREIGN CAREGIVER

The HENRI screen displays commands pretranslated in multiple languages; thus, foreign caregivers can easily understand patients' requests without guessing their meaning. In our experiments, the enhanced efficiency induced by the pretranslated commands in System II reduced the caregivers' response times as well as the patients' expression times. This improvement is beneficial considering the high number of foreign caregivers in Taiwan; the finding also highlights the importance of using appropriate translation functions in AAC systems.

Transnational workers are prevalent worldwide. Numerous countries have foreign workers whose native languages differ from the local languages used for communication at work (Chin and Phua, 2016). In general, such caregivers can easily learn the local languages. However, Taiwan uses traditional Chinese, which is not easy to learn; therefore, the communication obstacles caused by language differences are more severe in Taiwan compared with the aforementioned countries. Consequently, language obstacles (i.e., speaking and writing) could prevent the application of AAC systems to their full potential. We thus developed the HENRI system to provide voice commands in Chinese and English along with Indonesian and Vietnamese subtitles. This system can enable patients and foreign caregivers to communicate in real time, thus saving communication and guessing time while improving work efficiency and patients' quality of life. The proposed HENRI system is used by patients in their daily life activities, such a tacit understanding may be transferred to the system, thus possibly reducing patients' overreliance on caregivers. Consequently, caregivers could have some rest and improve their service quality.

5 COMMUNICATION WITH FOREIGN CAREGIVER

This study proposes an AAC system, namely HENRI, for enhancing communication efficiency between patients with ALS and their foreign caregivers or family members. This finding supports the results of previous studies (Felgoise et al., 2016; Caligari, et al., 2013; Körner, et al., 2013). The improved efficiency may prevent patients' overreliance on caregivers. Accordingly, the HENRI system can also improve the quality of life of patients, their family members, and

their caregivers. Overall, the HENRI system has multilingual functions and a user-friendly GUI and can improve the quality of communication between patients and foreign caregivers; the system thus responds to the call for a multilingual system that can meet the environmental needs of foreign caregivers (Chin and Phua, 2016; Tönsing, et al., 2018). The findings of this study provide a valuable reference for researchers in relevant fields to further explore the topic in the future.

6 ACKNOWLEDGEMENTS

We give thanks to all of the participants, including patients, family member, and caregivers of the ALS patients. This study was funded by National Taipei University of Tech Technological University Paradigm Taipei Tech project no: 7061401 and Taiwan Motor Neuron Disease. HENRI has been provided for use by Taiwan MNDA member patients.

Figure 1. Main HENRI screen: Buttons distinguish function categories by color.

Figure 2. Button actions: Magnetic control, animated concentric shrinking circles, and countdown timer.

Figure 3. HENRI result screen: Text displayed in three languages.

REFERENCES

B. Murray, 2006. Natural history and prognosis in amyotrophic lateral sclerosis. *NEUROLOGICAL DISEASE AND THERAPY 78* 227.

B. Tomik and R.J. Guiloff, 2010. *Amyotrophic Lateral Sclerosis 11*(1/2) 4–15.

C. Kumar, R. Menges and S. Staab, 2016. *IEEE Computer Society* 6–13.

C. Lai and H. Tseng, 2008. *Neuroepidemiology 31*(3) 159–166.

C. Lankford, 2000. *Eye Tracking Research & Applications* 23–37.

C. Lee, et al., 2013. *Journal of epidemiology 23*(1) 35–40.

C.W.W. Chin and K.H. Phua, 2016. *Journal of Aging & Social Policy 28*(2) 113–129.

D. Miniotas, O. Špakov and I.S. MacKenzie, 2004. *CHI 04 Extended Abstracts on Human Factors in Computing Systems* 1255–1258.

G. Barresi, et al., 2016. Focus-sensitive dwell time in EyeBCI: Pilot study, in *8th Computer Science and Electronic Engineering (CEEC)*. IEEE: Colchester, UK.

I. Käthner, A. Kübler and S. Halder, 2015. *Journal of neuroengineering and rehabilitation 12*(1) 76.

I.S. MacKenzie, 2014. *Human-computer interaction: An empirical research perspective*. Waltham, MA: Morgan Kaufmann.

J.-Z. Tsai, et al., 2008. *Journal of Medical and Biological Engineering 2008*(1) 39.

K.M. Tönsing, et al., 2018. *Journal of Communication Disorders 73* 62–76.

M. Caligari, et al., 2013. *Amyotrophic Lateral Sclerosis and Frontotemporal Degeneration 14*(7–8) 546–552.

P. Majaranta, et al., 2003. *Extended Abstracts on Human Factors in Computing Systems, CHI 03* (ACM) 766–767.

P. Majaranta, U.-K. Ahola and O. Špakov, 2009. *Proceedings of the SIGCHI Conference on Human Factors in Computing Systems*. ACM: Boston, MA, USA. 357–360.

R. Bates, et al., 2007. *Universal Access in the Information Society 6*(2) 159–166.

R. Menges, et al., 2016. eyeGUI: A Novel Framework for Eye-Controlled User Interfaces, in *Proceedings of the 9th Nordic Conference on Human-Computer Interaction*. ACM: Gothenburg, Sweden 1–6.

R. Spataro, et al., 2014. *Acta Neurologica Scandinavica 130*(1) 40–45.

S. Körner, et al., 2013. *Amyotrophic Lateral Sclerosis and Frontotemporal Degeneration 14*(1) 20–25.

S.H. Felgoise, et al., 2016. *Amyotrophic Lateral Sclerosis & Frontotemporal Degeneration 17*(3/4) 179–183.

Y.-H. Liu, S. Huang and Y.-D. Huang, 2017. *Sensors 17*(7) 1557.

Smart Science, Design & Technology — Lam et al. (eds)
© 2020 Taylor & Francis Group, London, ISBN 978-0-367-17867-3

A parametric interactive platform with tangible visualization for intuitive cognition and manipulation of urban information

Yu-Pin Ma*
Department of Architecture, National University of Kaohsiung, Kaohsiung, Taiwan

Wei-Chieh Chang
Institute for Physical Planning & Information, Taipei, Taiwan

ABSTRACT: Inspired by the concept of urban information modeling, this study focuses on the research of the information visualization of large-scale dynamic urban systems. In addition, in response to new generation learning modes and characteristics, this study uses the tangible user interface to develop an innovative design platform based on parametric computing tools to extract and integrate urban information for presenting a dynamic urban system. Such parametric interactive learning tools and platform using tangible visualization computations could provide a more intuitive way to communicate and manipulate urban information for inter-disciplinary integration and collaborative design.

1   INTRODUCTION

In order to convey the context and design concepts of cities, scholars and professionals have used digitally visualized two-dimensional image modes such as spatial statistics and quantitative analysis to transmit information and invoke perception (M. Carmona et al. 2010). Students have also learned the computing modes and development design (J. Beirão 2012, M. Berger et al. 2015) of urban areas through complex data and two-dimensional images. The traditional urban design tools are mainly CAD or GIS (J. Beirão 2012), which have difficulties with model integration. These tools also hinder the intuitive communication of GUIs in design, discussion, and teaching. Such interface tools are inappropriate for the new generation of students, who are deeply influenced by science and technology (M. Prensky 2001).

Aided by information technology (T. Stojanovski 2013, J. Gil et al. 2011), the urban information database and the spatial volume model are becoming larger and more complex. Different projects have different data formats due to different fields and methods of data processing tools, making the visualized and quantitative analysis of urban data time-consuming. To solve this problem, the logical concept of parametric computing has been introduced into the tools and a large number of tools have been developed, such as programmed modeling tools (J. Beirão 2012, Y. I. Parish and P. Müller 2001) and visualized

programming tolos (D. Rutten 2013, M. Berger et al.2015, C. Schneider et al. 2011, M. Bielik et al. 2012).

Mathematic computing with parametric adjustments can be used to visualize urban information. The advantages of parametric modeling lie in accurate lofting and urban data analysis. Therefore, urban development can be managed by accumulating urban information, and urban parameters can be extracted from the database for analysis, simulation, computing, and real-time observation of urban recessive information. This is why city information modeling (CIM) (J. Gil et al. 2011) and urban information systems (UIS) have drawn the attention of scholars and professionals (T. Stojanovski 2013, F. Salim 2014, G. Schubert et al. 2013).

Based on the concept of City Information Modeling (CIM), this study focuses on the information and visualization of large-scale urban dynamic systems. Given the differences between professional modeling and analysis tools, this study avoids the limitation of existing professional tools and explores the continuity and integration of informatization models in urban planning and design.

In response to the learning characteristics of digital natives and the rise of information-based modeling technologies, different programming languages and digital tools have been widely used in the analysis and simulation of urban planning and design (C. Schneider et al. 2011, J. Duarte et al. 2012, B. Hillier & S. Iida 2005, M. Roudsari et al. 2013). Digital design tool or platforms based on information modeling are

---

*Corresponding author: E-mail: yupinma@go.nuk.edu.tw

permeating the interaction between urban planning and design learning. This study intends to explore how to deepen the learning mode of digital natives and create an open and innovative new-generation learning platform, so that the complex and changeable results of urban issues can be presented through informatization-based, visualized, interactive, gamified, and tangible teaching aids (F. Salim 2014, C. M. Rose et al. 2015) to spark students' interest in deep learning (L. Vicent et al. 2015, F. Valls et al. 2017).

2 RESEARCH METHOD

This study focuses on design media that can intuitively assist the perception and computing of urban information. Starting from GIS data-aided urban information modeling, real-time urban information visualization, and platform-based interactive computing, this study adopts computing and design tools to visualize dynamic urban information. In order to increase the operability of the data, this study selects a Tangible User Interface (TUI) as the medium built in the space suitable for discussion, so that the tool interface can be free from the limitation of professional tools.

To implement real-time information visualization and give dynamic feedback, the tasks of selecting parametric design tools is as follows:

- The application and learning of urban planning and urban design are assisted by GIS and parametric computing tools.
- Real-time displays of recessive urban information provide assistance in decision-making and environmental information understanding.
- The development of design media through a tangible user interface (TUI) assists students in learning and perceiving urban information.

2.1 Tangible user interface

As mentioned in the literature review, any computer-aided tool is a window-based user interface. Through the mouse and screen, the user can convert the location information of an object into coordinates on the computer screen. However, complex software functions and interfaces are not intuitive or user-friendly. Unlike a window-based interface, a tangible user interface centers on transforming objects into an operation interface that allows objects to be capable of data input and output, (such as the touch of a hand or object detection). The computer collects and converts physical information through sensors that can read environmental data to trigger events intended by the users.

2.2 Parametric computing design

Grasshopper, a parametric design tool, is one of the visual programming tools used by this study.

Plugged into a Rhino interface, Grasshopper can check and correct planning and design processes by parametric computing logic. The merit of Grasshopper is the ability to extract numerous parameters (including environmental information parameters, physical information, and digital information). Since there are multiple evaluation criteria for the discussion of urban issues, different design criteria can be defined through the computing of parameters, such as shelter ratios, volume ratio restriction conditions, calculation of the best route, analysis of sunshine and shadow surfaces, and horizon analysis. Parameter computing design not only inspects the design criteria that have been set but also creates a dynamic urban model through the change of parameters.

This study uses motion sensing technology to extract features and assist the detection of physical objects so that objects can escape the restriction of barcodes. This study uses the user datagram protocol (UDP) to assist the software analysis and make up for the inadequacy of displays, in order to integrate development environment tools to develop a 2D table interface and achieve an interactive table design. UDP and visualized programming tools are used for data scaffolding to achieve real-time analysis of urban information models, the interactive computing of objects, and the display of dynamic models.

3 PLATFORM DEVELOPMENT AND INTERFACE DESIGN

In this study, parametric computing technology and a tangible user interface (TUI) are used as research methods, and visualized programming tools are used as the research tools. Aiming at the tangible perception of urban information and based on being human-centered, a set of Tangible City Information Perceptive Tables (TCIPT) is built to facilitate discussion and study urban planning and design. With TCIPT as the representative of this research table, the development process can be divided into two parts: hardware for interface design and software for platform development.

3.1 Platform development

Hardware facilities include infrared cameras, infrared lights, motion sensors, short-focus projection, and front projection screens. In order to operate urban information more intuitively, acrylic ① is defined as an interactive operational interface. Meanwhile, through short-focus projection②, the computer screen is projected onto an acrylic surface (in terms of design and operation, this is the base plane for discussion). Infrared light③ and infrared cameras④ are used to realize the intuitive operation of gestures and multi-point touch, and it is necessary to have the function of infrared image recognition. In addition, an infrared camera can be regarded as a low-precision sensor that can recognize the embryonic form of an object placed on the interactive interface, while a Kinect motion sensor⑤ is a high-precision sensor that can identify the colors and depth of an object (e.g. Lego) to generate the height of visualized models. A projection screen⑥ is used to display the perspective of a 3D model or urban information (data or statistical charts, etc.). The screen is used to observe the changes of the information before and after design. In order to intensify the interaction of the tables, the system is equipped with Arduino⑦ which, thanks to the characteristics of microcontrollers, allows the choice of different sensors according to interactive demands or the tangible information desired. As for the table-based platform used in this experiment, a variable resistor (8) is selected to generate digital shapes in the way of selective rotation and torsion.

3.2 Interface design

Due to the need to compile 2D CAD files (.dwg) generated by different geographic information and 3D models generated by different software, as well as to visualize the data for analysis of different topics including 3D information, weather data input assessment, and real data of tangible objects by inserting operational analysis formulas, this study uses Processing as a 2D user interface development tool, Rhino as a 3D display interface for volume and data visualization, and Grasshopper, a visualized programming tool, for computing to analyze and bridge different software. In order to connect with Grasshopper and draw graphics at the interactive end, such as the drawing on the shortest path, it is necessary to connect with the information of other software. Therefore, this study uses the User Datagram Protocol (UDP) as the connection method.

Visualized programming tools are capable of visualizing information through programming languages or simple operation modules. Grasshopper is a plug-in software for Rhino that features great flexibility in computing design because it can add numerous operation modules, such as the Firefly module for reading tangible information, the Mosquito module for reading open government information, and the Ladybug module for analyzing weather, light, and shadow. Visualized programming design tools are to parametric design what programming languages are to information engineering. Visualized programming design tools can use different modules, grammar, and logic to generate different design schemes. This study performs the following tasks. First, it inserts the Firefly module into Grasshopper, executes the Arduino microcontroller, and connects a variable resistor to expand the functions of the table. Second, the Ladybug module is inserted to perform tangible environment analysis on urban models. The results of the data visualization assist in understanding the impact of shelter volume on urban environments and in formulating the strategies of urban design. Third, the gHowl module is used to implement UDP and integrate the advantages of different software.

Overall, the hardware of TCIPT is defined as follows: A is a 3D data visualization display interface, which is used for displaying 3D model volume model or 3D data analysis; B is a 2D

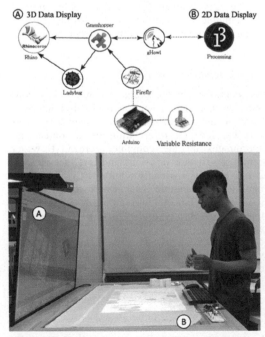

Figure 3. Parametric application framework for TCIPT.

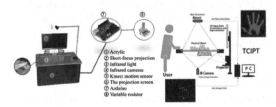

Figure 1. TCIPT hardware installation and system architecture diagram.

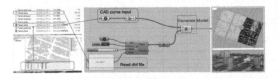

Figure 2. Using Grasshopper to compile 3D models in Rhino generated by 2D CAD files (*.dwg).

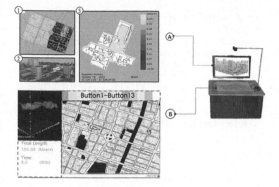

Figure 4. The hardware of TCIPT: A is a 3D data visualization display interface and B is a 2D data visualization interface.

data visualization interface that is used to collect different 2D data, such as plane configuration maps, related tables, statistical charts, urban planning maps, and cadastral maps. In addition, B is also an interactive terminal, which is the input or output interface.

4 CONCLUSION

Based on the urban information model, this study focuses on the informatization and visualization of a large-scale dynamic urban system. In response to the new-generation learning model and its characteristics, this study overcomes the limitations of traditional graphics user interfaces (GUI) and selects innovative design media constructed by a tangible user interface (TUI). By integrating four tools, including the Geographic Information System (GIS) widely used in large-scale sites, Rhino, Grasshopper, and (4) Processing, this study develops the prototype of a design media concept called a Tangible City Information Perceptive Table (TCIPT).

Regarding the need of real-time interaction, this study uses GIS to parameterize urban information and adopts Rhino to integrate 2D and 3D models generated by different software. Through Grasshopper, a parametric computing and visualized programming design tool that is plugged into Rhino is used to extract digitized GIS information. Operate and read climate information through Grasshopper plugin tools to provide real-time information visualization. For intuitive operation, TUI is used as the medium. Processing is used to integrate graphic information and chart information into the table operation interface. Finally, feature extraction and multi-point touch are used to enable users to manipulate the data more intuitively.

The prototype of the design media proposed in this study is mainly aimed at assisting urban information perception. The concept of the design media is proposed through simple instructions for design and operation. The current version is mainly applied to students' learning and the interactive discussion between students and teachers. The incorporation of larger scale models and more complex information, along with abundant user scripts, could allow this design medium to be applied to public discussion and professional work.

ACKNOWLEDGMENT

This research was supported by the Ministry of Science and Technology of Taiwan for her generous research grant, the coded number: MOST 107-2119-M-390-002.

REFERENCES

M. Carmona, T. Heath, T. Oc, S. Tiesdell, 2010. Public Places-Urban Spaces. Oxford: Architectural Press.

J. Beirão, 2012. CItyMaker: Designing grammars for urban design. TU Delft.

M. Berger, et al., 2015. CAD integrated workflow with urban simulation-design loop process. The Sustainable City X, WIT Press. Southampton. 11–22.

M. Prensky, 2001. Digital natives, digital immigrants part 1. On the horizon. 9.5: 1–6.

T. Stojanovski, 2013. City information modeling (CIM) and urbanism: Blocks, connections, territories, people and situations. In Proceedings of the Symposium on Simulation for Architecture & Urban Design. 12.

J. Gil, J. Almeida, and J. Duarte, 2011. The backbone of a City Information Model (CIM): Implementing a spatial data model for urban design. In 29th eCAADe Conference, Ljubljana, Slovenia. 143–151.

Y. I. Parish and P. Müller, 2001. Procedural modeling of cities. Proceedings of the 28th annual conference on Computer graphics and interactive techniques. ACM. 301–308.

D. Rutten, 2013. Galapagos: On the logic and limitations of generic solvers. Architectural Design 83.2. 132–135.

C. Schneider, A. Koltsova, and G. Schmitt, 2011. Components for parametric urban design in Grasshopper from street network to building geometry. In Proceedings of the 2011 Symposium on Simulation for Architecture and Urban Design. 68–75.

M. Bielik, S. Schneider and R. Koenig, 2012. Parametric Urban Patterns-Exploring and integrating graph-based spatial properties in parametric urban modelling. Proceedings of the 30th eCAADe Conference, Czech. 701–708.

F. Salim, 2014. Tangible 3D urban simulation table. Proceedings of the Symposium on Simulation for Architecture & Urban Design. 23–26.

G. Schubert, S. Riedel, and F. Petzold, 2013. Seamfully connected: Real working models as tangible interfaces for architectural design. International Conference on Computer-Aided Architectural Design Futures, Springer, Berlin, Heidelberg. 210–221.

J. Duarte, J. Beirão, N. Montenegro, J. Gil, 2012. City Induction: a model for formulating, generating, and evaluating urban designs. Digital Urban Modeling and Simulation. 73–98. Berlin Heidelberg: Springer.

B. Hillier & S. Iida, 2005. Network and psychological effects in urban movement. International Conference on Spatial Information Theory. 475–490.

M. Roudsari, M. Pak, A. Smith, M. S. Roudsari, M. Pak & A. Smith, 2013. Ladybug: a parametric environmental plugin for grasshopper to help designers create an environmentally-conscious design. 13th Conference of International Building Performance Simulatio Association. 3128–3135. Chambéry France.

C. M. Rose, E. Saratsis, S. Aldawood, T. Dogan, C. Reinhart, 2015. A tangible interface for collaborative urban design for energy efficiency, daylighting, and walkability.

L. Vicent, S. Villagrasa, D. Fonseca, E. Redondo, 2015. Virtual Learning Scenarios for Qualitative Assessment in Higher Education 3D Arts. Journal of Universal Computer Science. 21(8). 1086–1105.

F. Valls, E. Redondoa, D. Fonsecab, R. Torres-Kompen, S. Villagrasa, N. Martí, 2017. Urban data and urban design: A data mining approach to architecture education. Telematics and Informatics.

Smart Science, Design & Technology — Lam et al. (eds)

Evaluation of applying blockchain and cryptocurrencies in marine logistics

Mengru Tu* & Shenglong Kao
Department of Transportation Science and Center of Excellence for Ocean Engineering, National Taiwan Ocean University, Keelung, Taiwan

Chanmin Tsai
Department of Transportation Science, National Taiwan Ocean University, Keelung, Taiwan

Tsaifu Hu
Office of Library and Information Technology, National Taiwan Ocean University, Keelung, Taiwan

ABSTRACT: The global supply chain depends heavily on maritime logistics to transport a variety of goods, from raw materials, parts, to finished products. Maritime logistics has many advantages over air logistics concerning cost and cargo volume. Thus, many manufacturers choose maritime logistics to ship their parts or products to customers because it has lower transportation cost and higher cargo carrying capacity than that of air logistics. However, there are significant inefficiencies in shipping goods across countries due to the complexity of documentation and administrative procedures involved with multiple entities participating in maritime logistics. The rise of blockchain and cryptocurrencies may shed new light on improving efficiency and enhancing transparency for marine logistics processes. This paper explores the potential of applying blockchain technology and cryptocurrencies in marine logistics. A proof-of-concept blockchain (and smart contract) application based on the Ethereum platform is developed to assess the feasibility of the new blockchain-based approach and evaluate the new approach against the current marine logistics practice.

Keywords: marine logistics, blockchain, smart contract, cryptocurrencies

1 INTRODUCTION

Today's global industrial supply chain can extend over great distances crossing multiple geographical locations and may require international shipping to transport a variety of goods, from raw materials, parts, to finished products. For large industrial products and parts, such as cars, engines, and TFT-LCD modules, maritime logistics has many advantages over air transportation regarding cost and cargo volume. Thus, many manufacturers choose maritime logistics to ship their parts or products to customers in different countries owing to lower transportation cost and higher cargo carrying capacity of maritime logistics, as compared to those of air logistics. Despite the benefits of sea freight service for large industrial products, the industrial supply chain usually involves several entities over many production and logistics stages, which include a multitude of documents (e.g., invoices) and payments, thereby increasing the complexity of administrative procedures and human errors in order processing. Thus, maritime logistics still has many inefficiencies in shipping goods across countries in terms of administrative costs and order processing time. For example, customs procedures in each country and bank payments require many documents and processing time. The use of blockchain technology (BCT) for marine logistics might help reduce its complexity and increase its transparency. Using cryptocurrencies together with blockchain might also greatly reduce the time and cost of the financial transaction in international trade. This paper investigates the potential of BCT and cryptocurrencies in supply chain and evaluates the feasibility and benefits of applying them to improving marine logistics.

2 BLOCKCHAIN TECHNOLOGY

Blockchain is a distributed database encompassing a chain of blocks and acts as a global ledger maintaining records of all transactions on a blockchain network (A. Bahga et al. 2016). Blockchain comprises a physical chain of blocks that include 1 to N transactions, where each transaction added to a new block after validation is then inserted into the block

---

*Corresponding author. E-mail: tuarthur@email.ntou.edu.tw

(J. J. Bam bara et al. 2018). Blockchains allow non-trusting members to interact with each other on a distributed peer-to-peer network in a verifiable manner without a trusted intermediary (K. Christidis et al. 2016). BCT provides a solution for transferring value and ownership of digital assets without using any trusted third party, serving as a shared database where all asset transactions recorded in cryptographically chained blocks of data become immutable (S. A. Abeyratne et al. 2016). Thus, BCT naturally becomes an essential underlying infrastructure for many distributed trusted information technology and business applications. These applications include novel financial technology like the use of cryptocurrencies (e.g., Bitcoin) in trading, insurance claims processing, asset management, and many nonfinancial applications such as verifying product pedigree in the supply chain. In the evolution of BCT, smart contracts are incorporated in BCT as blockchain 2.0 technology. A smart contract can be seen as a "computerized transaction protocol that executes the terms of a contract" (N. Szabo 1994). A smart contract contains self-executing scripts residing on the blockchain that allow for the automation of multistep processes, which allow the user to have general-purpose computations occur on the blockchain (K. Christidis et al. 2016). Several blockchian platforms support smart contracts. The two major blockchain platforms for deploying smart contracts are Ethereum and Hyperledger. We will use Ethereum in this study because it has the cryptocurrency token Ether (ETH) that allows for transactions between participants in the Ethereum network; Hyperledger does not have cryptocurrency in its platform. In the supply chain and logistics management space, BCT is also increasingly gaining attention from academics (Bahga et al. 2016, S. A. Abeyratne et al. 2016, H. M. Kim et al. 2018, F. Tian 2017) as well as practitioners. Therefore, many experts believe that BCT has the potential to disrupt many conventional practices in a supply chain. As for the study of marine logistics, the application of BCT in this field has been investigated (A. Shirani 2018, R. Di Gregorio et al. 2018); however, few studies have examined this research topic so far, and we would like to explore the potential of BCT-enabled marine logistics in this research.

3 CASE STUDY FOR BLOCKCHAIN-ENABLED MARINE LOGISTICS

In this section, we first present a case study to explore the potential of applying BCT and cryptocurrencies in marine logistics. Then a proof-of-concept blockchain and smart contract application is developed based on the Ethereum platform, and the feasibility of the application is discussed.

We use an industrial supply chain as an example, where a TV manufacturer (importer) in one country purchases TFT-LCD modules (parts) from another manufacturer (exporter) in a different country by shipping. The as-is process model of maritime logistics

for the case study is described in Figure 1, which also shows all the parties involved in this logistics process. The process description is defined in Table 1. We then propose an alternative new maritime logistics process model using blockchain and smart contract, as illustrated in Figure 2 (process description in Table 2). To simply the analysis, we assume that third-party logistics service providers (3PL) take the role of the carrier for all transportation needs.

Based on the four order processing layers proposed by (S. R. Magal et al. 2011), a blockchain-driven supply chain can be decomposed by the aforementioned four layers (E. Hofmann et al. 2017), as order processing, shipping, invoicing, and payment layers. In

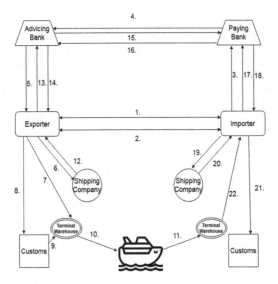

Figure 1. The as-is process model of maritime logistics for the case study.

Table 1. The process description of Figure 1.

1. Inquire and Quote	12. Acquire the B/L
2. Place an Order	13. Apply the export negotiation
3. Apply to open L/C	14. Advance payment from the export negotiation
4. Open an L/C	15. Mail the document under L/C
5. Advice an L/C	16. Reimburse the negotiation payment
6. Book the shipping space	17. Release documents against import payment
7. Effect Shipment	18. Clear the cargo receipts
8. Deliver the export customs document	19. Return the B/L
9. Inspection	20. Change the D/O
10. Load and effect shipment	21. Deliver the import customs document
11. Unload the cargo	22. Clear the Cargo

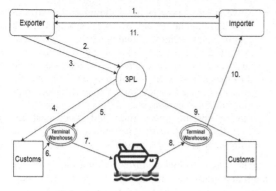

Figure 2. The to-be process model of maritime logistics for the case study.

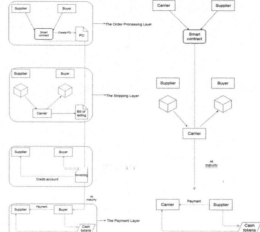

Figure 3. The blockchain-enabled supply chain layers for procurement and fulfilment process of the case study.

Table 2. The process description of Figure 2.

1. Importer Set a smart contract with Exporter	7. Loading and effect shipment
2. Exporter Set a smart contract with 3PL	8. Unload the cargo
3. Effect Shipment	9. Agent the import customs clearance
4. Agent the export customs clearance	10. Clear the Cargo
5. Effect Shipment	11. Payment (e.g. BTC)
6. Process the inspection	

this proposed layered blockchain-driven model, international trade documents must be digitalized and all participating parties in the supply chain can perform automatic verification against those documents stored on the blockchain. We further refine the blockchain process model in Figure 2 into two blockchain-driven process models, each related to a smart contract, as illustrated in Figure 3. Then, based on the two blockchain-driven process models, we shall design two smart contract data structures. One of them is shown in Figure 4, where we define the state variables and functions of a contract. Finally, in our proposed model, we suggest that all trading partners use cryptocurrencies to speed up the transaction and store that event in the blockchain.

We use the Solidity programming language with Remix IDE to develop a proof-of-concept blockchain prototype application with smart contract and Dapps on the Ethereum test environment. A Dapp allows users to access blockchain information. Dapp is a decentralized application running on a decentralized peer-to-peer Ethereum network. Some of the Dapp's user interfaces in this case study are shown in Figures 5, 6, and 7.

The logic (or rules of law) of a smart contract like the one in Figure 6 can be encoded in the execution steps. In this use case, the logic of the smart contract (Figure 6) is described in the following steps, and this execution logic is implemented on Ethereum using the Solidity programming language.

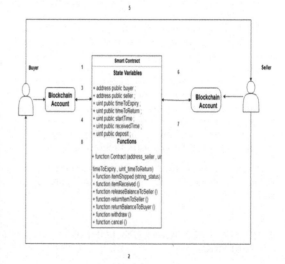

Figure 4. The smart contract design for the case study.

Transaction Information

[This is a Ropsten Testnet Transaction Only]

TxHash:	0x966b9c8613600485ad58eeee3afff2bed16d66070947c10821c2fc0353fa341d
TxReceipt Status:	Success
Block Height:	4644606 (25 Block Confirmations)
TimeStamp:	4 mins ago (Dec-17-2018 07:39:32 AM +UTC)
From:	0x60fd878c393412116f5cc92d1298db9bf66be78
To:	[Contract 0xee07bb05bfb26e61c69411119d941fbcb69746 Created]
Value:	0 Ether ($0.00)
Gas Limit:	977752
Gas Used By Transaction:	977752 (100%)
Gas Price:	0.000000001 Ether (1 Gwei)
Actual Tx Cost/Fee:	0.000977752 Ether ($0.000000)
Nonce & {Position}:	60	{13}

Figure 5. Setting up the smart contract running on Ethereum (step 1).

Figure 6. Seller updates item shipment information (step 2).

itemReceived   ∧

:   string

                    ⇄   transact

Figure 7. Buyer updates item upon receiving item (step 3).

1. Buyer sets up smart contract and makes a deposit in the contract account.
2. Seller ships item to buyer and updates item shipment information.
3. Buyer checks the delivered item and updates item receiving information.
4. If the buyer is satisfied with the item, the buyer releases the balance deposit to seller.
5. If the buyer is not satisfied with the item, the buyer can return the item.
6. Seller releases the balance deposit to buyer upon receiving the item returned by the buyer.
7. Seller can cancel the contract and return the deposit held in the contract account to buyer.
8. Buyer can withdraw the deposit upon expiry of the contract if the item is not received.

Some execution results and user interface implemented for this case study are shown in Figures 5, 6, and 7.

4 EVALUATION AND CONCLUDING REMARKS

Finally, we evaluate the new blockchain-based approach against the current marine logistics practice and give concluding remarks for this study.

Compared with the current marine logistics practice, the proposed approach, which uses blockchain and cryptocurrencies (such as Ethereum coin in this case) for marine logistics, clearly shows that it is not only feasible (as demonstrated by the proof-of-concept prototype) but also has many advantages over the conventional approach. The major benefits are summarized in the following points.

• Removing intermediary agencies to simplify and speed up the marine logistics processes. For example, the intermediary role played by financial institutions (banks) can be removed in the blockchain-based approach using cryptocurrencies, which greatly improves the efficiency of the payment layer.

• Lowering the complexity of administrative procedures and human errors in order processing. All supply chain entities, including the customs agencies of different countries, can perform real-time automatic verification against international trade documents stored on the blockchain. This greatly improves the efficiency for order processing, shipping, and invoicing layers.

• Saving administrative time and labor cost for overall marine logistics operations because of the aforementioned improvements.

The novel blockchain-based marine logistics approach presented in this paper provides a new avenue to overcome the current challenges facing marine logistics and to improve its international trade efficiency. However, this research is only the beginning, and prototype application in this study is very primitive and limited in use. The integration of IoT and blockchain might further enhance the efficiency of marine logistics, and it is an important research topic for us to pursue in the future. Many opportunities lie ahead for the application of blockchain and cryptocurrencies in marine logistics as well as in other fields. For example, integrating IoT and BCT might further enhance the efficiency and transparency of marine logistics, and it could become an important research topic to pursue in the future.

REFERENCES

A. Bahga, and K. Vijay Madisetti, 2016. Blockchain platform for industrial internet of things, *Journal of Software Engineering and Applications*, 9(10), 533.

J.J. Bam bara, P.R. Allen, K. Iyer, S. Lederer, R. Madsen, & M. Wuehler, 2018. *Blockchain: A practical guide to developing business, law, and technology solutions.*

K. Christidis, & M. Devetsikiotis, 2016. Blockchains and smart contracts for the internet of things, *IEEE Access*, 4, 2292–2303.

S.A. Abeyratne, & R.P. Monfared, 2016. *Blockchain ready manufacturing supply chain using distributed ledger.* UK.

N. Szabo, 2016. *Smart Contracts. [Online].* Available: http://szabo.best.vwh.net/smart.contracts.html.

H.M. Kim, & M. Laskowski, 2018. Toward an ontology-driven blockchain design for supply chain provenance, *Intelligent Systems in Accounting, Finance and Management*, 25(1),18–27.

F. Tian, 2017. "A supply chain traceability system for food safety based on HACCP, blockchain & Internet of things," In *Service Systems and Service Management (ICSSSM), International Conference* on pp. 1–6. IEEE.

A. Shirani, 2018."Blockchain for global maritime logistics" Issues in *Information Systems* Volume 19, Issue 3, pp. 175–183.

R. Di Gregorio, S.S. Nustad & I. Constantiou, 2018. "Blockchain adoption in the shipping industry".

S.R. Magal & J. Word, 2011. *Integrated business processes with ERP systems.* Wiley Publishing.

E. Hofmann, U.M. Strewe & N. Bosia, 2017. *Supply Chain Finance and Blockchain Technology: The Case of Reverse Securitisation*, Springer.

Cultural & creative research

Smart Science, Design & Technology — Lam et al. (eds)

Cultural value and significance of Nanyin Pipa

Ping Chen*
Jimei University, Xiamen, Fujian Province, PR China

ABSTRACT: Nanyin, one of the most ancient music forms surviving in China, is reputed to be the "living fossil" in the music history of China. In Fujian Nanyin music, Nanyin Pipa plays the role of a conductor, which is irreplaceable. Hence, this paper starts with Nanyin Pipa, explores its various features, and compares it with similar Pipas to highlight its historical and cultural value. Moreover, in the combination of tangible cultural heritage and intangible cultural heritage from both micro and macro perspectives, it grasps the cultural essence of Fujian Nanyin music and demonstrates the value and significance in the protection and inheritance of Nanyin music.

Keywords: Nanyin Pipa, development and evolution, comparative study, cultural value

1 INTRODUCTION

On May 20, 2006, Fujian Nanyin was included in the first batch of national intangible cultural heritage list; on October 1, 2009, Fujian Nanyin was officially recorded in the representative list of intangible cultural heritage of humanity by the UNESCO. Before and after the application for the list of world heritage, there have been experts and scholars studying and excavating Fujian Nanyin, which promoted the appearance of certain research findings quite mature and comprehensive in the form, playing methods, heritage, and development of Fujian Nanyin. However, there's a lack of research in the instrument. As we know, tangible culture and intangible culture complement each other; so do Nanyin Pipa and Fujian Nanyin. In Fujian Nanyin performance, Nanyin Pipa plays the decisive role of a conductor and carries a rich cultural message. It is crucial for the formation, spreading, and development of Nanyin. Therefore, this paper has summarized the research achievements of Nanyin Pipa by former artists and experts and made further exploration to highlight its unique charm as an ancient oriental art from the perspective of Chinese aesthetics and makes this traditional art carry forward in the land of southern Fujian.

2 SHAPE AND STRUCTURE FEATURES OF NANYIN PIPA

The mention of Fujian Nanyin reminds people firstly of its Pipa, which is different from the rest. Actually, Nanyin Pipa was not different from the rest at the beginning. It is just that in their historical development, other Pipas went through consecutive reformations and developed finally into the Pipa well-acquainted to us nowadays, while Nanyin Pipa maintains its ancient appearance in its long history. Therefore, Nanyin Pipa can be considered worthy of the reputation of "Jade."

Nanyin Pipa belongs to the Bent-neck Pipas. In fact, there is no difference between the Bent-neck Pipa and our most ancient indigenous Qin Pipa. Bent-neck Pipa was imported in Central China from Persia approximately in the Southern and Northern Dynasties, and to distinguish it from the Qin Pipa in China, it was called Bent-neck Pipa due to the features of its shape and structure (Zhang and Han, 1991). Bent-neck Pipa was quite fashionable in the Tang Dynasty, completely overwhelming the Qin Pipa, so that "Pipa" was used specifically to refer to Bent-neck Pipa. The earliest Bent-neck Pipa had a four-string-and-four-column, pear-shaped sound box, crooked nape, and short neck. It can be told from pictures in historical materials that the Bent-neck Pipa at that time featured a long body but a short neck, quite different from the Pipa commonly seen now. But after the Tang Dynasty, Bent-neck Pipa went through reformation, with its short neck gradually elongated and big sound box shrunk. The holding gesture of the player changed also from horizontal hugging to oblique hugging. However, compared with the introduction of Bent-neck Pipa into Central China, its introduction into southern China occurred somewhat later, thus the reformation in shape and structure was introduced while the original

*Corresponding author: E-mail: cp6180896@163.com

horizontal hugging gesture was kept. At the same time, the playing method was also inherited from that of the Tang Dynasty, with the finger playing method replacing the plectrum playing method. In the interviews of Nanyin performers nowadays, you can see that they still play pipa with fingers. In an overview of the development history of Nanyin Pipa, we can see it is closely connected to Bent-neck Pipa as well as in some way different from Bent-neck Pipa.

Nanyin Pipa as a whole can be called a piece of exquisite art ware. Viewed from the side, its neck and body form a curve line; viewed from the front, it is in the shape of a pear, with narrow neck, flat board, large string connection part, high collar, four phases and nine or ten frets, two crescent moon-shaped "phoenix eyes" on the two sides of the board, long neck, and Ruyi-shaped (S-shape) head. Many parts in the body are made with precious wood, animal bone, ox horn, and so on; even the tuning peg is inlaid with exquisite shells. Generally, it adopts nylon strings. Therefore, Nanyin Pipa features not only mellow and full tone, but also a pleasant appearance.

The Nanyin Pipa inherits the horizontal hugging gesture; but for a more specific definition, the holding gesture of Nanyin Pipa does not completely inherit from that in the Tang Dynasty, but inherits from the up-left oblique hugging gesture in the Song Dynasty. The holding gesture in and before the Tang Dynasty was oblique hugging gesture with head directed toward the lower left. The horizontal hugging gesture has been maintained until now, which makes Nanyin worthy of the title of living fossil in Chinese traditional music.

In general, Nanyin Pipa has not experienced too much change in shape and structure. Bent-neck Pipa gradually immigrated into Southern China after its change in shape and structure in the later Tang Dynasty. From the perspective of time, Nanyin Pipa began to develop after the reformation of Bent-neck Pipa, therefore, Nanyin Pipa inherits the shape and structure of the changed Bent-neck Pipa. Since the Tang Dynasty enjoyed a highly advanced economy and great national strength, its governors could spend more time in the study of arts. The art of music was highly developed in the Tang Dynasty; there was not only a variety of music forms, but also many outstanding artists. Bent-neck Pipa was developed into quite a mature status at that time; therefore, Nanyin Pipa has maintained its simplest appearance for several thousand years after its development.

3 TECHNIQUE CHARACTERISTICS OF NANYIN PIPA

3.1 *Difference with common Pipa in main techniques*

The most important technique for the playing of all Pipas is pluck and pick. Nanyin Pipa is not an exception, it just inherits some ancient techniques. Nanyin Pipa is different from the rest in string pressing; the player of Nanyin Pipa commonly uses index finger and ring finger to press the string, and sometimes uses the thumb, and never uses the middle and little fingers for string pressing.

The main techniques of Nanyin Pipa include Lian (quick pluck and pick), Diantiao (point pick), Quli, Jiaxian (tick fingering), Luozhi (turn-playing fingering), and other decorative fingerings.

(1) Lian fingering (quick pluck and pick): This mainly includes whole Lian, point Lian, rushing Lian, and passing Lian and other fingerings. The whole Lian technique is introduced in detail here. This technique refers to the consecutive point picking with the index finger and thumb of the right hand, somewhat similar to the rolling fingering technique adopted by contemporary Pipa, but it develops from heavy to light and from slow to fast; its last sound connects and forebodes the sound of flute, chord, and singing.

(2) The series of point and pick techniques mainly includes point, pick, Qudao, separate fingering, Caizhi, etc. These fingerings are closely connected with the time value of musical notes; during the performance, each fingering is set with fixed time value, generally with a half beat, some with one beat.

(3) Luozhi: Also called "turn-playing fingering," Luozhi consists of quick turn-playing fingering and slow turn-playing fingering. Compared with the turn-playing fingering adopted by common Pipa, the fall fingering is different in some degree. The turn-playing fingering adopted by the player of common Pipa generally follows the order of index finger, middle finger, ring finger, and little finger of the right hand; but the fall fingering of Nanyin Pipa starts from the little finger of the right hand, then comes to the ring finger, middle finger, and index finger for a pause, and then quickly picks up the string with the thumb. The quick fall fingering and slow fall fingering share the same performance method, differing only in speed.

(4) Decorative fingerings mainly include Danda (single-string striking), Shuangda (two-string striking), Moliu, Banfan, and Quanfan, etc.

3.2 *Comparison with the main techniques adopted by North Pipa*

In the long development history of Pipa, various schools have come up; North School Pipa is a branch among them. In its later development, North School gradually enriched Pipa's frets until they reached the current twenty-eight frets, and adopted the twelve-tone equal temperament scale, which allows free conversion of tones and increases the gamut to three and a half octaves. The instrument holding gesture changes from oblique hugging to vertical hugging, which completely liberates the right hand and contributes to the techniques of harmonics (including sao, fu, pie, hua), pai (make a beat sound), ti (make the string hit on the board), etc. Thus the techniques for the left and right hands amount to more than fifty types, and the performance style of Pipa is expanded.

In the playing of North Pipa, the thumb of the left hand is pressed on the back of Pipa, while the index finger, middle finger, ring finger, and little finger can flexibly move; the whole hand is in a semi-ball-cupping shape and can implement handle change flexibly. The main techniques of the left hand include glissando (including pushing, pulling, and rubbing) and striking, more than those adopted by Nanyin Pipa. As mentioned above, the middle finger and little finger of the left hand are never used in the playing of Nanyin Pipa. In the playing of North Pipa, a celluloid fingernail needs to be worn on the right hand, it is mainly divided into the pluck and pick system and the turn-playing fingering system. In general, the right hand techniques of North Pipa are richer than those of South Pipa; for the turn-playing fingering, the two types of Pipas are different, with North Pipa adopting the turn-playing fingering with index finger first and South Pipa adopting the turn-playing fingering with little finger first (Wang, 2002).

In playing techniques, Nanyin Pipa is not as rich as North Pipa, but this is because Nanyin Pipa has insisted on the tradition and does not have too much change. In this way, we can feel its original simplicity.

4 CULTURAL CHARACTERISTICS OF NANYIN PIPA

4.1 *Historical characteristics in articulation and tone*

The articulation of any instrument cannot be separated from its own resonance sound box; and each object has its own vibration frequency, namely the inherent frequency. Nanyin Pipa is no exception. Its resonance sound box consists of the instrument's front and back parts. Certainly, there is also a very important sound production object, namely, the strings. Nanyin Pipa has four strings. The stirring of strings will cause vibration; when the vibration frequency approaches the inherent frequency of the instrument, resonance is produced. Pipa has many frets; the pressing of different frets changes the effective length of a string and thus produces tones in different pitches. It is the unique model of Nanyin Pipa and its particularities with strings that create the gentle, mellow, full, and ponderous sound of ancient charm, and represent the ingenuity of each Pipa maker.

4.2 *Nanyin Pipa and the regional characteristics of Southern Fujian*

As an ancient music form, Fujian Nanyin has its unique notation method, namely the tonal discrimination notation method. It consists of three major parts—fingering, music score, and tune—and is basically spread in the way of handwritten notation,

a way unique to this music form. At the same time, Fujian Nanyin functions as the cultural tie among people in Southern Fujian and Southern Fujian people in Southeast Asia. It has the function of recreation by oneself, and also the properties of social practice.

Nanyin Pipa has four strings, the A String, d String, e String, and a String. Although Nanyin Pipa has four strings like other Pipas, only the finest string (a String) is used in the playing of Nanyin Pipa, and the other three strings are used for octave pause. The a String is the string of the highest pitch among the four strings, and most melody sounds are completed on this string; it can give a full play to the fine emotions typical of Southern Fujian. Most fingerstall stories in Fujian Nanyin are connected with love, such as Sima Xiangru and Zhuo Wenjun, and Wang Zhaojun marrying Xiongnu. Nanyin Pipa is the best instrument to describe such stories with its soft, reserved, and elegant tone, and it is also consistent with the reserved and fine aesthetic features of southern people.

4.3 *Inheritance mode of Nanyin Pipa*

Fujian Nanyin has been inherited through the mode of "private association," in which the master's oral teaching that inspires true understanding within can make the disciples learn more about the rhythm of Nanyin music. Although this inheritance mode is confronted with problems such as limited inheritance people and small spreading range, it effectively avoids the defects of inflexibility and stiffness that stand out in the mode of learning by notation. Now, since there are various media forms and the stave and numbered musical notation are widely adopted as the means of notation, even an outsider totally unfamiliar with Nanyin music can hum the melody generally based on the stave or numbered musical notation, but in this way, there will definitely be a loss in the unique charm of Nanyin music. Therefore, in the inheritance, we should maintain its original ecology as much as possible. In my opinion, the combination of two modes, namely the academy mode and private association mode, can integrate their specific advantages together, which might result in extraordinary effects.

5 UNIQUE ROLE IN MUSIC BAND

The music band of Fujian Nanyin consists of two parts, with one part belonging to the traditional stringed and woodwind instruments and the other belonging to traditional blow and beat instruments, which include the common string and chord instruments, string-pulling instruments, and percussion instruments. Besides, some percussion instruments that are typical of Nanyin music have been included. It is basically similar to a traditional music band in configuration. In the Tang Dynasty, Chang'an was the music center of Asia, and many instruments and the music band configuration there were quite mature. So when the music began to immigrate into Southern China, it somehow maintained its original form; we can ascertain from some data that traditional music band configuration had a great impact on Nanyin music bands.

Firstly, Pipa plays the commanding role in the opening and ending. Nanyin Pipa plays a crucial role in a Nanyin music band, the reason for which is that the opening and ending sounds are commanded by Nanyin Pipa in a Nanyin music band. When Nanyin Pipa carries out the consecutive point pick sound, other instruments and singing begin to follow in; in the ending part, Nanyin Pipa plays the sound of one octave lower to forebode the ending of prolonged sound, which can be explicitly distinguished by audiences; the progressing of one prolonged sound represents basically the end of a music phrase. It can be told from various aspects that Nanyin Pipa plays the commanding role in music bands.

Secondly, Pipa can perfectly combine the backbone tune with decorative sound. In traditional Chinese music, many have only backbone tune, in the playing of which the player can add some decorative sounds, etc. based on performance circumstances or the player's understanding of the tune to enrich it. Much Fujian Nanyin music also shares this feature. In a Nanyin music band, all instruments have their own role and are responsible for their own task. For instance, Pipa and Sanxian (a three-stringed plucked instrument) are mainly responsible for the backbone tune, while the Dongxiao (a vertical bamboo flute) and Erxian (a two-stringed plucked instrument) are used for the decorative sounds; the perfect combination of the two makes Nanyin music characterized by "simple playing method and rich sounds."

6 CULTURAL VALUE OF NANYIN PIPA

6.1 *Comparison of Nanyin Pipa with Japanese Pipa*

Nanyin Pipa gradually came into development after the maturity of Bent-neck Pipa and has mostly inherited the development results of Pipa in the Song Dynasty. In Japan, the Pipa has a variety of types. Since Le Pipa is the main instrument in the ceremonial court music of Japan, we mainly talk about Le Pipa here. The Le Pipa was introduced into Japan from China in the Tang Dynasty. Therefore, the two are somewhat similar in appearance, both with four strings and four columns, and two symmetrical half-moon-shaped sound holes on the soundboard. But the string peg in the Japanese Le Pipa is backward inclined, which is the same as with the Bent-neck Pipa of the early period, while exquisite patterns matching the ceremonial music appear in the middle of the soundboard of Le Pipa. The two are different not only in shape and

Figure 1. A Nanyin Pipa.

structure, but also in the selection of manufacturing materials. For example, the front board of Nanyin Pipa is made with natural-color paulownia wood, and its back is made with rosewood and decorated with flower pattern; while the soundboard of Le Pipa is mainly made with chestnut wood, mulberry wood, and paulownia wood, and its back is made with ormosiahenryi and rosewood, etc. In the aspect of strings, Nanyin Pipa uses nylon string, while Le Pipa employs string made with spun silk, thus creating a slight difference in tone.

In addition, in the aspect of playing, both Nanyin Pipa and Japanese Le Pipa adopt the holding gesture of up-left oblique hugging, but Nanyin Pipa is played with the fingers, while the Japanese Le Pipa is played with a plectrum.

6.2 *Comparison of Nanyin Pipa with Vietnamese Pipa*

Vietnamese Pipa is almost the same as Nanyin Pipa in terms of outside appearance, both with four strings and four columns, and a pear-shaped sound box. Some scholars consider that the reason for their similarity is that the Vietnamese Pipa spread into Vietnam from China later than the 14th century. But there is a difference in the designation of the four strings. In Vietnam, the four strings of Pipa are respectively called "soldier, book, map, and war", or "pine, orchid, plum, and bamboo." Vietnamese people cannot give a clear explanation for these designations, but it might be to remember clearly the sounds of the four strings and to facilitate its tuning.

In addition, in the aspect of playing, the phases and frets of Nanyin Pipa can fulfill playing functions, while for the traditional Vietnamese Pipa, only its frets can fulfill playing functions and its phases cannot be used in the playing. Later, to adapt itself in the music band, the Vietnamese Pipa as a solo instrument has expanded its own gamut by increasing frets, but still does not make use of the phases. Vietnamese Pipa shares the same holding gesture with the contemporary Pipas in China, namely the vertical hugging gesture, which greatly liberates the functions of the left hand.

6.3 *Comparison of Nanyin Pipa with North Korean Pipa*

As an article under the influence of the Chinese culture, Northern Korean Pipa was introduced from China, and added local features in its later evolution. To distinguish it from the Tang Pipa of China, it is called Xiang Pipa. The Southern Korean Xiang Pipa and Nanyin Pipa are quite different in shape and structure. As we all know, Nanyin Pipa is a kind of Bent-neck Pipa. However, Xiang Pipa belongs to Straight-neck Pipa, and it has five strings. The two have adopted completely different development routes toward two directions, thus their difference in shape and structure.

Comparison of Pipas from three countries has shown that they have similar parts and different parts. The Pipa in each region or country has carried out the inheritance and development in its local mode. Nanyin Pipa has inherited the essence of Bent-neck Pipa and also manifests the characteristics of Southern Fujian. In its long development history, Nanyin Pipa still keeps its traditional veil and greatly drives our exploration or study of traditional music, thereby higher historical value rises.

7 CONCLUSION

As an important instrument in Fujian Nanyin, Nanyin Pipa plays a significant role in the whole music band of Fujian Nanyin. It can be seen from the lineages described in this paper that Nanyin Pipa has inherited the essence of Bent-neck Pipa, integrated the unique characteristics of Southern Fujian, and survived a long history until now. Although it does not have the rich techniques and notes of North Pipa, it has been included in the world cultural heritage list with its exquisite craft, mellow tone, traditional holding gesture, and playing techniques as well as its integration into Fujian Nanyin music. Nowadays, Nanyin is not only sung in the Southern Fujian area, but is also widely popular in areas of Southeast Asia, from which you can ascertain the historical and cultural value of Fujian Nanyin music.

REFERENCES

Z. Zhang and S. Han, 1991. *The History of Pipa in China*. Shanghai: Shanghai Literature & Art Publishing House.
Y. Wang, 2002. *Fujian Nanyin*. Beijing: People's Music Publishing House.

Smart Science, Design & Technology — Lam et al. (eds)
© 2020 Taylor & Francis Group, London, ISBN 978-0-367-17867-3

The repurposing of temple reception halls - using Xiamen Rushi Hotel Group as a case study

Tien-Feng Hsu*
Ph.D Program of Mechanical and Energy Engineering, Kun Shan University, Tainan, Taiwan

Huannming Chou
Department of Mechanical Engineering, Kun Shan University, Tainan, Taiwan

ABSTRACT: The reception halls of temples go by such names as "shangke tang," "ke tang"; or "pilgrim lodges" are used in Taoist temples in Taiwan. As these names imply, these buildings serve as temporary accommodations for devotees who come to pay homage to Buddha or take part in some religious assemblies, and there is no charge for lodging traditionally. Instead, devotees generally make some donations to serve as lodge fees, whose amount is subject to their decision and is therefore a variable sum. In terms of management, the reception halls of temples are used mostly when temples hold some events, so they remain unoccupied most of the time and lack regular maintenance, except the pre-event period that temples usually recruit some volunteers to do the cleaning. As a result, the common lodging quality of the reception halls is poor. Nevertheless, thanks to the prevalence of Buddhism and Taoism, the number of temples in Taiwan with "reception halls" or "pilgrim lodges" is substantial, so we recommend these "reception halls" or "pilgrim lodges" be repurposed and reused by consulting the management of the Rushi Hotel Group in Xiamen. For the temples themselves, this kind of management can improve their management of "reception halls" or "pilgrim lodges" and increase their revenues at the same time. For the local regions, it can boost the development of the tourism industry.

Keywords: Reception hall, Pilgrim lodge, Tourism industry

1 FOREWORD

1.1 *Research background and motivation*

Currently, Taiwan has accumulated quite a lot of information on the repurposing of idle assets, but most of the research is focused on the reparation and repurposing of cultural heritage. According to the strict definition of the "Cultural Heritage Preservation Act," cultural heritage is defined as tangible or intangible assets with historical, artistic or scientific cultural value. There are nine types of tangible cultural heritage, including monuments, historical buildings, memorial buildings, settlement complexes, archaeological sites, historical ruins, cultural landscape, antiquities, natural landscapes and natural monuments. Intangible cultural heritage, on the other hand, consists of five types of assets, which are performing arts, traditional handicrafts, oral traditions, folklore, and traditional knowledge and practices. Based on this definition, even though there are several temple administration organisations that have been specified

as monuments, most temples do not meet the definition of cultural heritage as stated in the "Cultural Heritage Preservation Act," and there are very few discussions on the repurposing of temples.

In Taiwan, the number of registered temples on file at the end of 2015 amounted to 12,142. Classified by religion, the greatest numbers of temples are Taoist temples, accounting for 78.5%, followed by Buddhist temples, accounting for 19.3% (Ministry of the Interior's Department of Statistics, 2016). From the statistics, it is known that there are many temples in Taiwan, but most of the temples are idle, only visited by believers and pilgrimage tourists during specific folk festivals or when the temples organise assemblies or other religious activities. Many of these temples are also located by the mountains or the sea where the scenery is beautiful, or even neighbouring famous tourist spots; the idling of these assets are therefore a waste of these sites' sightseeing resources.

According to relevant news reports, the former Chiayi City Hall had been abandoned for more

*Corresponding author: E-mail: oceanseal@gmail.com

than 10 years, but with the efforts of cultural heritage groups, the land that was originally intended to be sold is now a provisional monument. However, half a year has passed, and the municipal government has yet to roll out any repurposing plans. Meanwhile, a tobacco factory that has been idle for 10 years and is as big as five football fields has been turned into a garden, but having been abandoned for 10 years, no actual evident actions have been taken. The hostel next to the old Chiayi Prison is similarly a monument preservation; the buildings are badly damaged, and in July a plan to rent it out in return for renovation was launched - anyone who is willing to pay and renovate it will be exempted from rent for five years, but till now barely anyone has enquired about it.

From the above reports, we can see the challenges that abound when it comes to the preservation and repurposing of cultural heritage. Both preservation and repurposing require investments in advance, and with Taiwan's central and local governments all in a state of financial difficulty, they are unable to invest into the preservation and repurposing of cultural heritage, and therefore hope for an injection of private funds. However, this of course involves an assessment of the repurposing plan's profit and loss, which means it must have considerable benefits to attract private funds. If any cultural activity relies on the government to provide the resources, the development is inevitably limited. If it can be converted into an economically profitable industry, on the other hand, it is bound to attract the investment of private capital and talents. The "industry" of the cultural and creative industry is a collective name that refers to agriculture, mining, manufacturing, business and other economic undertakings, and is also known as the commercial industry. As a project that contributes to the economy, it is necessary to establish the appropriate business strategy with profit as the main objective. Of course, the strategy must include cultural and creativity, or else it cannot be known as the cultural and creative industry. These two elements are the basis of the cultural and creative industry. Therefore, we need to study similar topics to provide new ideas for the repurposing of cultural heritage.

1.2 Research objective

The aim of this research targets Taiwan's repurposing of temple reception halls as well as developing innovative and effective marketing strategies to bring in the investment of private funds and talents. This research first explores the present conditions of the utilisation of Taiwan's temple reception halls. Then the reasons behind the inefficient utilisation and idling of the reception halls are identified. Finally, using the business operations model of the temple reception hall administrated by Xiamen's Rushi Hotel Group as an example, this research discusses if the group's operational strategy is effective and sufficient to resolve the current idling situation of temple reception halls. If it is effective, they can serve as the direction for the future repurposing of idle cultural heritage assets.

2 PRESENT CONDITIONS OF TAIWAN'S TEMPLE RECEPTION HALL UTILISATION AND REPURPOSING CHALLENGES

As described in the foreword, there are currently over 10,000 temples in Taiwan. Even though most Buddhist temples have a reception hall, the regulations for lodging are relatively strict. Of the larger Buddhist temples, unless there are large-scale activities such as seven days of Zen retreat, seven days of Buddhist retreat, Dharma Drum Mountain and Chung Tai Shan only allow volunteer believers to rest at their reception hall, and do not accept normal believers' accommodation request; only Fo Guang Shan has established two pilgrims lodges, Chao Shan Lodge and Ma Zhu Garden Hostel, which are open for bookings to the public from 2000 Taiwanese dollars per night. Each room can accommodate from two to eight, and the facilities are similar to those at hotels and B&Bs. Other than Fo Guang Shan believers, those that choose to stay at the two pilgrims lodges are mainly visitors of Fo Guang Shan and the Buddha Museum.

Compared to Buddhist temples, more pilgrim buildings attached to Taoist temples are open to the public, but in terms of actual management, each lodging's regulations are different. For the stricter ones, such as Dajia Jenn Lann Temple, an official application request in writing must first be submitted by pilgrimage groups, who can only stay at the temple upon approval by the temple; personal applications are not accepted. Most of the other temples accept personal applications, but commonly only pilgrimage groups are allowed to apply. As for the method of application, each temple has different regulations. Some accept only applications in advance, while some accept on-arrival registrations.

In terms of room facilities, there are different levels of standard. There are those that have been compared to hotels, such as Kaohsiung's Shun Xian Gong and Yilan Jiaoxi's SieTian Temple. There are also some that operate a model akin to B&Bs, such as Tainan Baihe's Shan Zhi Monastery which provides organic food and Kaohsiung Meinong's Yi Ben Dao Monastery that provides herbal baths. All of them have their own unique features, and even though they mainly provide lodgings for believers and pilgrims, in effect they have already attracted many tourists. For these temple reception buildings that are close to the level of regular B&Bs, each night's accommodation is priced from over 1000 to over 2000 Taiwanese dollars per night. In comparison to regular B&Bs, the cost is almost the same or slightly cheaper. Of course, this operational model is already similar to that of

B&Bs, and effectively reduces the idling of assets. Some of these pilgrim lodges such as Yilan Jiaoxi's SieTian Temple are fully booked on weekends or during the holidays. With such operational models, the incomes of these pilgrim lodges are self-sufficient, or even bring in more income for the temples. Therefore, there would not be any issue of repurposing.

However, most temples' reception halls and pilgrim lodges have relatively crude facilities currently, mainly because the establishment of these halls and lodges are primarily to serve the temples' believers, providing them with a place where they can rest temporarily when they come on pilgrimage or participate in all kinds of activities organised by the temple. Therefore, the temples do not put great emphasis on the room facilities. If it is cheap, then tourists that have a higher degree of price sensitivity, such as students, backpackers and etc. are likely to be attracted; however, most tourists are not able to accept such accommodation. However, to renovate the rooms up to the standard that regular tourists would be satisfied with will put the temples in a spot where they face the two huge challenges of finance and administration:

(1) Finance-wise: Except for certain temples that have large followings, most temples receive limited donations from believers, and renovating guest rooms is a considerable expense. And based on the function of temple reception halls and pilgrim lodges, most temples' management will not deem renovation a necessity. Therefore, a comprehensive financial plan that can reduce the temples' burden is required to obtain the willing cooperation of the temples in order to repurpose these temple reception halls and pilgrim lodges.

(2) Administration-wise: Hotel management is a specialized skill; even if there are no issues on the financial aspect, converting reception halls and pilgrim lodges to guest rooms when they are idle require the help of professionals such as sales, maintenance and administration to operate. From the perspective of the temples, these require funds and human power, but considering that is not the core activity of the temples, they are unlikely to cooperate with the repurposing of their accommodation unless an all-rounded service including marketing, maintenance and administration is provided.

3 THE REPURPOSING OF TEMPLE RECEPTION HALLS - USING XIAMEN RUSHI HOTEL GROUP AS A CASE STUDY

Targeting Buddhist temples' reception halls, Xiamen Rushi Hotel Group has developed a comprehensive marketing strategy that includes a collaboration model with temples, which helps to resolve the aforementioned finance and administration issues posed to the temples. Currently, the group has continuously collaborated with several temples to repurpose idle reception halls. The main solution is explained below:

(1) Finance-wise: If temples lack the funds to renovate guest rooms, the hotel group will advance all renovation costs, repaid by 50% of the guest rooms' income until the full amount has been settled. In addition, the temples are required to allocate 10 to 15% of the rooms' quota to the group for their own use or to sell in return for administration remuneration. Under this solution, temples do not have to pay any cash, the repayment is based on income and the administration fees are paid by providing partial rights to guest rooms - the temples are under no pressure in terms of funding.

(2) Administration-wise: The professional help provided by the hotel group is explained below: (a) Hardware: The group is responsible for performing the necessary renovations to the guest rooms up to the standard of 3-star hotels. However, there will be no TV so that guests can achieve peace of mind. The cleanliness and maintenance of hotel rooms are, on the other hand, the responsibility of the temples' volunteers in order to reduce operating costs. However, the volunteers will be trained and supervised by the hotel group to ensure that room service is up to standard; (b) Software: This part is where it is easiest to differentiate temple lodgings from normal hotels and B&Bs. Because of the environment of the temples, physical and mental conditioning activities are more suitable, and the arrangements should be focused on the unique features of temple life. Firstly, the rhythm of daily life should follow that of the temples': sleeping and waking up early, having three meals at regular times, and from the rhythm of daily life condition one's body and soul. In addition, collaborations with professional organisations can be arranged or speakers can be hired for short-term training camps, including fasting and detox, yoga, koudou, chadou, hanadou, guqin and other activities that are appropriate at the temples.

Based on this solution, and on the basis that each room requires 20,000 Chinese Yuan's expense in advance with each room charged at 250 Chinese Yuan per night, the hotel group estimates in terms of repayment that they will be fully repaid within the year if the hotel occupancy rate reaches 50% and above, within two years if hotel occupancy is 30% and above, within two and a half years if hotel occupancy is 20% and above, and within five years if hotel occupancy is 10% and above.

4 CONCLUSION

There are many temples in Taiwan, with many that do have reception halls or pilgrim lodges, only provide accommodation for believers, and also many that have reception halls or pilgrim lodges that are operated like B&Bs and are open to the public. However, the quality of stay is poor generally, hence these accommodations are mostly idle when the temples do not have any activities organised. If these reception halls or pilgrim lodges can be repurposed, the temples can increase their income, consumers have more options when it comes to accommodation, and the local tourism sector is also further developed. Our research shows that the challenges of repurposing reception halls are mainly related to two aspects, which are finance and administration. Based on Xiamen Rushi Hotel Group's collaborative solution, temples do not have to raise funds in terms of finance, with the renovation works being paid in advance by the hotel group, then repaid based on income with no repayment deadline, which takes pressure off the temple. In terms of administration, the hotel group provides professional help and support, and engages temples' volunteers in order to reduce operating costs. Therefore, we recommend using Xiamen's Rushi Hotel Group's operating model to repurpose these "reception halls" or "pilgrim lodges" in order to promote the development of the tourism industry.

REFERENCES

Ministry of the Interior's Department of Statistics, 2016. *2016 Week 28 Interior Statistics Report*. Taipei: Ministry of the Interior's Department of Statistics.

Smart Science, Design & Technology — Lam et al. (eds)

The public policy environment for innovative future: Two cases regarding human security

Wen-Ling Kung*
Department of Sociology, State University of New York at Albany, USA

Shun-Nan Chiang
Department of Sociology, University of California, Santa Cruz, USA

ABSTRACT: Human future is at stake due to the increasing intensity of climate change, global armed conflicts, and other social and ecological issues. Researchers, innovators, and policymakers frequently propose different types of innovations as potential solutions to these social and ecological issues and as the pathway to a better human future. Following the emerging scholarly attention to the role of the public policy environment in determining the direction and development of innovations, this article analyzes how the public policy context, embedded with the moral values and collective visions for the future of society, may impact the development and directions of innovations with two empirical cases regarding the reproductive health service in Taiwan and the school gardening in the Philippines. We point out that the legal and regulatory environment provides a framework that has shaped the development of the local/transnational ARTs market in Taiwan and affected the day-to-day practices of medical practitioners at the clinical level. We also reveal that two particular policy visions – the well-nourished community and youth as the future of farming lead to the policy support of school gardening and specific types of innovations. Overall, these two cases both demonstrate complex "innovation systems" that comprise of more than one kinds of innovations and other social arrangements such as legal status, eco-nomic potentials, or cultural values. Furthermore, the comparison of these two cases also suggests the importance of the policy environment, especially values and visions embedded in these policies, and challenges the assumption that innovation is merely a linear process of technological breakthrough.

1   INTRODUCTION

Human future is at stake due to the increasing intensity of climate change, global armed conflicts, and other social and ecological issues. Researchers, innovators, and policymakers frequently propose different types of innovations as potential solutions to these social and ecological issues and as the pathway to a better human future. Meanwhile, researchers and policymakers gradually pay more attention to the role of the public policy environment in determining the direction and development of innovations. A good example is the European Commission's (EC) policy decision regarding technology development. When the EC announced seven "Grand Societal Challenges" as the main issues for the human future, EC also proposed to implement a new regulatory framework titled Responsible Research and Innovation (RRI) to evaluate and prioritize the distribution of its funding (Robert Gianni et al. 2018).

Following this emerging scholarly attention and centering on issues about population crisis and food security, two intertwined issues at the center of challenges pertaining to human future, this article discusses the the relationship between the policy rationale and the direction of innovations with two empirical cases regarding the reproductive health service in Taiwan and the school gardening in the Philippines. We aim to show how the public policy context, embedded with the moral values and collective visions for the future of society, may impact the development and directions of innovations as well as how the process of problem-defining and scenario-setting interact with the development of innovations.

2   LITERATURE REVIEW

Social science scholars have discussed various ways that the development of innovation may interact with the idea of "future" (Verschraegen 2017; Brown and Michael 2003). Chiang (Forthcoming) proposes that there are at least three ways that the development of innovations interacts with the idea of "future" – "innovation in the future," "innovation for the future," and "innovation with the future." "Innovation in the future" refers to a realist perspective that treats the future as only the result

---

*Corresponding author. E-mail: wkung@albany.edu

by the development of innovations. It is possible to predict the consequence of current innovations. In contrast, "innovation for the future" refers to a prescriptive perspective that holds the belief for a certain scenario of the future and promotes specific types of innovations to achieve the future. Finally, "innovation with the future" refers to a performative perspective that assumes the co-production of the current technoscientific development and various visions of the future embraced by stakeholders surrounding the development of innovations. In other words, instead of treating "future" as a potential reality, the perspective of "innovation with the future" recognizes "future" as a discursive tool that possesses the power to impact advances of science and innovation in the present.

This article follows the approach of "innovation with the future" to study the connection between the development of innovation, the policy environment, and the impact of the vision for the future. Particularly, we employ the framework of "sociotechnical imaginaries" developed by Science, Technology, and Society (STS) scholar Sheila Jasanoff and Sang-Hyun Kim (2009; 2015). Built upon Jasanoff's previous work on the co-production framework as well as analyses of political culture and civic epistemology, Jasanoff and Kim develop "sociotechnical imaginaries" to synthesize and move forwards the discussion surrounding the use of technology (Jasanoff and Kim 2009, 2015). They define sociotechnical imaginaries as "collectively held, institutionally stabilized, and publicly performed visions of desirable futures, animated by shared understandings of forms of social life and social order attainable through, and supportive of, advances in science and technology" (Jasanoff 2015:6). Furthermore, sociotechnical imaginaries also requires efforts both from the institutional-cultural arrangement and technoscientific advances. Overall, Jasanoff contends that our understanding and creation of the society are co-produced with our normative perspective of how a society should be. This argument implicitly indicates the connection between the process of innovation policy-making and the discursive-materialistic construction of socio-technical future bridged by the institutionalized practices of innovators and policymakers.

3 CASE ONE: REPRODUCTIVE TECHNOLOGY AND SERVICE IN TAIWAN

The first case is about Assisted Reproductive Technologies and reproductive health security in Taiwan. Ever since the first successful "test tube baby" was born in 1985 in Taiwan, the number of licensed medical organizations, which can legally practice ARTs care services, has increased from 48 to 84 while the total number of annual IVF (in-vitro fertilization) cycles has soared more than four times within two decades (Ministry of Health and Welfare 2018). Along with flourishing innovations and rising demands of the ARTs emerged a new form of market

in Taiwan – a local and transnational ARTs market. Within this newly-formulated market, hospitals and clinics provide various types of reproductive care services including IVF with or without intracytoplasmic sperm injection (ICSI) procedure, gamete freezing, third party reproduction involving with gamete donation or gestational surrogacy services, and pre-implantation genetic screening/diagnosis (PGS/PGD).

Based on the analysis of secondary policy documents and in-depth interviews with policy-making stakeholders and medical practitioners, this empirical case study suggests that different moral values and visions of the future, embedded in the policy environment, have contributed to the applications of certain ARTs techniques and those extended care services in the clinical contexts.

First, moral values of the continuum of "sexuality-reproduction-motherhood," and the related social orders of family, kinship, life, and gendered roles in workplace and family have affected the establishment of Artificial Reproduction Act in 2007 and the three current amendment bills on the Act. Until now, according to the Act, only married couples that are clinically diagnosed as infertile (if they are unable to get pregnant over six months of sexual intercourse) can legally use IVF service in Taiwan. Unmarried people can only use gamete freezing service but not IVF service. The legal environment makes it possible for the medical organizations to adopt the cryopreservation technique and develop a potential market of social egg freezing service that targets single women. Meanwhile, medical organizations incorporate with transnational gamete transport service companies and overseas medical organizations to guarantee the feasibility to transport the frozen eggs to other countries if needed for their clients. This arrangement secures the future access to IVF service for single women under any circumstances, which is also possible due to the cross-border regulations according to the current Act (i.e., according to the Act, gametes or embryos can be transported to and from another country, only when the gametes or embryos are retrieved from the human subjects who transport them).

Second, the vision of a better national development through commodifying medical care has been the core assumption of the government's medical tourism policy ever since 2007. Along with the institutionalization of the third-party egg donation service in Taiwan, the law and policy environment has facilitated medical organizations to develop their transnational ARTs services for foreign citizens. In 2017, a news report on The Philippine Star interviewed with Dr. Maw-Sheng Lee. Dr. Maw-Sheng Lee is from one of the leading private ARTs clinics, Lee Women's Hospital, in Taiwan. The organization has conducted over one-tenth of the total number of annual IVF cycles in Taiwan in 2016. In the news, as Dr. Lee said, his clients have come from over 36 countries around the world. There have been more than 700 Filipino couples seeking IVF services in his organization in Taiwan, including a city councilor from Dagupan city in the

Philippine (The Philippine Star, Ching Alano 2017). According to another report released on May 7th, 2017, by a Japanese newspaper Yomiuri Shimbun, the number of Japanese women receiving egg donation in Taiwan has rapidly increased in the past few years. There have been 177 women traveling to Taiwan for receiving donors' eggs during 2014 and 2016 (Yomiuri Shimbun 2017).

Third, influential policies also include the pronatalist policy since 2006 and the relevant ARTs subsidy program for low and middle-income families since 2015 due to its lowest-low fertility rate in the past decade (Population Reference Bureau 2016). The government of Taiwan recognizes the low fertility and aging population as a "national security crisis" in the near future and intends to expand the amount of subsidy for IVF service. Medical practitioners, in general, support the policy though requesting for more financial assistance from the government.

Indeed, the legal and regulatory environment provides a framework that has shaped the development of the local/transnational ARTs market in Taiwan and affected the day-to-day practices of medical practitioners at the clinical level. However, the condition that different stakeholders envision the future of the ARTs in different ways may continue to foster social controversies. For example, the unresolved controversies about the legalization of gestational surrogacy service in Taiwan remain a heated debate among stakeholders, which may either facilitate or impede the adoption of technical innovation and the development of ARTs market.

4 CASE TWO: SCHOOL GARDENING IN THE PHILIPPINES

The second case is about the issue of malnutrition in the Philippines. Malnutrition has been a significant issue in the Philippines for more than a century. The government and the nutritional community have focused on different types of malnutrition in different historical periods. Currently, the challenge is particularly complicated because of the coexistence of various types of malnutrition including energy deficiency, micronutrient deficiencies, and obesity. According to the 2015 National Nutrition Survey, while one-third of children are still undernourished, one-third of adults are overweight.

Among different social or technological innovations to address the issue of malnutrition, school gardening stands out to be a favored measure and is promoted by the Department of Education. In 2007, the Department of Education announced the Gulayan sa Paaralan (School Garden) program and mandated that all public schools should set up gardens to support the school feeding program, which was also a national program implemented by the Department of Education. At the end of 2018, the Department of Education announced to re-implement "School inside a Garden" (SIGA) program, which

indicates the continuation of the policy trend promoting school gardening.

Then, why, among different potential measures, has school gardening been popular and recognized as a viable solution to the issue of malnutrition in the policy context? Moreover, how does this policy preference impact the development of innovations regarding malnutrition? To answer these research questions, we collect data from semi-structured interviews, participant observations, direct involvement of certain components of the project, and secondary data including official reports and news articles. Overall, the promotion of school gardening as a solution to malnutrition is associated with two particular policy rationales.

The first policy rationale refers to the vision that "school gardening" could catalyze the change of the entire community. Policymakers and promoters argue that school gardens could not only provide fresh produces to the school feeding program and offer educational opportunities for good nutrition practices, but also serve as a platform to promote home gardens in the community and deliver nutritional knowledge to all the community members. Eventually, children will be fed both in schools and their homes, and school gardening will help the country to keep children well-nourished year-round. This policy vision is particularly foregrounded by the Southeast Asian Regional Center for Graduate Study and Research in Agriculture (SEARCA). SEARCA launched its own "School Plus Home Gardening Program" (S+HGP) in 2016 and has held several regional and international events advocating this vision (Blesilda Calub et al. 2019).

The second policy rationale concerns the vision for the future of farming in the Philippines. Setting up gardens within schools has never been a new idea. In the Philippines, as early as the U.S. colonial period of 1907, there were discussions and proposals to create gardens in public schools. However, the U.S. colonial government treated school gardening as training for future farmers. The same rationale serves as essential background for the promotion of school gardening in recent years. Encountering the trend of urban migration and aging population, the Philippines government envisions the shortage of farmers as one crucial challenge. Since the government still acknowledges the agricultural sector as one crucial pillar for the Philippines economic development and the foundation of the Philippines culture, there are different calls for making agriculture more interesting to youths. School gardening thus fits well into this agenda and supported by different stakeholders. This is also why the Department of Education encourages teachers to incorporate school gardening into the curriculum so that agriculture may become more appealing to students.

Based on these policy visions, the government and different nonprofits began to promote school gardening; this further leads to different types of innovations associated with school gardening. The list of major innovations is as below: 1) Organic farming. Knowledge and techniques of organic farming are the foundational components of school gardening since all the children will have regular contact with the garden. Farming methods suitable for small-scale agriculture gain particular attention and promotion because of the limited space in schools. Major aspects include solid waste management, organic compost making (Vermicompost), and pest control. 2) Bio-Intensive Gardening. Bio-Intensive gardening is a particular type of farming system. The primary purpose of the system is to reduce the input and achieve the goal of sustaining of the garden. Thus, this innovation is particularly suitable for school gardens so that there is no need to secure external funding to maintain the garden. 3) Infrastructural innovation. Infrastructures also play an essential role. This includes the design of the entire garden. School gardens require an excellent design to maximize productivity and accommodate different kinds of educational needs. For example, some schools also incorporated aquaculture and livestock in their school gardens. Moreover, based on prior knowledge, a significant problem of the existing school gardens was that schools could not produce good quality seedlings for transplanting due to the changing weather patterns in the Philippines. Thus, the government and some nonprofit projects also designed greenhouses with a rainwater collection system as an adaptation strategy. This innovation enables schools to produce good quality seedlings year-round which sometimes were also shared to parents for their home gardens.

To conclude, the case of the policy promoting school gardening in the Philippines demonstrate how particular visions of problems at stake and a better future may lead to the policy support and development of specific innovations.

5 DISCUSSION & CONCLUSION

We present two distinct cases regarding population crisis and food and nutrition security in two countries. Juxtaposing these two cases further reveal several key points worthy of attention. First, it is important to reconsider what "innovation" encompasses. These two cases both demonstrate complex "innovation systems." These innovation systems comprise of more than one kinds of innovations and other social arrangements such as legal status, economic potentials, or cultural values. In other words, the degree of sophistication (i.e., high tech or low tech) may not be the main differentiation between these innovation systems. It is more important to pay attention to all the components included in the entire innovation systems.

Furthermore, following the framework of sociotechnical imaginaries, these two cases also reveal the importance of the policy environment, especially values and visions embedded in these policies. These analyses challenge the assumption that innovation is merely a linear process of technological breakthrough and suggest the significance of broadening our understanding about the innovating process and the interplay between public policy environment and the development of innovations.

REFERENCES

B.M. Calub, L.S. Africa, B.M. Burgos, H.M. Custodio, S.Chiang, A.G.C. Vallez, E.I.N.E. Galang, and M.K.R. Punto. 2019. The School-Plus-Home Gardens Project in the Philippines: A Participatory and Inclusive Model for Sustainable Development. *SEARCA Policy Paper 2019*. Southeast Asian Regional Center for Graduate Study and Research in Agriculture (SEARCA). College, Los Baños, Laguna, Philippines.

C. Alano, 2017. Taiwan's Fertility Doctor Brings Affordable In Vitro Treatment. *The Philippine Star*, October 30.

G. Verschraegen, ed. 2017. *Imagined Futures in Science, Technology, and* society. Routledge Studies in Science, Technology and Society 34. London; New York: Routledge, Taylor & Francis Group.

Ministry of Health and Welfare, 2018. *Yearly ARTs National Summary Report 2016*.

N. Brown and M. Michael. 2003. A Sociology of Expectations: Retrospecting Prospects and Prospecting Retrospects. *Technology Analysis & Strategic Management 15*(1) 3–18.

Population Reference Bureau. 2016. *2016 World Population Data Sheet*.

R. Gianni, J. Pearson and B. Reber, eds. 2018. *Responsible Research and Innovation: From Concepts to Practices*. Abingdon: Routledge.

S. Chiang, Forthcoming. Future in the 'Making:' Multiple Ways of Engaging the Future. *Science, Technology, and Society*.

S. Jasanoff, 2015. Future Imperfect: Science, Technology, and the Imaginations of Modernity. In *Dreamscapes of Modernity: Sociotechnical Imaginaries and the Fabrication of Power*, edited by Sheila Jasanoff and Sang-Hyun Kim, 1–33. University of Chicago Press.

S. Jasanoff and S. Kim, 2009. Containing the Atom: Sociotechnical Imaginaries and Nuclear Power in the United States and South Korea. *Minerva 47* (2) 119.

S. Jasanoff and S. Kim, eds. 2015. *Dreamscapes of Modernity: Sociotechnical Imaginaries and the Fabrication of Power*. Chicago: University of Chicago Press.

Y. Shimbun, 2017. "3年で100人 子ども誕生." *Yomiuri Shimbun*, May 7, 12.

Smart Science, Design & Technology — Lam et al. (eds)
© 2020 Taylor & Francis Group, London, ISBN 978-0-367-17867-3

The influence of culture on iconic recognition and interpretation: Using Russian as a case study

Mariia Ominina* & Siutsen Shen
Department of Multimedia Design, National Formosa University, Hu-Wei, Yunlin, Taiwan

ABSTRACT: The paper studies cultural differences in the interpretation of pictograms. The online survey was conducted via Google forms and Facebook to collect the data during December 2018. The survey asked Russian participants the meanings of six preselected globalized signs and six localized signs from Japan and Taiwan. There were 98 respondents from 30 countries where the main group was from Russia (41%). The survey observed the distinction of the recognition of the global and local signs among people with various cultural backgrounds. The results demonstrate the connection between cultural background and understanding of localized signs. The issues of the need for localized and global signs were discussed and were compared with the results of the survey. More than half of the localized signs did not achieve 50% of the recognition rate for the Russian group. Meanwhile, half of the local signs were misunderstood by all participants. Furthermore, the globalized signs had the highest recognition rates, and some of them had 100% recognition rate for the Russian participants. Based on the results, factors of the effectiveness of the signs, such as sign prevalence and abstractness, were discussed. Overall, the survey has shown that there is a need for more research to be done with more meticulous selection of the signs. It is hoped that, in the future, further iconic/sign testings with higher recognition rates can be validated for Russian people in Taiwan.

1 INTRODUCTION

1.1 *Localisation*

The term localization is mostly used in digital and cultural fields. Localization involves taking a product and making it linguistically and culturally appropriate to the target locale (country/region and language) where it will be used and sold (Esselink, 2000). Thus, the processes of globalization in the business and digital environment contributed to the appearance of the localization phenomenon.

First, products were designed for international markets and then adapted for local users considering their preferences and cultural background. For achieving 'culturally appropriate' products, not only should linguistic aspects be considered, but also colors, typography, images, and layout.

Localization 'also involves more than just making the product readily available in the form and language of the target market. It must speak to the target audience, based on its cultural norms and their worldview' (LISA 2005). Translation cannot replace the localization process fully; it is only a part of localization.

1.2 *Semiotics and abstractness*

Semiotics, also called semiology, is a study about signs and users' behaviour about the signs. According to Aristotle's definition of the three dimensions of signs, recognition of the sign is the part based on the psychological, social, and cultural backgrounds of a referent (Sebeok 2001). Each sign conveys information to a concrete audience according to their languages and cultural backgrounds. A sign or icon must convey the intended value of the sender to the recipient of the message; that is, the selected pictogram must refer to the participants at both ends of the communication channel.

According to Peirce's classification of signs as icons, indexes, and symbols, an 'icon is a sign that is made to resemble, simulate, or reproduce its referent in some way' (Sebeok 2001). It is impossible to see or listen to a sign without thinking about this object at the same time. Consequently, it is impossible to recognise a sign without knowing its meaning.

One of the cognitive features used for describing and sign rating is semantic distance or semantic closeness. 'Semantic distance is a measure of the closeness of the relationship between what is depicted in an icon and the function it is intended to represent' (Ng, A. W., & Chan, A. H. 2008).

Abstraction can be defined as the word opposite to the word concreteness, by which is meant the minimalistic and schematic image of an object from the material world. 'This degree of simplicity

---

*Corresponding author. E-mail: mariaominina@gmail.com

corresponds to the kind of abstraction associated with representational images serving a symbol or icon function' (Zender 2006).

1.3 *Hofstede's cultural dimensions*

Geert Hofstede's theory of cultural dimensions is one of cross-cultural communication issues in societies. It describes the impact of cultures on the values of members. Hofstede builds this theory based on the global online survey conducted between 1967 and 1973. The phenomenon of culture was defined by Hofstede as 'the collective programming of the mind that distinguishes the members of one group or category of people from others' (Hofstede 2011). Culture is a complex of factors which affect humans' unconscious behavior. In childhood, the impact of culture has deeper roots than integration to other cultures in adulthood.

There are six dimensions in Hofstede's theory: power distance, uncertainty avoidance, individualism versus collectivism, masculinity versus femininity, long-term versus short-term orientation, and indulgence versus restraint.

For this study, this system was used to compare the cultural dimensions of Russia, Taiwan, and Japan. Based on these cultural variations, authors may find connections between culture and recognition rate.

1.4 *Related works*

The largest tests of the understanding and effectiveness of pictograms were carried out in ISO 9186 (Public Information Signs), based on the choice of participants and judges from a number of answers. 'According to the ISO standard, a symbol is accepted if 67% of the users understand it in an unquestionable way or almost so (ISO 9186-1989). In the United States, the pictogram must be understood by 85% of the users in order to be standardized (ANSIZ 535-1987)' (Tijus, C. et al. 2007).

Shen et al. (2006) tested icons created by the author on the international group of participants. Zender (2006) did research about icon design for communication based on a research study sponsored by Procter & Gamble.

Cho et al. (2007) performed a survey of cultural differences observed in pictogram interpretation by participants from Japan and the United States. Tijus et al. (2007) investigated the understanding of pictograms and their usage in daily lives.

Schröder et al. investigated using 48 icons for 12 mobile phone functions, some of which were original ones as well as icons specifically designed for experimental purposes. They were testing age groups and speed of reaction, but they also tested icon concreteness (abstract vs. concrete) and icon complexity (visually simple vs. visually complex).

2 RESEARCH PLAN AND PILOT STUDY

2.1 *Problem introduction*

The survey was conducted to find out which group of signs (localized or globalized) were easier to interpret among people from various cultures, and if so, why.

It should be determined which type of signs were more user-friendly and easier to recognize through the online pilot study. Also, the qualitative indicators of signs like semiotic distance and icon complexity were taken into account.

2.2 *Participants*

The total amount of participants was 99 people. There were two groups of participants. These 99 participants were divided into two groups: the international group consisted of people from 35 countries where 31% were from Europe, 7% were from the Americas, and 18% were from Asian countries. The main group of research was Russians (46%). It should be mentioned that Russia is located close to Europe and Asia, thus Russia has mixed culture from both East and West.

The gender balance was not equal for the International group (67% female, 33% male). For the Russian group, there were 83% female participants compared to 17% male participants.

2.3 *Pilot study*

The online survey was conducted via Google forms and Facebook to collect the data during December 2018. The survey was a part of the research related to the localized and globalized pictograms.

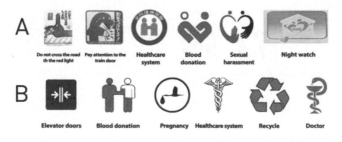

Figure 1. The set of the tested pictograms.

During the survey, 12 preselected signs were checked. They were divided into two groups, such as localized and globalized signs where 'A' was localized and 'B' was globalized signs.

The main localized group for testing included four Taiwanese local signs, since it was planned to create a sign system in Taiwan in the future. The Japanese group was formed to check signs used in countries with a culture similar to Taiwan's.

The goal of adding Japanese signs was to find the connection between the recognition rate and similarities of the cultural backgrounds. Moreover, Japanese signs and visual culture are known for their originality and high level of localization. The reason was the cultural variation and contrast in the interpretation of signs.

The International group was formed from worldwide known signs but was not related to a particular country.

The Health systems signs were added based on data from a previous survey conducted in August 2018, which demonstrated that medical signs were the most difficult to recognize for foreign participants in Taiwan. All signs were mixed in one part of the survey so that participants could not get hints. In this part of the survey, participants answered multiple-choice questions. They chose from three options where only one was correct.

3 RESULTS AND ANALYSIS

'Ultimately a pictogram's effectiveness should be measured primarily in terms of people's ability to understand it' (Tijus, C. et al. 2007).

3.1 Localised signs

There were four common Taiwanese localized signs picked by the author (from the left side of the table). They had high levels of abstractness and low semiotic distance.

The 'Night watch' sign had the lowest rate for all localized signs (14%).

The 'Healthcare system' and 'Blood donation' signs had almost the same amount of correct answers. The wrong answers for both signs were 77% ('Family center' was the same option in both of the answers).

The correct answer for 'Sexual harassment' was recognized by 33% of the participants, the highest among Taiwanese signs.

Meanwhile, Japanese signs had a higher recognition rate than Taiwanese. 'Do not cross the road to the red light' had 85%, and 33% of the participants recognized 'Pay attention to the train doors'.

Overall, it can be seen that localized signs had a low recognition rate, and some of them did not achieve even 40% recognition, which is a frontier for acceptance.

3.2 Globalised signs

There were six globalised signs with different levels of abstractness and spread throughout the world. These were the 'Healthcare system', 'Elevator doors', 'Pregnancy', 'Blood donation', 'Doctor', and 'Recycle' signs.

The lowest recognition rate was for 'Healthcare sign' because the sign was accidentally adopted from the area of law.

Night watch	Blood donation	Healthcare system	Sexual harassment	Pay attention to the train door	Do not cross the road th the red light
14%	30%	31%	33%	33%	83%

Figure 2. A table with the correction rate results of the survey.

Healthcare system	Elevator doors	Pregnancy	Blood donation	Doctor	Recycle
76%	90%	93%	98%	100%	100%

Figure 3. A table with the correction rate results of the survey.

The rest of the globalized signs were recognised by more than 90%. 'Pregnancy' and 'Doctor' have a high level of prevalence in Russian. Even a sign such as 'Recycle' was recognized by 100% of Russian participants, despite the fact that until 2019 Rusia was not a recycling country.

The global signs had a high level of distinction for all signs despite their abstractness and semiotic distance level. Perhaps the extent of a sign's prevalence in the world had a greater effect on the recognition of the sign than abstract metaphors.

3.3 Analysis

Our aim was to make signs effective for foreigners and non–Chinese-speaking people in Taiwan. However, in this study we recruited Russian participants as a studying point to represent the definition of foreigners.

Localized signs had lower recognition level than globalized signs, but overall Japanese once had better performance than localized signs. It is important to take into consideration that Taiwanese society is conservative and collectivistic rather than individualistic based on Hofstede's theory. According to Hofstede's theory, Russia is more individualistic than Taiwan (Russian [39] vs. Taiwan [17]).

The 'blood donation' and 'healthcare' signs were used in both groups of preselected signs (globalized and localized). Based on the similar design concept of the 'blood donation' sign, the globalised 'blood donation' sign had more than three times the recognition rate (98%) than the localised one (30%). The reason for this result was the Taiwanese version's abstractness. The localized 'Healthcare system' sign did not have anything in common with the globalized signs of 'Medicine' or 'Doctor' (Caduceus and bowl of Hygeia). Interestingly, the localized 'Blood' donation and 'Healthcare system' signs used the silhouettes of people in the images ('heads' and abstract forms imitating 'bodies'), which could be another reason that the 'Blood donation' and 'Healthcare' signs had higher levels of erroneous results in that most participants recognised them as 'Family center' (67%) or 'Young family center' (70%).

For the 'Neighbourhood watch' sign, even the caption 'safe' in English text did not help Russian participants understand its correct meaning (14% correct answers). The answer preferred by Russians was 'Social housing' (71%) due to the harmonious colour and silhouettes used in this sign.

In terms of the heart shape of the 'sexual harassment' sign, Russian participants recognized it as a disease symbol or 'Cancer awareness center' (48%).

Therefore, the researchers have learnt that difference in cultural background carries different metaphoric cognition, which leads to various results in this research.

4 CONCLUSION

This study has shown anticipated results that cultural background and abstractness of the signs do make the interpretation of signs different. The preselected signs chosen by the researchers proved to be biased and may be subjective, such as the combination of Taiwanese and Japanese signs. The original idea of mixing both signs in one group as localised signs was based on the opinion that these countries have a similar culture, which led to confusion for Russian participants. For future work, we need to make more revisions on the next questionnaire such as the selection of a set of localized signs and pictograms in Taiwan mainly for Russians. Potential target groups shall include non-Chinese speakers, tourists, and short-stay residents who feel more confident, and help them in way finding. Currently, there is not any unified system of local signs and icons in Taiwan. Most daily services such as Seven-Eleven, HSR, and public transport system generally use globalized signs. There will be more testing in order to achieve higher recognition of our intended iconic/sign design. Each stage will be iterative and refined carefully based on further validations. We hope to implement this visual iconic system for Russians and even Slavic users who live in Taiwan.

REFERENCES

Declercq, C. (2011). Advertising and localization. In *The Oxford Handbook of Translation Studies*.

Esselink, B. (2000). *A practical guide to localization* (Vol. 4). John Benjamins Publishing.

Hiippala, T. (2012). The localisation of advertising print media as a multimodal process. In *Multimodal Texts from Around the World* (pp. 97–122). Palgrave Macmillan, London.

Hofstede, G. (2011). Dimensionalizing cultures: The Hofstede model in context. *Online readings in psychology and culture, 2*(1), 8.

LISA (2005) Lisa Forum Cairo: Localization: Perspectives From the Middle East and Africa. Available at www.lisa.org

Ng, A.W., & Chan, A.H. (2008, March). Visual and cognitive features on icon effectiveness. In *Proceedings of the international multiconference of engineers and computer scientists* (Vol.2, pp. 19–21).

Sebeok, T.A. (2001). Signs: An introduction to semiotics. University of Toronto Press.

Schröder, S., & Ziefle, M. (2008, July). Effects of icon concreteness and complexity on semantic transparency: Younger vs. older users. In *International Conference on Computers for Handicapped Persons* (pp. 90–97). Springer, Berlin, Heidelberg.

Tijus, C., Barcenilla, J., De Lavalette, B.C., & Meunier, J.G. (2007). The design, understanding and usage of pictograms. In *Written documents in the workplace* (pp. 17–31). Brill.

Zender, M. (2006). Advancing Icon Design for Global Nonverbal Communication: or What does the word bow mean? *Visible Language, 40*(2), 177.

Smart Science, Design & Technology — Lam et al. (eds)
© 2020 Taylor & Francis Group, London, ISBN 978-0-367-17867-3

The negative effect of smartphone use on the happiness and well-being of Taiwan's youth population

Siu-Tsen Shen*
Department of Multimedia Design, National Formosa University, Hu-Wei, Yunlin, Taiwan

Stephen D. Prior
Aeronautics, Astronautics and Computational Engineering Design, The University of Southampton, Hampshire, UK

Chin-Huang Wang
Department of Applied Foreign Languages, National Formosa University, Hu-Wei, Yunlin, Taiwan

Liang-Yin Kuo
Department of Multimedia Design, National Formosa University, Hu-Wei, Yunlin, Taiwan

ABSTRACT: The pursuit of happiness is considered to be the proper measure of social progress and the goal of public policy. Human well-being is more than wealth. United States Senator Robert F. Kennedy famously pointed out that GDP "measures everything, in short, except that which makes life worthwhile," and today GDP is increasingly recognized as an insufficient measure of quality of life. With good reason: we have gotten richer – but not happier. We have therefore failed, it seems, to convert wealth into well-being. In the latest report, Taiwan was ranked 26th out of 156 countries, with a score of 6.441, while Finland was rated the highest (7.632) and Burundi the lowest (2.905). Five Nordic countries appear in the top ten ranking (1–4, and 9), and this leads us to reflect on why this is the case. In this survey, almost two-thirds of the participants (65%) stated that they considered themselves addicted to their smartphones. Although generally happy with their lives (scoring 7.0), the 161 participants showed moderate anxiety, with an average Nomophobia score of 68.

1 INTRODUCTION

Thanks to the rapid growth of available apps, smartphone users are spending more and more hours in the digital world and are becoming addicted (see Figure 1). According to a recent report from App Annie, the amount of time people spend online had increased by 30 percent. The average person uses 40 apps per month and spends three hours a day using apps (Annie 2018). Statistics show that 98 percent of 18- to 29-year-olds in the USA access the internet regularly.

Online harassment is becoming an issue for many users, and this includes offensive name-calling, online trolling, and physical threats. A recent study found that 25 percent of users in the USA across all age groups have been called offensive names online, with 28 percent of males and 22 percent of females having experienced this type of behavior (Statista 2018).

Furthermore, data indicates that the age range of 18 to 24 has the highest penetration rate with 95.7 percent using smartphones (eMarketer 2018). In order to protect vulnerable students, some schools have now banned smartphones during the school day as well as asked parents to prohibit their use during weekends (Hymas 2018).

2 SURVEY OF HAPPINESS

Increasingly, happiness is considered to be the proper measure of social progress and the goal of public policy. The rising use of so-called smart devices, with people reporting using these for up to eight hours per day on average, raises concerns on whether these devices are liberating or inhibiting people's productivity.

An initial study investigated smartphone use and levels of happiness. This was conducted in three phases, with Part 3 being a questionnaire to elicit the Nomophobic value. A further study is planned to investigate the level of happiness of Taiwanese young people with and without their social apps on smartphone devices. The aim of this study will be to measure the level of happiness before and after not using the participant's

---

*Corresponding author: E-mail: stshen@nfu.edu.tw

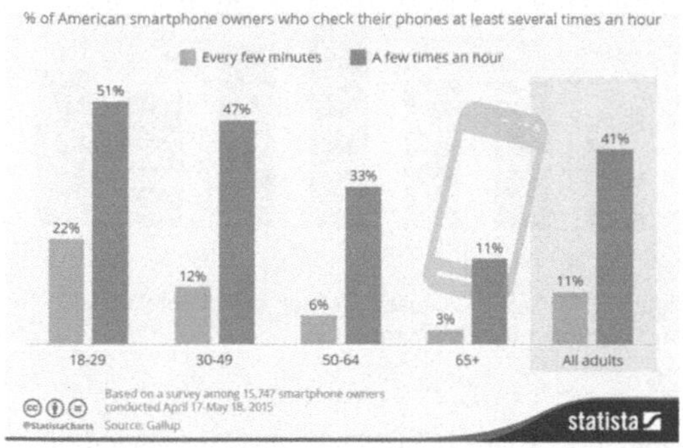

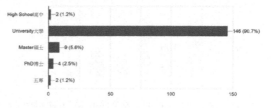

Figure 1. America's growing smartphone addiction.

social apps on smartphones for a period of at least one week and up to one month.

The hypothesis is of participants' initial unhappiness (similar to drug withdrawal), followed by a gradual improvement in happiness when participants realized the benefits of reducing their reliance on social media apps.

The initial user trial consisted of approximately 161 individuals (81 males and 80 females) within the age range of 15–28. After being interviewed, selected participants agreed to share personal information on their smartphones for this research (General Online Questionnaire, 13 questions; Online Happiness Index, six questions; and NMP-Q after Yildirim and Correia (2015), 20 questions).

The happiness index was carefully measured and analyzed weekly (Online Happiness Index, six questions). This helped to rethink the impact of social media on young people. This research conducted a series of formative studies, including three online questionnaires and interviews, to observe and record users' cognitive behavior and preferences on smartphones over a one-year timeframe.

After being interviewed, the selected participants who agreed to share personal information on their smartphones for this research were carefully documented.

Users with iOS were asked to download the Moment App for free, or they had their default setting App Screen Time, e.g., iOS12. Moment allowed the users to see how much time they were spending on their phone and which Apps they were using the most. Users with Android were asked to download the Quality Time App for free to measure the user's screen time in the same way. Users were asked to take a screenshot daily and send it back to the researcher within their trial period of time (at least a week, but possibly up to one month).

After an initial analysis of the users' feedback, selected members of each group were called back for further interviews; this was an opportunity to discuss their feelings and overall quality of life before and after the research.

Figure 2. Analysis of the participants by educational level.

3 RESULTS

The initial survey (Part 1) obtained 161 responses. These participants were classified as 18% Beginner/ Novice, 66% Intermediate, and 18% Expert. Roughly 50% were male and 50% female. Almost 60% of the participants were aged 20 or 21. Ninety-eight percent described themselves as students, with 91% having a University degree.

In terms of their smartphone usage, 45% had over 12 years of experience, and a further 30% had between 11 and 12 years. When asked about smartphone use per day, 30% stated that they were using it more than 12 hr/day. A further 15% stated that they were using their smartphone between 11 and 12 hr/ day. Another large group (27%) used their smartphone between 3 and 4 hr/day.

Almost 78% of the participants took their smartphone to bed at night. By far the biggest group (40%) used an Apple iPhone, with HTC, Sony, and Samsung in second place with around 11–15%. When asked about their operating system, however, 60% used Android and 40%, iOS. The participants tended to have relatively new smartphones, with 40% (1–2 yrs), 29% (<1 yr), and 24% (3–4 yrs). When asked about the number of social media apps on their smartphone, a large majority (57%) had between 1 and 5 apps, with the next highest category (22%) having 6–10 apps. The most common App was Line (87%), followed by Instagram (61%), Facebook (57%), and Messenger (39%).

The Part 2 – Happiness Index questions related to overall life expectation and feelings of happiness. Almost two-thirds of the participants (65%) stated that they considered themselves addicted to their smartphones. In terms of overall happiness with their life, the majority described themselves as happy, with a score of 7/10 (27%).

When asked about the extent to which they felt the things they do in their life were worthwhile,

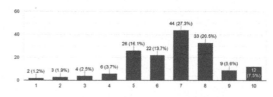

Figure 3. Degree of happiness with life (scale of 1 to 10).

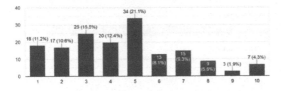

Figure 4.   Overall, how anxious did you feel yesterday?.

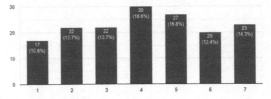

Figure 5.   Anxiety caused by not being able to keep in touch with family via smartphone.

there was a mixed reaction, with a range of responses from 5/10 to 8/10 (ranging 16–24%). When asked how happy they felt yesterday, the majority of the participants were positive, with scores of between 5/10 and 10/10 (averaging 15%). When asked how anxious they were yesterday, the majority of the participants had low scores, 1/10–5/10.

The final question asked the participants to name three things in life that gave them happiness. There was a range of responses, varying from family and friends to eating, sleeping, playing, and music.

The Part 3 Nomophobia questionnaire repeated a study conducted by Yildirim and Correia (2015). In response to these 20 standard questions, the Taiwanese youth expressed a clear view of being moderately anxious without access to their phones, with an average score of 68 (161 participants).

To put this into perspective, the Nomophobia scale is shown below.

The exception to this was Q.13, which related to anxiety caused by not being able to keep in touch with their family.

4   CONCLUSION

In this paper, the authors surveyed Taiwanese youth to elicit information on smartphone use and general happiness. There is no doubt that, in general, Taiwanese youth are both happy and enjoying life.

By a factor of 2:1, however, they recognise that they are addicted to their smartphone devices. This in itself may not be a big problem; however, when this relates to a need to be liked together with a lack of personal communication skills, there may be a longer-term issue which may not be revealed until later in life. The lifestyle implications of playing online rather than in the real world may be storing up health issues for the future.

With an average Nomophobia score of 68, Taiwanese youth are officially classified as moderately anxious. Whilst this is indicative of a modern, technologically sophisticated country, it does create some alarm as to the reliance of youth on smartphones and social media apps.

REFERENCES

App Annie, 2018, The average smartphone user accessed close to 40 apps per month in 2017, retrieved December 5, 2018, from https://www.appannie.com/en/insights/market-data/apps-used-2017/.

eMarketer, 2018, US smartphone user penetration by age, 2018 (% of population in each group), retrieved December 5, 2018, from https://www.emarketer.com/Chart/US-Smartphone-User-Penetration-by-Age-2018-of-population-each-group/219283.

C. Hymas, 2018, Secondary schools are introducing strict new bans on mobile phones, *Telegraph*, London, UK, Telegraph Media Group Limited.

Statista, 2018, Share of adults in the United States who use the internet in 2018 by age group, retrieved December 5, 2018, from https://www.statista.com/statistics/266587/percentage-of-internet-users-by-age-groups-in-the-us/.

C. Yildirim and A.-P. Correia, 2015, Exploring the dimensions of nomophobia: Development and validation of a self-reported questionnaire, *Computers in Human Behavior* 49(Supplement C): 130–137.

Table 1.   The Nomophobia Scale (Yildirim and Correia 2015).

Score	Nomophobia Level
NMP-QScore = 20	Absent
21 ≤ NMP-QScore < 60	Mild
60 ≤ NMP-QScore < 100	Moderate
100 ≤ NMP-QScore ≤ 140	Severe

Smart Science, Design & Technology — Lam et al. (eds)
© *2020 Taylor & Francis Group, London, ISBN 978-0-367-17867-3*

The artistic characteristics and social significance of the production and installment of the "King Gesar" stone relief

Yinkun Wang*
Fujian Yinli Stone Decoration Co., Ltd., Fujian Province, PR China

ABSTRACT: The "King Gesar" stone relief was produced and installed at the King Gesar Square in Shangri-La, Diqing Tibetan autonomous prefecture, Yunnan province. The relief gives an all-around manifestation of the world's longest historical epic that has been passed on for thousands of years on the Tibet Plateau of China, guiding people to witness the full and unique glamour and glory emitting from the exceptional cultural heritage of the Tibetan ethnical group.

Keywords: Shangri-la, red sandstone relief, King Gesar, Tibetan ethnical group.

1 INTRODUCTION

Shangri-La – a land of beauty and wonder with an average altitude of 3,280 meters above the sea level and a place where diligent, wise, and true Tibetan people have lived and thrived for generations. Looked down by the magnificent Mt. Shika and Mt. Meili, the people here have built delicately designed and decorated temples by the edge of high cliffs, contributing to a Tibetan cultural breeding ground that is immersed in peace, tranquility, eternity and mystery.

2 THE ARTISTIC CHARACTERISTICS OF THE PRODUCTION OF INSTALLATION OF THE KING GESAR STONE RELIEF

The King Gesar stone relief consists of two scrolls to the east and west respectively, "The Coronation of King Gesar upon Birth" and "Fight by King Gesar for the People's Benefits". The two scrolls are altogether 218 meters long, 5 meters high on average, totaling to more than 1000 square meter. The construction period was from May of 2017 to December of 2017. It was finished with saffron red sandstones mined in Yunnan and designed and constructed by Fujian Yinli Stone Materials Decoration Company whose work ranged from the drawing design to the carving, production, and installment. With the improved taste of the market, red sandstones have become of significance in the decoration work rather than just in the construction industry, as it is involved in the people's life. Under its decorating nature and given the work of the carving art, red sandstones have become more aesthetically pleasing (Yang, 1999; Laurence Brahm, 2008).

During the course of production of the red sandstone relief, the writer was engaged in the historical materials collection (for as long two months) and gave a full statement of King Gesar's life from his birth to the point he united various tribes, big or small. As the reincarnation of Padmasambhava in the Tibetan legend, King Gesar fought all his life in punishing the evil, carrying forward the kindness, advancing the Buddhism, and dissimilating the culture, becoming the epic hero of which Tibetan people are proud. King Gesar was born on A.D. 1038 and died on A.D. 1119, during which he vanquished all kind of demons and monsters, eradicated the violence, took care of his people, traversed the north and the south, and united over 150 tribes of different sizes, keeping the territory of Ling Kingdom intact as shown in Figures 1 and 2.

With laying off of the King Gesar stone relief, the 1:1 work produced as per the design drawing. Carving: first finish the initial carving (for a rough silhouette) and then the detailed carving by the carving team and artistry artist led by the writer. Polishing: once the carving work is done, process the details and polish the work with coarse and fine abrasive cloth. Relief installation: finish the steel skeleton as per the site and dimensions and install the keel of box iron and angle iron processed by hot-dip galvanizing in a dry-hang manner. Once the fastenings or sub keel are welded, use stainless steel bolts to connect the stainless steel hanging pieces. The positions of stainless steel pins and T-shape stainless steel

---

*Corresponding author. E-mail: 1346525023@qq.com

Figure 1. A portion of the design of "The Coronation of King Gesar upon Birth".

Figure 2. A portion of the design of "Fight by King Gesar for the People's Benefits".

Figure 3. Laying off of the King Gesar stone relief.

hanging pieces can be adjusted by the bolt holes of hanging pieces. The next step of work may continue only when the plate is confirmed to be perpendicular and the concealed work successfully pass.

The stone relief of "The Coronation of King Gesar upon Birth": it was a time of natural calamities and man-made misfortunes where demons and evil spirits polluted the ground and people suffered. As the merciful gods were discussing whether to send a son of god to the world to free the mass from the misery, the son of god Tuibagewa volunteered to come down to Tibet and be the king of the black-hair Tibetan. Later, Gedanlamu, a diligent, smart, kind, and forthright girl conceived in a dream on the grassland and gave birth to Gesar from her left armpit after ten months of pregnancy. Coincidentally, the cow of the Gedanlamu's gave birth to a foal named Jiangtongsi at the same time. By three months old, Gesar shot through the straw with a bow and arrow made from willow branches; by three years old, he shot down two birds with sandalwood bow and bamboo arrow; by four years old, he started to study scriptures; by twelve years old, he won in the tribe's horse racing event. With Gesar sitting on the throne, the marvelous horse Jiangtongsi neighed three times that made the ground shake and the treasure gate of the crystal mountain open, through which gods appeared with a white helmet, a piece of bronze armor, and a red-vine shield. Then Gesar, who was surrounded by gods and fully suited, appeared: a girdle inlaid with Mamao soul stones, a quiver attached with the soul of the god of war,

a leopard-skin bow case possessed by the soul of Weierma, longevity underclothes, longevity belt that belonged to the god of war, and a pair of battle boots by which the world was awed.

The stone relief of "Fight by King Gesar for the People's Benefits": when Gesar claimed the throne, he was determined to eradicate all demons ravaging the world and change the divisive landscape where tribes of different sizes and small countries established scattering regimes of their own and fought and raided each other. Armed with his divine bow and arrows, King Gesar rode on Jiangtongsi and headed eighty brave, resourceful, courageous, and valiant commanders who were the reincarnation of mahasiddhas on a march towards the demons that were situated eighteen snow mountains away. From then on, he exerted his full diving power, fighting and conquering, and eventually subdued the northern demons that invaded Ling Kingdom, beat the Lord of Baizhang of Huoer Kingdom, Lord Sada of Jiang Kingdom, Lord Xinchi of Menyu, Lord Nuoer of Dashi Kingdom, Lord Chidan of Kaqiesongershi, Lord Tuogui of Zhugu, Nine-soul Lion of Ladake Kingdom, captured Gongzanchijie, king of Songba Kingdom, conquered the seven demons of Ali Kingdom, slayed Sengezhaba, king of Minu Kingdom and Nimazanjie, king of Mugu-luo Kingdom, extradited Asailuocha, saved the wandering souls of the three evil destinies and the three goods destines, and defeated tens of "Zong" successively. Ultimately, King Gesar brought unity and peace to the world by clearing the demons, punishing the evil, supporting the weak, and saving lives, and the mass was once again living in peace, stability, abundance, and happiness. King Gesar, having brought down the demons and evils of the world and become fully accomplished, summoned the mass, men and women, the young and old, from neighboring countries and tribes to Senzhudazi Palace. During the joyous event, He informed the people that would return to the gods. With tears running, the present wished to him to stay, but only to witness the body of King Gesar, along with that of the two princesses Zhumu and Meisa who stood beside him, rose slowly towards the black

Figure 4. The stone relief of "Fight by King Gesar for the People's Benefits".

and white halo in the sky. Almost immediately, rainbows started to appear in the sky, fragrance filled the air and the rain of flowers begun. Surrounded by gods, King Gesar and the two princesses returned to heaven.

3 THE SOCIAL SIGNIFICANCE OF STONE CARVING ART OF THE KING GESAR STONE RELIEF

The King Gesar stone relief bears a milestone significance in the history of art development and Tibetan stone carving. Reliefs derive from sculptures and paintings, a method that compresses the objects and expresses 3D dimensions using perspective and other elements that are shown in one or two faces. Reliefs are often done on surfaces, hence they are used more frequently on buildings as well as articles and utensils. In recent years, they are taking greater importance in urban beautification. Reliefs are as diverse as circular carving in terms of contents, forms, and materials.

The King Gesar stone relief was finished under the traditional relief technique of Huian. The writer is well-adept to express the static in a dynamic manner, demonstrate mercy with rage, and compassion with resentment. Because of the long-time professional experience of the writer and the carving artists and the increasingly finer Huian stone carving artistry, plus the ingenious conception of the writer, the relief strikes the viewers as vivid and refined and an irreplaceable work. The relief also underscores the dynamism, sense of wildness, and braveness in the aspect of artistic design.

The King Gesar red sandstone relief is a sheer reflection of the ardor for real life, the consistent pursuit of beauty, the open expression of the vigor of life, rather than the symbol of the intangible next world. Artistically, the relief is brimmed with a refreshing, vigorous, and wild vibe that immensely resonates with real life, and emits an encouraging and inspiring power.

The King Gesar red sandstone relief was the representation of the progress of the advanced Tibetan culture back in the time when it was finished and stood in contrast with the culture of the ruling monk and noble class. In the new millennium, new century, and new historical background, we need to move forward to contribute to a new Tibetan culture under socialism. To create our own culture, we must understand the nature of the advanced culture and place a firm grip on the direction along which the advanced culture progresses. The China Western Development program has brought substantial opportunities to minority groups in far areas that were never seen before. The progress of the material civilization drives the development of the spiritual civilization, which is bound to propel and push forward the development of the material civilization. Material civilization and spiritual civilization are complementary to and benefiting each other that will facilitate the prosperity and development of the Tibetan economy.

REFERENCES

E. Yang, 1999. *The Life of King Gesar*. Shenyang: Chunfeng Literature and Art Publishing House.
Laurence Brahm, 2008. *Searching for Shangri-La*. Beijing: New World Press.

Smart Science, Design & Technology — Lam et al. (eds)
© 2020 Taylor & Francis Group, London, ISBN 978-0-367-17867-3

An exploration of architectural woodcarving decoration art

Yangchun Xu*
Zhangzhou Gu Xiang Yi Diao E-commerce Co., Ltd., Zhangzhou, Fujian Province, PR China

ABSTRACT: Back in the time of Clannish society 5,000 years ago, our ancient people were already aware of decorating buildings and houses to improve their living environment and quality. Wood, as a major building material, was used for portraying animals and plants or a person's working and living scenarios, and woodcarving came into being as a result. After the Qing Dynasty, the woodcarving technique was widely used for architectural decoration. In the Ming and Qing Dynasties, apart from architectural decoration, woodcarving was also applied in furniture, playthings, and household utensils. The decoration not only enriches the creators' artistic values, but also embodies the dignity of buildings. Architectural woodcarving art is popular due to its profound and rich connotations, and has been handed down from age to age as one of the traditional folk arts and one of the Chinese cultural treasures. Located in Minnan Delta and in vicinity of Chaoshan, Zhao'an County has a long history of using woodcarvings for architectural decoration, in particular, in the bucket arches, beams, rafters, and mortises and tenons in ancestral halls and temples. The themed woodcarving decoration enhances both stability and aesthetics of buildings, and thus is a folk technique that integrates aesthetics and practicality.

Keywords: Zhao'an woodcarving, decoration of ancestral halls and temples, heritage, innovation and development

1   CHARACTERISTICS OF ARCHITECTURAL WOODCARVING ART OF ZHAO'AN

Located in Minnan Delta and in vicinity of Chaoshan, Zhao'an has its heritage and folkways substantially influenced by Chaoshan, and its architectural woodcarving art, originating from Chaoshan, is in close connection with Chaozhou's woodcarvings (China Intangible Cultural Heritage Image Net, 2019). This perfect technique, which started in mid-Tang Dynasty and flourished in the Song, Ming, and Qing Dynasties, has given way to the special Zhao'an woodcarving art, thanks to Zhao'an craftsmen's continual efforts and innovations. The key features of the art include the following (Figure 1):

(1) Famed for Refined Detailing: Drawings should be made prior to the production of any piece of work, whether the parts of the ancestral hall, temple, and pailou; the doors and windows; or the furniture, household utensils, and playthings. And woods are chosen according to decoration requirements. The pliable but tough camphorwood is the best, with fir wood as an alternative, while mahogany is mainly used for furniture. The technical process includes raw material chiseling, fine carving, and painting or gilding. Craftsmen would choose relief, sinking carving, round carving engraving, hollow carving, and other techniques.

(2) Meticulous in Material Selection and Technique Implementation: Craftsmen of Zhao'an have been good at learning more carving techniques from past to present, and there are many different architectural decoration styles and a huge coverage for woodcarvings. As a result, different requirements may lead to various selections of decorative materials. The perennial fir wood is the best for the parts of the ancestral hall, temple, and pailou, while camphorwood is primarily used for portraying mountains, waters, people, flowers, birds, or stories due to its flexible texture, dampness resistance, and insect repellence. The camphorwood is durable and does not crack or deform. The carvings are normally composed in a Z shape, with a clear route and contrasting arrangement. Skillfully arranged in reasonable order, patterns like flowers, birds, fishes, and insects can be remarkably true to life.

(3) Wide Coverage of Themes: Stemming from Chaozhou woodcarving, Zhao'an woodcarving has taken its shape thanks to the craftsmen's dedicated creations themed from folktales, ancient operas, and classic literature. With a wide coverage of themes, Zhao'an woodcarving covers nature, animals, people, and even gods. People-themed carvings are mainly inspired by classic stories from the Chinese

---

*Corresponding author: E-mail: 1097073474@qq.com

Figure 1. Xu's Ancestral Hall, Longkeng Village, Zhao'an (Built late in Tang Dynasty).

Four Great Classical Novels and ancient Chinese poetry, e.g., "Three Brothers Fight Lu Bu," "Taking Oath in the Peach Garden," "Lu Zhishen Uproots a Tree," and "Lin Chong Goes up Liangshan Mountain." Landscape-themed works are reputed for special local scenic spots like Wang Yang Tai and Xiang Lin Tower. In many cases, the carvings on the decorative parts of ancestral halls and temples well echo the porcelain carvings on the main buildings.

(4) Profound Grassroots Feature: Most of the woodcarving craftsmen have been absent from the classroom of arts, and few of them have been highly educated. They originally started their artistic career to make a living. Taking root in Zhao'an, the hometown of calligraphy and painting arts, the woodcarving craftsmanship features a profound folk style and strong grassroots flavor that renders everything in the world in a simple and obvious manner. In the decorative structural parts of ancestral halls, temples, and pailous, the woodcarving skills feature bold lines, strange shapes, and color contrasting, showcasing the dignity of the architecture under the feudal etiquette. However, the treatment on furniture, doors, windows, and other playthings is characterized with delicate lines and gentle touches, similar to the technique used in graphic arts, and is welcomed by the people. The craftsmen need to keep reflecting and improving their skills to respond to the people's higher aesthetic demands.

2 ACHIEVEMENTS IN THE PAST FEW DECADES

I have been interested in woodcarving since my childhood. At the age of 15, I learned woodcarving from my master Mr. Shen Zhaogui in the downtown area of Zhao'an, and I was appreciated and carefully enlightened due to my hard studying. When I was 22, I started to independently contract the woodcarving projects of ancestral halls and temples in Zhao'an and Chaoshan. My works mainly include mascots, e.g., lion, elephant, pixiu, otter, dragon, phoenix, qilin, and flying fish-shaped bracket, as well as landscape, flower, bird, and legendary story. I am capable of flexibly combining round carving, hollow carving, and relief in the creations. Over the past two decades, my achievements in ancestral halls and temples include the following (Table 1):

(1) From 1994 to 1995, with the entire woodcarving project of Tiandi Temple built in Fengzhong Village, Waisha Town, Longhu District, Shantou City, it was the first time I contracted a large project. I independently made 600 pieces of work including dragon, phoenix, flower basket, lion, flying fish, people, and story, and developed, designed, drew, and carved each of them on my own. I poured a lot of effort, time, and soul into the works. The completion of each work made me happy and drove me to more carefully make the next one. I still remember how I struggled when required by the temple management committee to make two woodcarvings of people. I visited all the temples and ancestral halls, as well as all the bookstores, in the Shantou area. It took me 2–3 months to complete the two carving works of "Three Brothers Fight Lu Bu" and "Taking Oath in the Peach Garden," inspired by the two stories in "The Romance of the Three Kingdoms."

(2) Among the woodcarving works I created over the past decades, the Wen Xian Shi Jia (an ancestral hall of nine families in Zhao'an county) was the most outstanding achievement that I made. Originally built in the Ming Dynasty, the building is a relic site under provincial-level protection nowadays. As part of the renovation, the woodcarving structure was required to feature the ancient processing technique of Qing Dynasty. I undertook the project from 2010 to 2012. On the over 400 pieces of woodcarvings, most of the patterns were arranged and organized in a skillful way so as to give full expression of delicacy of lines and beauty of forms. In the hall, the civil themes on the left were presented in contrast

Table 1.   The woodcarving projects of ancestral halls and temples in Zhao'an and Chaoshan.

Name	Theme	Character	Material	Craftsmanship
Temple	3 upper beams with 5 brackets	Bucket arch and detailing	Camphorwood, fir wood	Hand-made
Ancestral hall	3 upper beams with 5 brackets	Bucket arch and detailing	Camphorwood, fir wood	Hand-made
Residence	Chinese decoration	TV background wall and flower windows	Mahogany	Both machined and hand-made

Figure 2. Shen's Ancestral Hall, Zhao'an (Renovated from 2010 to 2012 with fir wood and camphorwood).

Figure 4. A camphorwood carving named "Flowers bring fortune, bamboos presage peace," 6 cm in thickness and 50 cm in diameter, collected by Fuzhou University.

with the military themes on the right. The works of "Bai Ren Tang" and "Qi Xian Shang Jing" featured the vivid presentation of people as well as otter, lion, and flying fish. Other patterns included bucket arch, bracket, bent plate, phoenix, corner-trays, etc. The entire ancient woodcarving project of the ancestral hall was completed in three years, and it was widely appreciated as a masterpiece of elegance, magnificence, and excellence (Figure 2).

As part of China's artistic treasures, the woodcarving art represents a form of Chinese traditional folk art. In our current era of peace and prosperity, people are in pursuit of material culture, as the living standard keeps rising. Many wealthy families choose furniture with aesthetic woodcarvings, while decorating their homes with woodcarving craftworks. Therefore, I established the Ancient Woodcarving Decoration Studio in 2007 with an aim to combine woodcarving (root carving) with calligraphy and painting arts. Dedicated to satisfying people's demands in material and cultural life, we lead the decoration industry with innovative industrial arts.

3 INSPIRATIONS AND PROSPECTS

Art comes from life and goes beyond it. Influenced by Chao'shan woodcarving technique, the Zhao'an woodcarving decoration art is one pearl of Chinese traditional culture. As a woodcarving craftsman, I have devoted myself to architectural woodcarving decoration for several decades. I have the following inspirations:

(1) Never stop improving woodcarving craftsmanship by learning from professionals, skilled craftsmen, and grassroots workers from all over the world and absorbing artistic essence from the existing works. At the same time, I should take every opportunity to visit and examine architectural complexes throughout the country, with the goal of combining the carving styles (tile carving, stone carving) of other places with the local woodcarving art.

(2) Give priority to infusing the brilliant Chinese culture and positive energy of arts into every piece of work, since the government is attaching importance to the traditional culture (craftsmanship) this year. I should try hard to make more and more elaborate works by combining people's aesthetic interpretations and the cultural DNA of the Hometown of Calligraphy and Painting.

Figure 3. Chinese Decoration in Residence (Mahogany and camphorwood).

REFERENCES

China Intangible Cultural Heritage Image Net, 2019. *Chaozhou Woodcarving*. Retrieved 2019/2/2, from http://www.ihchina.cn/Article/Index/detail?id=14022

Smart Science, Design & Technology — Lam et al. (eds)

Aesthetic and educational value of Fujian wood carving art

Guangmin Yang*
Jimei University, Xiamen, Fujian Province, PR China

ABSTRACT: Fujian wood carving belongs to an artistic heritage that integrates the culture and nature of Fujian Province, and delivers unique ideas and concepts in nature, life, culture, and aesthetics. In contemporary society, it functions as the records of history, and also as a future aesthetic education resource.

Keywords: Fujian wood carving, aesthetics, education, value

1 OVERVIEW OF FUJIAN WOOD CARVING

As early as in the Tang and Song Dynasties, there has been commendable practice for Fujian wood carving. In the Ming and Qing Dynasties, Fujian wood carving has developed into an independent folk handicraft, and from the Republic of China period until now, Fujian wood carving has continued to develop and grow. In Fuzhou, Quanzhou, Putian, Zhangzhou, Xiamen, Sanming, Western Fujian, Northern Fujian, and other areas, many wood carving relics produced after the Ming and Qing Dynasties have been kept, and there are still many practitioners of wood carving now. As a kind of mature folk handicraft, Fujian wood carving is characterized by six aspects: firstly, wide application of timbers. The materials include camphorwood, phoebe zhennan, China fir, boxwood, longan wood, tea plant root, litchi wood, rosewood, sandalwood, teakwood, taxuschinensis, michelia alba, cryptomeriafortunei, pomegranate wood, dalbergia odorifera, jujube wood, lithocarpus glaber, wenge wood, etc.; secondly, perfect techniques. The carving techniques include line carving, intaglio carving, relief carving, hollowed-out carving, circular carving, semi-circular carving, penetrated carving, embedded carving, etc.; thirdly, precise working procedure. The process covers making bricks, finishing, polishing, coloring, waxing, inlaying, etc.; fourthly, rich themes. The themes of figures, animals, plants, utensils, landscapes, and abstract patterns are all included; fifthly, profound connotation. The stylized combined patterns represent people's hope for good life, with themes like persuading people from practicing vices to practicing virtues, advocating respect for the old and cherishment of the young, filial piety to parents, and love and loyalty to the country and country leader, or wishing longevity to old people, healthy and happy life, and conjugal love between wife and husband, etc. It shows a reflection of the characteristic that "the pattern must have a connotation, and the connotation must be auspicious"; sixthly, various categories, which include building wood carving, furniture wood carving, gold lacquer wood carving, puppet head carving, and artistic wood carving, etc.

Fujian wood carvings, such as building wood carving, furniture wood carving, gold lacquer wood carving, and puppet head carving, bear profound traditional features. Moreover, they are mostly function oriented with practicability, and can be categorized into the broad heading of craft wood carving. For independent wood carvings made with valuable hardwood and tree root, those of small volume can be displayed on desks while those of large volume can be displayed in spatial environments; the function of these carvings inclines to appreciation, and thus they can be categorized under the broad heading of artistic wood carving. It is concluded from a comprehensive evaluation in terms of local features, proportions and influences, etc., that building wood carving, artistic wood carving, and puppet head carving can be thought of as representative of Fujian wood carving (Guo, 2004; Miao, 2008; Lin, 2006).

2 BUILDING WOOD CARVING: BEAUTIFICATION OF LIFE

Wood carving is a kind of decorative method that has been used in ancient Chinese buildings for a long time, and it was the most important category of Fujian wood carving in the early period. In the long period of agricultural society, most dwellings were wood

*Corresponding author. E-mail: xmygm@163.com

structures, which provided an excellent platform for wood carvings. You can find the traces of wood carvings in traditional dwellings, temples, ancestral halls, drama stages, and other public folk constructions in Fujian. Wood carvings will appear in the beam, square-column, bracket set, queti (sparrow-shaped brace), window and door, sunk panel, etc., for decoration and beautification. The custom of using Fujian wood carvings in buildings had been popular until the late Qing Dynasty and the early Republic of China. A considerable number of exquisite wood carvings can be found in traditional dwellings in many areas, such as Fuzhou, Quanzhou, Putian, Liancheng in Western Fujian, Taining in Northern Fujian, and Fu'an in Eastern Fujian. Nowadays in the renovation or reconstruction of temples, ancestral halls, and other buildings in various places of Fujian, the tradition of using wood carvings as decoration has also been kept. In what follows, the building wood carvings in Quanzhou are taken as an example to explore the traditional building wood carvings of Fujian (Figure 1).

In the ancient buildings of Quanzhou, the beam, column, purlin, and other large-sized wood components that bear weight directly are generally not carved, or at least carved no better than decorative carving on architrave. The secondary load-bearing components, such as bracket, Guatong, and Shizuo, are mostly carved with bas-relief. Connection components, such as pensile flower column, Shuchai, Doubao, Tuomu (bolster), Shusui, Tongsui, and Menzan (decorative cylinder), mostly adopt the form of penetrated carving. The door leaf oriented toward the courtyard is often decorated with small exquisite wood, sometimes with various patterns patched by mortise and tenon, or patterns carved by an entire wood block. Most wood carving decorations are distributed in the Aoshou (concave space) of Xialuo (first row of rooms), Xiating (front lobby), Dating (main lobby), and Jutoujian (wing rooms), and those carvings are generally combined with painting.

As for carving method, Yuanguang and Shizuo are often decorated with front carving, which can be described as "single-side carving" in the jargon of carving craftsman; if it is decorated with carving on both the front and back, it is described as "double-side carving." Tuomu (bolster) is carved into the shapes of fish with dragon head, dragon and phoenix, flower, and grass; Shuchai is carved into shapes of immortal, lion, and tiger; arch is carved into flying immortal, man of unusual strength, and water dragon, etc.; Shuzai is often decorated with open calligraphy and paintings or entangled water dragon with scroll grass pattern.

The wood carvings in the former residence of Yang Amiao are representatives of wood carvings decorated in Quanzhou buildings in Ming and Qing Dynasties. Most brackets, short columns, sparrow-shaped brace, and other wood components used in the former residence of Yang Amiao are multilayer hollowed carvings decorated with different patterns. These carvings feature sophisticated carving techniques, ingenious conceptions, and fine lines; combine anima in simplicity; and make people admire the ingenuity and rich imagination of the builder. With these exquisite carvings, this ancient residence recovers the elegance of old times and helps visitors feel a thick classical atmosphere and magnificent artistic charm. The components of the round-ridge-roof square kiosk in front of the lobby, atrium, and eastern guest room share the same style and crafts with the bracket, short column, and sparrow-shaped brace of the gate, and are painted with gold powder, which creates a resplendent and magnificent world. In the gold-plated spindle nose, carvings of figures from various dramas and legends are true to life, and those reliefs feature complex and dense layers. The carving characteristics of the Qing Dynasty are explicit and should belong to 2.5-dimension multilayer stereoscopic work. But the layout of single work is scattered and planar. This style makes the building decoration features of the work become the example to certify that a work possesses the traditional exquisite carving craft that is unique to the ancient folks of Quanzhou.

3 ARTISTICAL WOOD CARVING: EXPRESSION OF THOUGHTS AND FEELINGS

Artistical wood carving takes up quite a proportion of contemporary Fujian wood carvings. An artistic wood carving that is well documented in the history of Fujian might be the carving *Lord Guan* made by Qing Dynasty woodcarver Liao Xi and selected to be exhibited in the Panama Pacific International Exposition held in the United States. Among the folk, *Lord Guan* is generally worshipped as the God of War, and is of a property similar to that of religious statue. The exhibited *Lord Guan* of Liao Xi is more of a property for aesthetic appreciation. Nowadays, the wood carvings of Avalokitesvara,

Figure 1. The traditional building wood carvings of Fujian.

Maitreya, Lord Guan, and Bodhidharma made by Fujian woodcarvers are favored by collectors. These works have lost their significance in idol worship and thus have become objects for appreciation. It is worth mentioning that there are a considerable amount of root carvings in the folk wood carvings of Fujian. In the conception and production of those root carvings, the carver takes advantage of the original shape of the material and then carves it skilfully. Although the procedure is different from that of other traditional wood carvings, their function is more for display and appreciation. The tree root carvings involve many images, including figure, animal, plant, etc., and have become a distinctive category among Fujian wood carvings. In Fujian, there are many practitioners of artistic wood carvings; Fujian is also the cradle for many outstanding artists and works of influence nationwide. In the following paragraphs, several representative works of Zheng Guoming, a Chinese master in arts and crafts, are appreciated to highlight the charm of Fujian artistic wood carvings.

First is the *Meditation on the Past in Red Cliff*. The poem *Meditation on the Past in Red Cliff* written by Su Dongpo is unrivaled in the field of poetry and wins universal praise. This poem was often taken by artists in past dynasties as the prototype for their creations and transformed into verses, calligraphy, and paintings. This is the only painting that describes this poem and exists now. As we all know, it is hard to deliver a poetic conception with a paintbrush. In this work, the creator conveys the meaning of the poem with carving and makes the viewers feel personally on the scene, and the bold innovation and crafts of this creator are really worthy of admiration. In this work, the creator combined natural wood and stone materials, with natural stone as cliff and wood carved into the ship, figure, river, and trees, which are exquisite and precise; the carving of figures in terms of their identity, personality, and expression is especially concise and vivid. This work perfectly realizes innovation on the basis of tradition (Figure 2).

Second is *Be Sworn Brothers*. The story of Liu, Guan, and Zhang swearing to be brothers in Peach Garden has become a literary quotation, and the righteous spirit reflected from the story of the three

Figure 3. Wood carving *Be Sworn Brothers* (Zheng Guoming).

brothers can be reputed as the moral ideal of the Chinese nation. Basically, this work takes on the form of a mountain, with the three brothers standing high in a line shoulder to shoulder like a mountain towering into the clouds and also like an impregnable fortress. Freehand brushwork is adopted in the carving of figures to highlight their broadness and heroism. Their bearing with heads raised toward the vastness of heaven is a psalm to their utter specific devotion and braveness. This is a monument for excellent national tradition (Figure 3).

Third is the *Grasp for the Moon*. The monkeys grasp for the moon, which might only be hope in vain; but, their perseverance and efforts in pursuing dreams are awe-inspiring. The creator describes the scene full of fantasy and fun on a section of withered bamboo; the joint, wall, and surface of the bamboo are made the best of to give a full display of the images of monkeys, water, and moon. You can tell from this work the creator's ingenuity and artistic skills. This small artwork combines fun with noble character, and can bring people both afterthought and surprise.

Fourth is *Master Hong Yi*. In his later years, Master Hong Yi carried forward Buddhism in Southern Fujian, and finally passed away in Southern Fujian. It can be said the Buddhism connection between the Master and Southern Fujian is very deep. The reverence and understanding of Zheng Guoming for Master Hong Yi can be told from this work. The ingenuity of this lies in the fact that it expresses the complex unity of Master Hong Yi, monk, and Buddhism. With a hand holding a bamboo cane, a haggard complexion, and a somewhat tired and weak body, the Master has entered his later years as a human; his eyes are directed into the distance with measureless kindness and mercy, indicating the dignified Master's affection to all beings in the world. In his expression, there looms some joy and transcendence, representing his satisfaction and delight after reaching high-level Buddhism cultivation. In its creation of external features, it mainly adopts the realistic description method, while in the handling of detail, it applies comparison between freehand brushwork and realistic description, complexity and simplicity, abstraction and concretization. This work not only embodies the romantic charm of traditional wood carving, but also absorbs the ideas of modern

Figure 2. Wood carving *Meditation on the Past in Red Cliff* (Zheng Guoming).

art, stresses the exploration of the object's rich spiritual connotation, and realizes perfect unification of materials and theme.

These works reflect the pursuits of contemporary wood carving artists in Fujian:

Firstly, they take the core notion of traditional national spirit as their purpose and aim to expand the theme scope of traditional wood carvings. To improve the spiritual and cultural value of wood carvings, they no longer limit their attention to the traditional themes and forms handed down from generation to generation; instead, they go deep into the grand library of the whole traditional culture to search themes that are inspiring, and open the passage between wood carving and the whole cultural tradition. *The Red Cliff Ode*, *Farewell to My Concubine*, and *Be Sworn Brothers* all are the results of such purpose-clear explorations.

Secondly, in work forms, they break through the inherent mode of taking wood carvings as building accessories and religious belief symbols to carry out artistic exploration in the creation of artistic conception. *The Red Cliff Ode* expresses a particular circumstance, in which rocks, trees, ships, figures, river, and other images form into a stereoscopic space. This is an artistic space coming from the tradition independently, and exists as an entity not relying on other buildings and beliefs. *Be Sworn Brothers* is also a scene, which forms an enhanced spiritual space. It is not to highlight a certain hero, but the common spirit among the heroes.

Thirdly, in application of materials, they break through the tradition of only using wood for carving, and make comprehensive application of natural wood-hugging-stone, root carving, collage, inlay, etc. They choose craftsmanship and carving methods based on the features of the material; on the basis of the natural and primitive form of the used materials, after conception and careful implementation of tools, they shape the materials into a work integrating natural and artificial efforts, and highlight a novel artistic charm.

Fourthly, in carving techniques, they are no longer restricted to traditional techniques like relief carving, penetrated carving, flat cutting, and round cutting; instead, they take the artistic purposes of theme and conception as the emphasis and choose the proper methods. In many works, to express intense and strong affection, they often adopt flat carving in the creation of circular type, and jump cutting in stereoscopic description; this method, integrating the square and circular, and the relief carving and penetrated carving result in familiar and fresh effects.

4 PUPPET HEAD CARVING: DRAMATIZED LIFE

The puppet show is a comprehensive artistic form integrating literature, folk arts, music, fine arts, and many other art forms. It can be said that the puppet is the most important prop in the performance of puppet show. The puppet is a prop behind the scenes, but is an "actor" on the stage. Among the folk, the performance of puppet shows and the production of puppet show props belong to different industries and are undertaken by different personnel. The production of puppets is generally undertaken by wood carving artists. The flourishing of the puppet show and its collision with the excellent wood carving traditions in Fujian result in outstanding puppet head carving crafts.

In terms of puppet production crafts, puppet head carving is the core. The folk craftsmen for puppet head carving in Fujian are divided into the school of Jiang Jiazou in Quanzhou and that of Xu Niansong in Zhangzhou. Though the size of a puppet head is small, its production crafts are not simple. For example, the production process adopted by Jiang Jiazou, the famous artist in puppet head carving, includes cogging, shaping, detailed carving, paper mounting, polishing, space filling, clay brushing, powder mounting, face carving, and wax covering—a total of ten working procedures, with each working procedure having high technical requirements. Excellent puppet head carving artists can make a work vivid with small changes; puppet show performers stage touching stories with these works of art.

On the basis of tradition inheritance, Jiang Jiazou has continued to make innovations with the combination of daily life experiences and integration of many cultural factors from local customs and proverbs. He has created more than two hundred and eighty puppet head images of different personalities, developed more than ten different hair buns and hair braids, and created more than ten thousand pieces of carved and family rose puppet heads. His works show an intense Quanzhou style and thus are widely popular among the people sharing the same language and culture in Southern Fujian. Among his works, the representative work "Woman Matchmaker" is created on the basis of traditional "Quelao" and is very vivid. By analyzing the personal character based on the behaviors and languages of common people in life, Jiang Jiazou defines the "Woman Matchmaker" as a woman around forty years old, of rich life experience, with emaciated complexion, some wrinkles on the cheeks, and in need of sticking plaster to get refreshed from a lack of sleep. Since she acts as a go-between for betrothal, she has a tongue in her head; thus it is carved with thin lips, an obsequious smile, and a clean and tidy Suzong bun. The most obvious characteristic is the "matchmaker mole" (a nevus on the nasolabial fold) beside her mouth. The puppet style gives full play to the character traits and occupation characteristics of a figure; audiences could have direct feeling and psychological identification with the role characteristics in the drama through one glimpse of this puppet image. In another work, "Traitor Minister Head," a salient long hair is depicted on the face, and a pattern of something entangled by some thin lines is depicted on the cheek. According to the custom in Southern Fujian, a long hair appearing in the eyebrow will make the man look especially fierce, and Jiang Jiazou exaggerated this hair, which makes the audience grasp the personality of the Traitor Minister Head figure upon seeing it. For the pattern on the cheek, there is a saying among Southern Fujian people about greedy people, "Your face is covered with coin-string pattern," which highlights that greedy people show their greediness on their faces. Thus, Jiang Jiazou depicted it on the face of the traitor minister. In this way, the "fierceness" and "greediness" of the traitor minister are expressed using the styling concepts particular to customs and proverbs in Southern Fujian. The cultural and psychological assimilation makes local audiences sharing the same local language willing to accept this artistic model bearing cultural characteristics unique to the Southern Fujian area. His works also realize such excellent artistic attainments as combining realistic description with vividness, and endowing the puppet with life.

Jiang Jiazou not only describes the personality of a character from appearance, but also expresses the psychological activity of a role in the drama from the complicated inward world. His works not only stress the realistic description of a puppet head, but also its artistry in modeling. "BaiKuo" is one of his representative works, namely a carving of an old man about eighty or ninety years old. Jiang Jiazou has carved it into a man amiable, honest, healthy and humorous with a wide brow ridge and a large chin as well as silvery eyebrows and beard, very vivid and lovely. However, in fact, old people in the minds of Jiang Jiazou generally have one foot in the grave, with haggard facial muscles, sloppy skin, drooping eyelids, and lifeless sight. Jiang Jiazou stresses the artistry of modeling, so he is reluctant to put the image of an old man in his later years on the stage, and believes that audiences prefer a healthy and optimistic image. The works of Jiang Jiazou not only reflect the character traits of a role, but also the artist's healthy and optimistic view of life. They show the fact that the creator understands the aesthetic and psychological characteristics in art works.

Nowadays, the puppet show is declining among the folks, but there is still space for the existence of the puppet head carving crafts. The puppet heads carved by the inheritors of the school of Jiang Jiazou including Jiang Chaoxuan, Jiang Bifeng, Huang Yiluo, and Wang Jingran, and those of the school of XuNiansong including XuZhuchu, etc., are very popular among museums and private collectors. The puppet head, a traditional performance prop, is gradually being oriented toward the functions of display and appreciation.

5 EDUCATIONAL VALUE OF FUJIAN WOOD CARVINGS

5.1 Beautification of life and teaching from artistic works

Wood carving is widely and frequently used in traditional folk society. From the perspective of its content, it has an important educational function in traditional folk society and exists in nearly all aspects of folk life, such as buildings, drama stage, furniture, and religious and custom activities. The historical characters, drama roles, and various auspicious patterns presented by Fujian folk wood carvings also appear frequently. Living in an environment accessible to wood carvings, the people are influenced by what they constantly see and hear, and can learn much knowledge from the stereoscopic textbook – wood carving; thus wood carving exerts its function of shaping the thoughts and ideas of people and influencing unconsciously their behaviors and manner.

5.2 Craftsmanship spirit and artist self-consciousness

The traditional appellation of woodcarver is craftsman, which is an occupation among all folk crafts. In a traditional society, craftsmen would not pursue an official career, had no estate and farmland, and depended solely on their crafts for living. On the way to becoming a craftsman, for purposes of surviving at the beginning, it is unavoidable to have some economic considerations. The profession of the craftsman has been handed down for many generations among the people in Fujian, and many well-known craftsman families and schools have been established; among them, there is no lack of emotional sustenance, and this also reflects a kind of professional integrity. Traditional craftsmen are patient; they can bear an indifferent post, and work with great care; this is the craftsmanship spirit, which includes rich emotions and spiritual ideas.

In modern society, practitioners in the wood carving industry have gradually diversified. In addition to the traditional inheritance from father to son, and from master to apprentice, there are many wood carving artists who learn from a master at the beginning and then receive modern art education. Those practitioners with a relatively diversified education background create possibilities for the future development of the wood carving industry. Moreover, some commendable practitioners possess the "self-consciousness of an artist," and they not only personally engage in the creation of wood carvings, but also conclude their personal and industrial experiences by further transforming them into valuable text, which expands the way for the spreading of experiences.

No matter whether it is the traditional craftsmanship spirit or the contemporary self-consciousness, people's devotion to and affection for artistic activities make the wood carving art more attractive. The devotion and affection of practitioners toward their industry are commendable and worth learning, and the spirit of the people becomes an important education resource.

5.3 Reverence for nature, cherishment of belongings, and application of proper crafts

The traditional materials for Fujian wood carving derive from local timber trees; longan wood, litchi wood, and other representative woods of Fujian. These functioned as the main materials of Fujian wood carvings in early times. With the logistics growing more convenient, many precious woods now enter Fujian and become important raw materials for Fujian wood carvings. In the application of these materials, all the wood carving artists make use of the "artful carving" principle to keep the primitive texture of these materials and give full play to their features based on their characteristics, form, and color. From this, you can tell the artists' reverence for nature and cherishment of belongings. It often takes decades or even a hundred years for the hardwood to mature, thus this kind of material is precious in the minds of woodcarvers; the full use of such a material is the highest reverence for it. One might say that it is because the hardwood is imported, rare, and expensive, and then the craftsmen have economic considerations in application of the hardwood. However, the application of common tree roots in the field of "root carving" can give better play to the craftsmen's capability in choosing proper crafts, and their reverence for nature and cherishment of belongings.

REFERENCES

F. Guo, 2004. *Wood Carving Arts of Fuzhou*. Fuzhou Hai-chao Photography Art Press.

H. Miao, 2008. *Wood Carvings*. Beijing: China Society Press.

W. Lin, 2006. *Folk Wood Carvings in Fujian*. Beijing: Writers Publishing House.

Smart Science, Design & Technology — Lam et al. (eds)
© 2020 Taylor & Francis Group, London, ISBN 978-0-367-17867-3

Research on the transforming of green energy culture and cognitive science

Juiwen Yu*
Ph.D Program of Mechanical and Energy Engineering, Kun Shan University, Tainan, Taiwan

Huannming Chou
Department of Mechanical Engineering, Kun Shan University, Tainan, Taiwan

ABSTRACT: Green energy generation such as hydropower, wind power, and solar power generation equipment at the edge of sea, in mountainous areas, and in the country has formed a landscape that is unfriendly to both the natural environment and the urban architectural landscape and even affects the ecological environment. Products for saving energy such as air conditioners, thermal pumps, and variable-frequency motors with high energy efficiency ratio also have been developed. Power management systems and power transmission through intelligent grids are becoming more important in cities, and electric motorcycles and vehicles are also getting more popular in the city. The use and development of green energy and products that save energy have changed people's daily living habits and have caused industrial changes and employment moves that have impacted the economy and society, including oppositional opinions for the use of renewable energy, which have resulted in transformation of the traditional culture into the green energy culture. This cultural variation is in fact a conversion process explained by cognitive science. This paper analyzes the turning point of green energy culture from the perspective of cognitive science, and believes that through the education and development of cognitive science, noise from the development of renewable energy will be eliminated and the green energy culture will develop into an important cultural model that will change global human life and civilization.

Keywords: renewable energy, green energy, green energy culture, cognitive science

1 INTRODUCTION

Due to heavy use of fossil fuels and related chemicals by mankind, large amounts of carbon dioxide and other heat-retaining gases are being emitted, and the global temperature is warming up and rising due to the greenhouse effect, causing climate change, threatening human health, and endangering birds, fish, and other ecosystems. The danger of biological extinction could happen in the future. The world has increased the use of renewable energy (green energy) to reduce greenhouse gas emissions. Governments also provide relevant promotion policies. The amount of green electricity generation has been rising year by year. Various types of high-efficiency electricity-consuming products such as electric vehicles and motorcycles have come out in succession, while gasoline vehicles and motorcycles are being gradually prohibited or alternated. However, some forms of green power generation such as wind and solar power will cause changes in urban and environmental landscapes and have affected cultural,

economic, social, and political factors. Green technology has become an effective recipe to save the planet from the threat of climate change, and the green energy culture has sprouted into small trees and continues to grow rapidly throughout the world.

Culture covers a wide and complex range. Generally speaking, culture is a group of people's common lifestyles and attitudes which involve ways of survival, morals, systems, norms, religions, and customs, that is, the behavior patterns of a certain group of people living together. Alfred Louis Kroeber (1876–1960) and Clyde Kluckhohn (1905–1960) argued that culture is all design that has been made for survival in the history. It may be explicit or implicit; it may be rational or irrational, or even without reason. The core of culture consists of a set of traditional concepts—especially the value system. This statement has gained widespread popularity among social scientists. The formation of a green energy culture is the alienation effect caused by changes in this set of value systems that mankind

---

*Corresponding author. E-mail: edmundyrw@hotmail.com

follows. In this study, cognitive science is the main clue to explore the basic driving force of green energy culture and the source of the power itself. The perspective of cognitive science is used to explore the turning point for the development of green energy culture, and further make it a global or national desire to enhance the use of green energy in order to solve the problem of climate change as soon as possible.

The objectives of the study include collecting the current status of global greenhouse gas emissions, green energy generation, and the origins of green energy culture, exploring the turning point for cognitive science to influence cultural change, and elaborating that cognitive science is actually the cornerstone on which to form green energy culture and have a sustainable development for the earth.

2 COLLECTION OF INFORMATION AND ANALYSIS

Use the Internet to collect data, charts, regulations, measures, or other information relevant to each unit for analysis and application. Data collected will be statistically analyzed or plotted using computer to show the extreme or trend of event development and help to judge and analyze the event development.

3 RESULTS AND DISCUSSION

3.1 *Global greenhouse gas emissions*

Data from the International Energy Agency in Figure 1 shows that energy-related global greenhouse gas emissions have increased from 15.1 billion tons in 1975 to 32.1 billion tons in 2015 of carbon dioxide equivalent per year, doubling emissions over 40 years (International Energy Agency, 2016). However, emissions have slowed down since 2013, even showing a flat trend form 2014 to 2015 with the same amount of emissions, and may have the opportunity to start reducing the amount of emissions by 2016. Figure 2 (Wikipedia, 2017) shows the world's top 20 single countries or regions in the world in terms of their emissions as

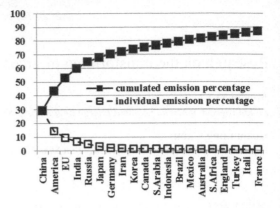

Figure 2. The percentage and the cumulated percentage of the greenhouse gas emissions of individual countries or regions relative to global emissions in 2015 (Wikipedia, 2017).

a percentage of the whole world's emissions and the cumulative rates of emissions by countries in 2015. China ranked first in the world for 29.28% of the world's emissions, followed by the United States 14.23%, the European Union (EU) 9.55%, India 6.76%, Russia 4.85%, and Japan 3.45%. The top 20 countries accounted for 86.96% of the world's total emissions, and these countries have built renewable energy power generation units to generate electricity to reduce the use of fossil fuels and emissions of greenhouse gas from thermal power generation.

3.2 *Global green power generation electricity*

Since 1988, the United Nations Intergovernmental Panel on Climate Change (IPCC) has been trying its best to promote and invite all countries to commit themselves to a reduction of greenhouse gas emissions. Countries all over the world have started gradually to use renewable energy to generate electricity and hope to reduce their greenhouse gases emission. Figure 3 shows the changes in total generation, fossil

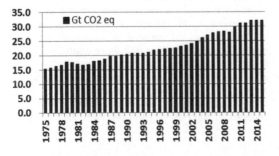

Figure 1. Global energy-related CO_2 emissions (International Energy Agency, 2016).

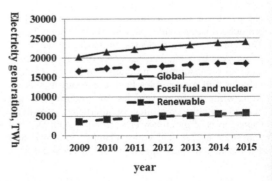

Figure 3. The electricity generation for global, fossil fuel and nuclear, and renewable energy.

fuel and nuclear power generation and renewable energy generation in the world from 2009 to 2015. The world's electricity generation increased gradually from 20,183 TWh in 2009 to 24,097 TWh in 2015. Fossil energy and nuclear power generation did not increase after 2013. On the other hand, in the same duration the generating capacity of renewable energy increased from 3,633 TWh to 5,711 TWh, an increase of 57.2%, which is larger than the increase of global generating capacity of 19.4% over the same period. Among the countries, the generation of renewable energy in China increased from 615 TWh in 2008 to 1,425 TWh in 2015, an increase of 131% (Wikipedia, 2018a). The United States renewable energy power generation increased from 400 TWh in 1998 to 609 TWh in 2015, an increase of 52% (Wikipedia, 2018b). Renewable energy generation in EU (28 countries) increased from 310 TWh in 1990 to 925 TWh in 2015, an increase of 198% (Wikipedia, 2018c). Japan's renewable energy power generation increased from about 100 TWh in 1990 to about 150 TWh in 2013, an increase of 198% (Wikipedia, 2018d).

3.3 Green energy culture formation

The use of green power generation is currently the most effective method to reduce carbon emissions; however, it gradually brings about various economic, environmental, social, and political changes in the development process. For example, the purchase of electricity-saving appliances such as heat pumps, high-efficiency air conditioners, and electric vehicles and motorcycles has caused changes in the economy and consumer behavior. Disruption of social and political arguments and opposition aspects are caused by the deactivation of petrochemical fuels or nuclear power plants. The use of wind or solar power has changed our landscapes and urban scenes, and noise and ecological protection issues have caused objections to be raised to the use of wind power or solar power generation. The prevalence of low-carbon-footprint products has led to changes in raw materials, manufacturing processes, and buying behavior, and even caused international trade obstacles. That is to say, these products have caused social and cultural problems in living, industrial, and political areas such as food, clothing, housing, transportation, education, and music. The global green energy culture has gradually formed, and it is actually a result of cognitive system changes.

3.4 Cognition and cultural change

Cultural change may come from the living environment or social reasons; it may come from alienation within the culture body itself or from the interaction between two cultures followed by an acculturation effect and then the occurrence of alienation. Social force is actually a product deriving from and clinging to a collection of living entities. It has influences on individuals by providing a set of social value norms that can be recognized by the individual and forms an external force to plasticize the individual's behavior. However, the reason why social value norms have the external force to shape an individual's behavior is that he or she internalizes those norms through cognitive ability and develops the phenomena of socialization. Therefore, cognitive ability with the power of internal plasticization is the driving force of cultural transformation (Yu & Chou, 2011).

The research team led by Nisbett (1996), a social psychologist at the University of Michigan, found that when people change their culture, they learn to use new ways of perception. The conscious brain is active and always adjusting itself (Doidge, 2007). When people change their culture, it involves two levels of representation. First, members of society change the intrinsic cultural cognitive representation in order to adapt to the external culture. That is, members of society, through their socialization course, change their own cultural cognitive representation. Second, members of society should use their own cultural cognitive signs to influence the change of external culture, which also must be achieved through the operation of social forces. Strictly speaking, for the cultural changes, the interaction between social force and cognitive force is not easy to separate out. However, in either case, the conscious brain is always active. The conscious brain, on the one hand, changes one's intrinsic cultural model release through the initiative of cognitive ability and on the other hand, influences or changes the external culture together with the cultural model release of others.

In the process of the interaction between the intrinsic cultural release of the individual and the external culture, the conscious brain is not only active, but always adjusting itself. In other words, the cognitive power generated through mental model is dynamic, and it is self-adjusting. Therefore, from the perspective of motivation, cognitive ability is the basic driving force of cultural change, and cognitive ability is the overall demonstration of mental function. In fact, this cognitive ability is actually generated by the integration of the seven subconscious minds and the mental body, the eighth consciousness (Xiao, 2003).

4 CONCLUSION

Human beings have become aware of the hazards from greenhouse gas emissions. The world's major countries have made an adjustment and reduced the amount of fossil fuel power generation in order to reduce greenhouse gas emissions from fossil-fuel-fired power generation and to increase the use of renewable energy power generation year by year with the hope to change the Earth's environment and achieve sustainable development and human survival. Scientific evidence and international public opinion as well as other social forces are a kind of external force for change. However, the cognitive power of human beings makes them adjust behaviors dynamically and initiatively, to start utilizing green energy to generate electricity, thus gradually resulting to the formation of the green energy culture. Consequently cognition is the basic motivation for the transformation of a green energy culture.

REFERENCES

International Energy Agency, 2016. *CO₂ emission statistics*.

J.W. Yu & H.M. Chou, 2011. *Journal of Kun Shan University 8*, 141–163.

N. Doidge, 2007. *The Brain That Changes Itself: Stories of Personal Triumph from the Frontiers of Brain Science*. Penguin ISBN 9781101147115.

P. Xiao, 2003. *The Shadow of the Lamp*. Buddhist True Enlightenment Practitioners Association, Taipei.

R. Nisbett, 1996. *Culture of Honor: The Psychology of Violence in the South*. Westview Press.

Wikipedia, 2017. List of countries by carbon dioxide emissions.

Wikipedia, 2018a. Renewable Energy in China.

Wikipedia, 2018b. Renewable Energy in the United States.

Wikipedia, 2018c. Renewable Energy in the European Union.

Wikipedia, 2018d. Energy in Japan.

Smart Science, Design & Technology — Lam et al. (eds)
© 2020 Taylor & Francis Group, London, ISBN 978-0-367-17867-3

Develop innovative design education with critical thinking and interdisciplinary practices—based on Parsons School of Design

Xiangyuan Zeng*
School of Design, Fujian University of Technology, Fuzhou, Fujian, PR China

ABSTRACT: With the education mission of cultivating creativity, Parsons School of Design enjoys a reputation of being the world's first-class design university, with the cultural gene of critical thinking permeating its campus. Its introduction of interdisciplinary practice into design education in campus culture, academic research, teaching organization, and professional learning is innovative, highlighting the distinctions of an education that is innovative, diversified, and open, and also constitutes its sustainable power to lead the development of design education in the world.

Keywords: Parsons School of Design, creativity, critical thinking, interdisciplinary practice

1 INTRODUCTION

In *The QS World University Rankings* (Quacquarelli Symonds Limited, 2019) by Art and Design disciplines, Parsons School of Design at The New School (hereinafter referred to as "Parsons") has been high on the first, the top of the world's top five, referred to by the media as a "palace level Design institute," attracting numerous students and designers from all over the world. How did this design university in lower Manhattan become one of the most prestigious art and design universities in the world? The author had the honor to study in Parsons as a visiting scholar for one year, and personally experienced its education practice in many aspects, such as school philosophy, strategic planning, academic research, curriculum system, class teaching, management service, and so on. In my personal opinion, Parsons leads the world in the development of design education with the education feature of cultivating creativity based on critical thinking and interdisciplinary practice in the global context.

1.1 The source of creativity development of "art + industry"

Parsons originated from Chase College, founded by the American impressionist painter William Merritt Chase in 1896, which was originally aimed at gathering progressive art groups and pursuing a more-free and personalized artistic expression. In 1898, Chase College was renamed New York School of Art. Frank Alvah Parsons, who joined the faculty in 1904 and quickly became head of the school, predicted that art would be closely associated with industry in the new industrial revolution. This vision created a series of American "firsts" for the school, such as the first American fashion design, interior design, advertising, and graphic design majors. And, on the basis of the creation of art and industrial connection theory, Parsons created the modern design education curriculum system. Many students who graduated from Parsons became successful designers. And they were hired as full-time or part-time faculty to bring their design experience in an industrial society back to the School of Art, and they further adapted the teaching content, created the perfect modern design course system, and made Parsons the birthplace of outstanding design knowledge and creativity education. In 1970, the School became part of The New School for social research, which expanded and strengthened Parsons' mission of designing education, namely "Creativity, Innovation and Challenge the strangers" (The New School, 2019a). Now Parsons has five colleges: School of Constructed Environments, School of Art and Design and History, School of Art, Media, and Technology, School of Fashion, and School of Design Strategies. Today's "art + industry" is interpreted as "art + technology," while the mission of creativity education remains the foundation of Parsons and the source of its development.

1.2 Research foundation of "design + society" creativity

Parsons is the only design school in the United States to be established at TNS (The New School), which has rigorous liberal arts. It relies on the other four

---

*Corresponding author. E-mail: 34407123@qq.com

colleges, shares resources, and promotes the other colleges. These colleges are Eugene Lang College of Liberal Arts, College of Performing Arts, New School of Social Research, and College of Public Engagement. The core values of "academic freedom, diversity, inclusiveness and exploratory experiments" defined by TNS complement Parsons' creative mission. Today, with the rapid changes in the global economy, society, and environment, unprecedented new problems and trends are emerging continuously. The design of education needs to pay attention to the changes of the times from a broader and deeper perspective, and graduates need to be able to choose more flexible and diverse career paths in the face of future complex social problems. Thus, Parsons gives priority to Humanity and Culture in the design system, and uses the academic research method of "design + society" to integrate the cultivation of creativity into the process of designing education. Here, the research of "design + society" promotes the creativity needed to solve the problems of the times, such as research on social democracy, urbanization, technological innovation, economic empowerment, sustainable development, immigration, globalization, and many other social themes, which can reveal the deep collection of problems behind the design. With the branch and merger, crossover, and integration of the research field of "design + society," the perspective of design continues to enter new fields. Parsons' students work freely in an interdisciplinary environment with artists, designers, economists, sustainability experts, policy analysts, management experts, psychologists, linguists, social innovators, journalists, and writers to develop new insights and abilities, release individual potential, and enhance design creativity with the participation of design thinking and artistic creation methods.

1.3 The creative change dimension of "local + global"

"Design a resilient word" (The New School, 2019b), it is Parsons' creative focus on design practice. Among Parsons' five strategic planning goals, "Puts student success first" – which Puts: student success; Academic Programs and Quality; Global Education; External Profile; Infrastructure (The New School, 2019c). Parsons has set up thirteen undergraduate programs, twenty-two graduate programs, and five associate degrees programs. Based on Parsons' localized and international student groups and academic and cultural foundation, these programs develop and construct students' creative ability to explore local and global issues, which reflects the flexible nature of cultivating students' creativity and is an important symbol of students' future success. The so-called localization requires students to study regional needs and constraints in the localization context; discusses the contradictions and constraints of design, manufacturing, sales and other elements; and creates solutions that can promote environmental and

social sustainability and demonstrate the lasting relevance of art and design. The so-called internationalization is to guide students to adopt the design principle focusing on universal needs in the context of globalization and propose designs that can be implemented across national boundaries and cultures. For example, in the course planning of the Masters of Industrial Design, the core course Design Studio series is: Form Design – Local Design – Global Design – Professional Thesis. This "local + international" curriculum system helps students understand how to balance economic production, material consumption, sustainability, social and environmental improvement, local manufacturing, and global supply chain competition in the design and to explore the integration and coordination of the power of these factors. It also trains the student to pay attention to the vertical and horizontal, and elastic design thinking, with a final design practical action to prove his understanding of the present society and the environment. Such training gives students more reflective ability and more responsible thinking dimension, "helping students apply the innovation creativity of design to the challenges of making the earth more just and livable after graduation, and becoming socially aware manufacturers, strategists and industry experts" (The New School, 2019d).

2 TAKE CRITICAL THINKING AS THE CULTURAL GENE

Critical thinking is a thinking method to develop innovative ability. It is not only a kind of thinking skill and thinking tendency, but also a kind of personality or temperament; not only can it reflect the level of thinking, but it can also highlight the humanistic spirit. As the school's vision puts forward, it can "help them become critically engaged citizens, dedicated to solving problems, and contributing to the public good" (The New School, 2019d). Parsons attaches great importance to the cultivation of students' critical thinking, which is regarded as Parsons' cultural gene, penetrating Parsons' campuses and inheriting from Parsons' faculties and students.

2.1 Critical thinking culture in media communication

The school homepage (https://www.newschool.edu/) when first opened displays the catch phrase of bell hooks, a Parsons alumna, author, and social activist: "I believe that salvation is critical thinking of my life, I believe I saved my life is, in fact, critical thinking. - bell hooks)." In the static state of the page, every ten seconds, another Parsons alumnus's motto is displayed, such as the words of Ana Sui, a famous Chinese fashion designer and alumnus of Parsons: "We don't need more of the same, we need better ideas." On the homepage, quotes of a total of ten famous alumni of Parsons are shown, all of which illustrate the importance of critical thinking from different perspectives. In the layout design, the area of aphorisms takes up more than 80% of the layout (The New School, 2019e), which reflects the school's concept of giving prominence to the development of students' critical thinking and innovative ability. At Parsons, it is the norm for students to self-organize a large number of BBS activities on critical themes. Students post information about intensive activities on ubiquitous bulletin boards on and off campus, such as the activities on crisis of democracy, social justice and labor rights, gender equality, environmental protection, urban development, community survey, design of disaster relief, the pioneer of fashion and art, technology, workplace challenges, and so on. Parsons transmits the meme of critical thinking through various forms of campus media.

2.2 *Develop critical thinking skills in class discussion*

In Parsons, apart from a few lectures, the teaching method of most courses is small-class seminar (12–15 people). In class, the teacher guides students to state their opinions, explain the plan, encourage criticism and discussion, and finally express opinions. In the process, every student often asks or is asked "what" and "why." Therefore, the ideas of explanation, analysis, evaluation, inference, and explanation around the theme burst out from the collision of critical thinking. Coupled with the teacher's evaluation, students can find self-calibration of ideas and promote the improvement of design thinking and design ability. Taking the Product Design Program as an example, in the course planning of the first spring semester, it not only arranged the Integrative Studio 1, but also arranged the Integrative Seminar 1, for 3 credits respectively. The former carries out design practice around the design theme, a series of visual modeling analysis, prototyping skills, and interdisciplinary cooperation training, while the latter conducts concept discussion around the field related to the theme, requiring the sharing of ideas and concepts through reading and writing so as to acquire the ability of "intellectual dialogue." The former develops the habit of critical thinking through oral expression, while the latter internalizes critical thinking skills deep into the heart through written expression.

2.3 *Academic freedom creates an atmosphere of critical thinking*

Parsons' academic freedom is manifested in the freedom of academic research and teaching methods, and it is this free academic atmosphere that creates conditions for the development of critical thinking. The reason why Parsons attracts scholars with truly independent academic thoughts to teach in the

school is inseparable from the core value of academic freedom. Professor Robert Kirkbride, dean of the School of Constructed Environment, has been studying the interaction between memory and the built environment. He, through the interpretation of cultural relics, buildings, literature and background investigation, and other unique research trains of thought, has put forward such academic views as how the physical nature of thought and multi-perception can shape insight in human survival experience. He applied these academic achievements in teaching and opened a course called *Poetics of Design*. Professor David Brody of the School of Art and Design and History is an expert in material culture, visual culture, and design studies. He was once asked by an interviewer, "What attracted you to Parsons?" He replied, "It's the school's openness to new ideas and its unparalleled location … Here, I don't have to stick to a specific goal, I can start new courses, bring together students from different subjects."

Academic freedom is embodied in the flexibility and diversity of teaching methods. In addition to independent lectures, professors often invite guests from inside and outside the university who are related to the subject to participate in class lectures, Sometimes classes are "moved" directly to museums, communities, galleries, or design firms, or to other research sites in New York City, such as field investigations, field visits, community discussions, creative workshops, exhibition defenses, and so on. Most of the guests come from design companies, research institutions, social organizations, or other universities. They express their views on the content of the course. Their critical thinking is often novel and unique, which is called the "outer brain" of the course learning, providing students with channels for social participation. In addition, the mechanism of students' independent course selection also reflects the academic freedom of teaching methods. Many courses accept students from different professional backgrounds. For example, Robert Kirkbride's course *Poetics of Design* not only receives students of the School of Constructed Environment, but also is open to other schools' students. In class discussions, the atmosphere for critical thinking and creative energy is often strengthened due to different subject backgrounds.

3 TAKE INTERDISCIPLINARY PRACTICE AS INNOVATIVE ACTION

Interdisciplinary practice refers to the practice of giving up a single disciplinary method and cooperating with other majors to solve problems in various fields of background knowledge. As an interdisciplinary subject of science and technology and culture and art, design has the characteristics of multi-angle, multi-method, and strong practicality in solving problems. Parsons unleashes the creativity of teachers and students by taking interdisciplinary practice as innovative action.

3.1 *Interdisciplinary practice in academic research*

Parsons is equipped with seventeen Research Labs, such as CDA (Center for Data Arts), Center for New York City Affairs, Center for Transformative Media, DESIS (Design for Social Innovation and Sustainability), India China Institute (ICI), Visualizing Finance Lab, etc. The main mission of these research laboratories is to challenge existing paradigms and promote emerging academic practices with an interdisciplinary approach to design. For example, CDA (Center for Data Arts) brings together artists, designers, data scientists, and other academics from the MIT Media Lab, led by Ben Rubin, and works in large-scale media installations, open source software, original academic research, and multi-mode experiences to visually perceive hard-to-understand data, and implements art and design projects in the private and public sectors. The CDA project team draws on talent from TNS programs in anthropology, economics, journalism, strategic design, music, cognitive science, and public policy. Working with scientists, humanists, policy makers, and artists, the CDA project team extends the boundaries of data representation by using methodologies from data economics, cognitive and learning sciences, and information aesthetics; pioneers radical new techniques for turning complex information into meaningful narrative experiences; and rethinks creative problem solving in large-scale data exploration for product and interface designers, architects, urban planners, graphic artists, writers, and fashion designers. For example, in intensive care units, family homes, or trading halls, CDA learns how information enters and flows through the environment. At each point, CDA examines who is controlling the collection, access, and use of data, and proposes changes to data and ways of making decisions that may have an impact on culture, economy, and institutions.

3.2 *Interdisciplinary practice in professional learning*

At Parsons, "Students must be interdisciplinary in the way they learn and act" (The New School, 2019b). Interdisciplinary practice in professional learning is mainly reflected in three aspects: firstly, major setting is interdisciplinary; secondly, course planning is interdisciplinary; thirdly, learning methods are interdisciplinary. In spite of the cross-disciplinary characteristics of Design majors, Parsons has set up special majors from the perspective of cross-disciplines. For example, the School of Design Strategies has disciplines such as Trans-disciplinary Design, Integrated Design, Design and Urban Ecologies, etc. The School of Art, Media and Technology has Data Visualization, Design, and Technology majors. Setting up interdisciplinary majors not only accords with the reality of subject integration in the era of creative economy, but also builds a platform to attract experts and scholars from different disciplines to teach in the university.

In terms of curriculum planning, both undergraduate and postgraduate majors take Design Studio as the core course, which is also called Integrative Studio. These studio courses are generally 3–6 credits and are based mainly on design projects to integrate knowledge of other disciplines and carry out practical design training. Different students choose design projects in different fields according to their individual academic interests. In the process of guiding the implementation, teachers of the course will invite teachers of the major, foreign majors, and experts in and out of the university to discuss and criticize in the interdisciplinary context so as to help students repeatedly optimize the design project and bring students the creative power of interdisciplinary knowledge and technology.

In terms of curriculum learning, teachers seldom teach knowledge and skills, but promote students' independent interdisciplinary practice in a task-driven way. In the process of independent learning, students will actively explore interdisciplinary knowledge and skills. For example, they will actively seek to cooperate with experts, scholars, or classmates in relevant fields to achieve the learning goal of a certain course, such as ethnographic knowledge in user research, behavioral method in demand survey, supply chain and value chain theory in manufacturing analysis, and engineering and technical knowledge in sample production.

4 CONCLUSION

Parsons, with a history of more than 100 years of design education, is committed to the mission of "challenging the status quo, releasing and improving creativity" and adheres to core values of "academic freedom, tolerance and experiment." Its innovative education philosophy and diversified and open education methods have shaped the alumni group with extraordinary achievements and made it a world-class design university. The memes of critical thinking and the innovative action of interdisciplinary practice are the best interpretation of Parsons' advanced educational concepts and methods, and have provided a significant model for the design of education in the contemporary world. Instilling the critical thinking and interdisciplinary practice indispensable for cultivating design creativity into every link of design education is expected to provide a reference point for China's design education, which has the largest scale in the world.

FUNDING

This research is funded by the key project of social science planning of Fujian Province (FJ2016A023).

REFERENCES

Quacquarelli Symonds Limited (2019). QS Top Universities – Art & Design. https://www.topuniversities.com/universities/subject/art-design.

The New School (2019a). History. https://www.newschool.edu/Parsons/history/.

The New School (2019b). About. https://www.newschool.edu/Parsons/about/.

The New School (2019c). Strategic-plan. https://www.newschool.edu/about/university-resources/strategic-plan/.

The New School (2019d). Mission-vision. https://www.newschool.edu/Parsons/mission-vision/.

The New School (2019e). Home. https://www.newschool.edu/.

Green technology & architecture engineering

Smart Science, Design & Technology — Lam et al. (eds)
© 2020 Taylor & Francis Group, London, ISBN 978-0-367-17867-3

Exploring the perceptual cognition of green design related to cultural and creative products

Yi-Ren Chiu*
Ph.D. Program of Mechanical and Energy Engineering, Kun Shan University, Tainan, Taiwan

Huannming Chou
Department of Mechanical Engineering, Kun Shan University, Tainan, Taiwan

ABSTRACT: While maintaining the fundamental needs for product functionality, quality, and durability and at the same time taking into consideration the need to meet environmental requirements throughout the product life cycle, green design for cultural and creative products should be able to create an implicit sense of emotional perception, cultural cognition, and a social value system in consumers of the products. The main objective of this research is to promote use of the Qualia cognitive method in the research and development for green design of cultural and creative products, and through exploring the cognitive sense's operation to initiate the mental cognition of the underlying subconscious and the deeper levels of the consciousness, in order to design cultural creative products that touch the hearts of the consumers while complying with green product requirements.

Keywords: cultural and creative products, green design, Qualia, cognition sense, subconscious

1 INTRODUCTION

1.1 *Research background and motivation*

Due to large-scale consumption of mother nature's natural resources that has disrupted the earth's biological environment, as well as the extensive pollution resulting from the technological process of converting resources into products, many developed countries have started to rethink their actions.

In 1987, the General Assembly of the United Nations (UN) officially proposed a definition for sustainable development, suggesting that sustainable development should be able to balance developmental needs in three main aspects, i.e., economic, environmental, and social, at the same time. The UN Environment Programme (UNEP) defined "green economy" as "an economy that can improve human well-being and social equity while significantly reducing environmental risks and ecological scarcities, and can also be thought of as one that is low-carbon, resource-efficient and socially inclusive." The 2016 European Environment Agency report included circular economy in the scope of green economy. The focus of a circular economy is placed on resource efficiency, while a green economy's broadest focus is human well-being. Therefore, through the promotion of the circular economy, any polluting industry can reduce waste and pollution,

allowing abandoned by-products to return to the resources recycling system in order to be regenerated, recycled, and reused, creating a green manufacturing system that contributes to a sustainable environment with high added value (Huang, 2017).

Good design can be touching. Therefore, enterprises should carry out "soul project" designs that are both complicated and delicate in order to give birth to products requiring operations with sensible techniques, which can more accurately and sensitively strike consumers' sensory neurons. "Sensibility" is humans' ability to form a direct emotion and feeling for a certain object through the senses, as opposed to the concept of rationality. John Locke is one of the earliest Western philosophers that suggested the concept of sensibility. He believed that human beings, when they are first born, are empty like a blank canvas. Denying the existence of inherent rationality, Locke advocated that human beings accumulated knowledge and gained experience of external things through the senses. "Sensibility Engineering" is, then, a technique that transforms human beings' desired perceptions or abstract imagery into concrete design elements (Nagamachi, 1995).

To promote the development of the cultural and creative industry, and construct a rich cultural and creative social environment, the green design of

*Corresponding author: E-mail: chucyr@yahoo.com.tw

cultural and creative products is necessary. Under the premise of maintaining the product's function, quality, and lifespan, as well as emphasising environmental compliance requirements throughout the entire lifetime of the product, green design produces emotional perception, cultural cognition, social value system, and other implicit meanings towards the product in consumers (Qi Guo Innovation International Co., Ltd., 2014).

Based on the format of the research problem and research issues, this research uses the research method of content analysis and review of the literature to organise relevant information for the discussion.

1.2 Research objectives

The so-called "Qualia" is a sense of life and emotional satisfaction – a touching and romantic feeling that touches the deepest part of consumers' hearts. Qualia products or services are the manifestation of the sensibility value's five main Qualia elements – Attractiveness, Beauty, Creativity, Delicacy, and Engineering – beyond the basic premise of quality. Qualia establishes unique characteristics and increases the added value of products and services, and in turn leads to consumers' pleasure and sense of poignancy on an emotional level.

The aim of this research targets the use of the Qualia cognitive method during the promotion of green design in the product development process, and through exploring the cognitive sense's operation to initiate the mental cognition of the underlying subconscious and the deeper levels of the consciousness, in order to design cultural and creative products that are touching and meet green product requirements.

2 LITERATURE REVIEW

2.1 The supply chain relationships of cultural creative designed product industry

Product design services serve as an important part of the entire industry chain. The upstream supply chain can be considered as the product development stage. Operators provide demand, and the design technology applied in the research and development phase includes dynamic design, mechanical design, electronic design, IC design, software design, and material design.

The midstream supply chain can be considered as the product commercialization stage. The actual output includes product appearance design, graphic design, mechanism design, and industrial and commercial packaging design, while the design application in the production and manufacturing phase involves industrial engineering design, mold design, and automation engineering design.

The marketing sector in the downstream industry chain utilizes brand design, electronic product design, exhibition design, and the like (Chiu and Chou, 2016).

2.2 Green design standards and methods

The design method of green design is a type of modern design thinking (Elkington and Hailes, 1988). Foreign expert Brewer once mentioned in a Green Design article that the highest standards of a product's ecodesign considerations include product recycling, waste reduction, increased product durability, product's ease of dismantling and assembly, material adaptability, choice of the least-polluting materials and processes, energy efficiency, etc. The concepts of a method that minimalises environmental impact are as follows: 1. Design for recyclability; 2. Design for reuse; 3. Design for remanufacturing; 4. Design for disassembly; and 5. Design for disposal (Sarkis, 1998).

2.3 Technical evaluation of cultural and creative products' five main Qualia elements

According to Yan and Lin (2012), it is shown that: (1) There is a close relationship between cultural attributes and the design of cultural products. (2) Successful products meet the five main Qualia elements: Attractiveness, Beauty, Creativity, Delicacy, and Engineering, while successful cultural and creative industries meet the aesthetic economy's business model, i.e., exquisite "art" and "culture," "creativity" complements "design," and "the industry" creates "the brand." Whether Attractiveness, Beauty, Creativity, Delicacy, and Engineering are equipped with technical qualities can be evaluated based on the cognition of the five main Qualia elements.

2.4 Cognitive sense's operation and the mental role cognition plays in the underlying subconscious and the deeper levels of the consciousness

"The Basic Driving Force of Cultural Change – Cognitive Science Oriented Exploration," the "cognition sense" originated from the "mental schema" is dynamic and automatically self-adjusts. So, from the source of the driving force, it can be said that "cognition sense" is the basic driving force of cultural change, and "cognition sense" is the overall representation of mental effects. The schema of the mental system must contain perceptual consciousness itself. It must also contain underlying subconscious factors such as emotions, fear, and unconsciousness. It must even include the function of giving rise to awareness and consciousness as well as the subconscious mind, and the "original mental entity" which stores all the operational results.

In cognitive science, the term "cognition" refers to all psychological activities related to the mental effects of animals, including reasoning, judgment, sensation, vision, language, emotion, memory, transformation, learning, and physical skills. Therefore, it involves disciplines such as neurophysiology, psychology, computer science, linguistics, mathematics, philosophy, and other interdisciplinary integrations among the aforementioned disciplines.

"Cognition sense" is, therefore, the unified effect of the mental system and must include the dome, hypothalamus, amygdala, hippocampus, limbic system, the cerebral cortex, and other physiological bases of the nervous system, as well as the three levels of mental effects that are the consciousness, subconscious, and the original entity of mind. Each of them has a different level of function that contributes to the completion of mental effects, and through mental effects shows "cognition sense" to complete the memory – the formation of memory is in fact the result of the operation of "cognition sense."

The original mind entity, which stores memories and brings forth active cognition sense, has long been described in Buddhist literature. It is called the eighth consciousness – ālayavijñāna. It uses the function of the cognitive sense that derives from itself to form and store memories – a treasure trove of memories. The operation of cognition sense requires the integration of eight different mental consciousnesses, which are called the eye consciousness, ear consciousness, nose consciousness, tongue consciousness, body consciousness, mental consciousness, manas, and ālayavijñāna. "The ālayavijñāna operates within the dense forest of all aggregates within which manas takes the lead. The conscious mind analyses all states including forms and others. Depending on the sense faculties, the five vijñānas manifest and discriminate the external state. All states grasped are nothing more than the ālayavijñāna." The statement "The ālayavijñāna operates within the dense forest" describes the eighth consciousness – ālayavijñāna, which is the mental schema, and its integrated operations of the five aggregates and eighteen elements originated from it, and its cognitive effects in completing the entire mental system. During the process of the entire mental system, the "mental faculty" is the lead and also the initiator, hence it reads "manas takes the lead." In Mahāyāna, the "mental faculty" is also called "manas" or "the seventh consciousness," which is similar to the cognitive unconsciousness known in cognitive science. However, the scope it encompasses is even broader than that known in cognitive science. "The conscious mind analyses all states including forms and others" describes the cognitive unconsciousness, i.e., the manas, once it initiates the cognitive mechanism and transmits information to the mental consciousness highland of cerebral cortex; thereby the conscious mind then determines various states such as form, sound, smell, taste, touch, and mind. At this time, the eye consciousness, ear consciousness, nose consciousness, tongue consciousness, and body consciousness, in line with their respective sense faculties (roots) – the eye faculty, auditory faculty, olfactory faculty, gustatory faculty, and tactile faculty – discriminate the five sense-objects of form, sound, smell, taste, and touch. The cognition of such information is in collaboration with the operation of the cognitive unconscious and is therefore described as "Depending on the sense faculties, the five vijñānas manifest and discriminate the external state." After this, the mental faculty transmits the cognitive information to the cortex brain for thought analysis by the conscious mind, following which the information is again transmitted back to the mental faculty to be determined if the memory will be stored. The cognition of all of these apprehended states, including the eighteen elements themselves, are the manifestation of, and originated from the eighth consciousness – ālayavijñāna, the mental system's schema.

3 ANALYSIS OF GREEN CULTURAL PRODUCT QUALIA'S DESIGN CASE STUDY AND QUALIA ELEMENTS

This research cites the case study of Meng's Handmade Paper products to discuss trendy handicraft's Qualia elements as described in the research of Chen et al. (2013). Meng's Handmade Paper products are largely heartwarming handmade paper experiences that focus on interaction with consumers. Through the environment-friendly LOHAS recycle and regenerate paper concept, Meng's Handmade Paper recycles and converts paper into reusable resources, satisfying the standards and ideology of green design. Postcards are the entry-level representative product of the brand, allowing consumers to create a handmade paper postcard (Figure 1), called "There's a Place," imbued with blessings from far away for their beloved. Their products also include dinner table lighting (Figure 2), which uses photopermeable material to represent a touching handmade product made with love from the bottom of the heart. Meng's Handmade Paper's product line even extends to all types of paper coasters, plant fibre boxes, manually bound notebooks with handmade paper covers, and other such green cultural product designs.

Figure 1. "There's a Place" product. http://mengs-hande made-paper.blogspot.tw/.

Figure 2. Dinner table lighting product. http://mengs-handemade-paper.blogspot.tw/.

4 CONCLUSION

Combining all the above discussion, in the world of design, the mental model represents the symbolic meaning of a certain object in a person's heart, and shows the importance of mental effects' "cognition sense" towards design. Once sensory stimulation reaches the brain, it secretes endorphins and produces a feeling of happiness; so once the cognitive unconscious – the mental faculty – initiates the cognitive mechanism and transmits the information to the cerebral cortex's consciousness, and when past experiences evoke the memory of a special feeling, one would want to feel it again. The wonder of the taste, the depth of touching feeling, imbue the sense-objects with the highest value that is unparalleled. Even if one uses the five sense faculties to experience the wonder, the mental consciousness affects the sensory organ's direct judgement. This means that through the body's experience of the five senses, due to the internal thought organ – the crossover and integration of the mental consciousness as well as the underlying subconscious – the mental faculty initiates memories and past experiences, is what truly prompts such

strong and vivid feelings in us (Qiu, 2016). Therefore, in promoting green design during the product development process, the "Qualia" cognitive method – through the dynamic operation of the cognitive unconscious, with the cognition of the five main Qualia elements in order to initiate the mental effects' role on the underlying subconscious and the deeper levels of the consciousness, and in line with an experience of the cognition and judgment of the eye consciousness, ear consciousness, nose consciousness, tongue consciousness, body consciousness, and mental consciousness – can be used to initiate the design of green cultural Qualia products.

REFERENCES

H.Y. Yan and R.T. Lin, 2012. "A Study of Value-added from Qualia to Business Model of Cultural and Creative Industries," *Journal of National Taiwan College of Arts 91*, 127–152.
J. Elkington and J. Hailes, 1988. *The Green Consumer Shopping Guide from Shampoo to Champagne-High Street Shopping for a Better Environment*. London: Victor Gollancs.
J. Sarkis, 1998. *European Journal of Operational Research 107*(1), 159–174.
J.W. Qiu, 2016. *Quality with Sense*. Taipei City: Taiwan Creative Design Centre.
M. Nagamachi, 1995. *International Journal of Industrial Ergonomics 15*, 3–11.
Qi Guo Innovation International Co., Ltd., 2014. *Cultural and Creative Industry's Practical Design*. Taichung City: Qi Guo Innovation International Co., Ltd.
Y.J. Chen, et al., 2013. *The Research on Qualia of Modern Crafts*. Taipei: National Taiwan University of Arts' Crafts and Design Department, Taiwan.
Y.R. Chiu and H.M. Chou, 2016. *Proceedings of International Conference on Innovation, Communication and Engineering*.
Y.Z. Huang, 2017. *Circular Economy*. Taipei City: Common Wealth Magazine.

Smart Science, Design & Technology — Lam et al. (eds)
© 2020 Taylor & Francis Group, London, ISBN 978-0-367-17867-3

The application of grey sculptures and color painting in the large residential architecture in Anzhen walled-village

Jianqing Weng*

Landscape Management Center, Sanming, Fujian Province, PR China

ABSTRACT: Grey sculptures and color painting is a distinct decoration technique that was seen in the south China architecture during the Ming Dynasty and Qing Dynasty. The paper drew an instance from the large residential architectural Anzhen walled-village and analyzed the decoration of roof, walls, and grey sculptures, and the expression of the theme message, technical features, and colors, in a view to summarize the relation between the artistic style and the architecture.

Keywords: Anzhen walled-village, Grey sculptures and color painting, Decoration technique

1 INTRODUCTION

Anzhen wall-village, also known as "Chiguan ciy", situates in Yangtou village of Huainan town, 110km to Yongan city of Fujian province. Chi Zhanrui, a then country gentleman led its construction in the 11th year of Guangxu emperor's reign in the Qing dynasty (A.D. 1885) and finished 14 years later. The village sits on a land of approximately 10,000 square meters with a floor area of 5,800 square meters and 350 rooms. Facing the east, the village has a square-shape front and a semi-circle shape rear. It clings to the mountain and is featured by consistently increasing heights that appears distinctly layered and grand. Categorized as a walled and earth building, the village centers around the hall and is separated into three courtyards that sit along the central axis in a symmetrical manner, making it a rarely seen large scale rammed earth folk architecture constructed in China (Zhang, 2001; Feng, 2004; Hu, 2007; Liu, 2010). Since it was built in a turmoil period, the village was built under the purpose to fight off bandits and is complete with 90 watcher's openings and 180 shooting openings, making it a superior military defense work. The eaves, doors, windows, beams, supporting brackets, window frames, structural brackets, and walls of the building are decorated with meticulously and ingeniously carving. The color painting on grey sculptures and wall paintings specifically, are of extremely high artistic and aesthetic value and scientific exploration value, and are hailed as the wonder in the domestic architectural design history. The village was titled the province-protected key cultural relic in 1991 and the state-protected key cultural relic in 2001.

2 PRODUCTION TECHNIQUES OF GREY SCULPTURE

2.1 Raw materials: Lime, straw, handmade Yukou paper, sand, brown sugar, glutinous rice flour, etc.

The reason the special technique for making grey sculptures are praised as "breathing" is because the raw materials are usually found in people's everyday life and that are natural and green.

2.2 Ratio of materials for making grey sculptures

Grassroots lime: made by combining the lime with straws fermented with whitewash. The material is a green product that is free of the contraction caused by the changing weather, because straws are tube-shaped and soft that can be flexibly shaped under different weather.

Paper lime: made by combining the lime with bamboo-made paper (Yukou paper). Subsequent to the fermentation of bamboo fibers with lime, the product is of a fine texture, favorable plasticity, small fibers, and high density, and has a harder texture than grassroots lime. It is rigid outside and soft inside and durable against rainwater, hence the "breathing" material. The dew it absorbs at night may evaporate in the daytime in the sun, enabling it unaffected by any problem brought by material contraction and granting grey sculptures a long-time performance.

In making grey sculptures, adding salt may facilitate the fermentation of lime, adding glutinous may

---

*Corresponding author: E-mail: cxyy2005@163.com

generate softer lime, while adding brown sugar may help the lime cover and seal the surface. Only with the proper ratio of materials can the lime product perform stably and longer.

3 DECORATION OF GREY SCULPTURES

3.1 Conception

The first and most vital step of making grey sculptures is conception where the sculptor needs to determine the conception and means of expression based on the main building design, the place of sculpture installation, and dimension requirements.

3.2 Determination of position and shape

Once there is a clear conception, the dimension, size, position, and shape of the work need to be determined. For high relief and 3D circular carving, skeleton frames need to be prepared first and leave enough space for the stacking of grey sculpture.

3.3 Basic form

Stack to produce the basic shape of the grey sculpture on the fixed skeleton with grassroots lime. The thickness of each stack shall not exceed 5cm. Whenever a thicker stack is to be done, the new stack shall only be constructed after the first stack is dried for several days. Haste makes waste.

3.4 Plastering

Once grassroots lime is applied, use paper lime to carry out detailed processing on the surface. During the plastering process, natural pigments may be added, stirred, and mixed in the paper lime to enable the color more permanently, a similar method used in dough modeling.

3.5 Coloring

Natural mineral pigments are used for coloring grey sculptures. Coloring shall be finished on the surfaced applied with paper lime and shall be applied only when grey sculptures are 70% to 80% dry. Because the paper lime is still in some extent moist, a state favorable for the pigments to permeate the lime and blend in the material for going through oxidation reaction together. Under this method, the colors may remain fresh and durable against time.

3.6 Shaping

The last step is air drying after the previous processes are finished. The grey sculptures, once dried are of high strength, strong weather resistant, and bright colors.

4 CONCEPTION OF GREY SCULPTURE WORKS

The conception are mainly derived from the welcomed traditional cultural essence that covers opera characters and plots, folk legends, auspicious animals, flowers and trees, auspicious patterns, utensils and talismans, as well as symbolic geometrical elements, in a view to express the spirit and aesthetics.

As shown in the grey sculpture – Brimming Delight and Peony of Opulence, themed by flowers and birds found in Anzhen walled-village (Figure 1), the patterns consisted of animals and plants are used to deliver an propitious message. The high relief of the chirping magpie is obvious enough for people to detect the happiness it stands for. Brimming Delight is one of the traditional propitious patterns in Chinese culture, which means the arrival of Spring and joyous events, and an elated spirit. Peony, known as the "King of flowers" and "Flower of opulence" is featured by its flamboyant color, wide-spread petals, elegant postures, and overwhelming fragrance, and extensively used as folk house decorations because people believe it is the icon of splendor, grace, wealth, auspice, and prosperity.

The grey sculpture captures a historical story where Liubei, the first emperor of Shuhan regime, relentlessly visited Zhu Geliang, a wise man who was practicing a secluded life, to become his military counselor and finally succeeded in persuading Zhu to serve him (Figure 2). The sculpture demonstrates the story in a vivid and lifelike way and brings back the details of how Liubei treated his subordinates with manner and virtue and put them into good use. The idea behind the sculpture is that one can only

Figure 1. The grey sculpture – *Brimming Delight* and *Peony of Opulence*.

Figure 2. The grey sculpture – *Brimming Delight* and *Peony of Opulence*.

Figure 3. The myth of Eight Immortals.

succeed in career only if he or she treats others with genuineness, aspire for the talent, and take in talents.

The myth of Eight Immortals. The story of the eight celestial xian is a tool to propagate Taoism philosophy that one could grow wings and ascend to heaven if he or she extricates the mass from various world suffer ings and that the kind will be rewarded and the evil are bound to face retribution. These myths serve as the base with which folks use to realize their wishes. The work is an absolute demonstration of the diverse and rich artistic creativity of artisans (Figure 3).

5   DECORATIVE FEATURES OF GREY SCULPTURES

Grey sculptures are knows for their rich colors, three dimensional views, and techniques that integrate clay sculpture and color painting. They are made with exquisite skills chosen from a wide range of themes that are easily understood by the mass, which is usually consists of legends, classical stories, characters, land-scape, flowers, animals, auspicious creature, propitious patterns, etc. Grey sculptures provide function and decoration to the architecture in a subtle way (Figure 2).

The grey sculptures of Anzhen walled-village form a 30–40cm wide decorative band between the ridge tile, window, roof tile of the first floor, and the walls of the second floor (Figure 4). The color difference between the grey tiles and white walls instill dynamism into the building instead of just having a simple range of colors. In the aspect of visual layers, the grey sculptures utilized multiple carving techniques that include low relief, high relief, 3D circular carving, hollow-out carving and more, and succeeded in creating a multi-layered 3D visual effects. In terms of patterns and layout, the decorative band

is a inclusive and genius adhesive of architecture and patterns of stones, flowers, grass, characters, animals, and auspice, giving the band distinct categorized contents and auspicious patterns that connect all the other elements. The perspective effect is well managed that grants a fine effect of close-range, medium-range, and long-range view.

In terms of functionality, the embellished decorative band also serves to guard against wind and rainwater, keep the tiles in place, and ensures that window corners and pedestals are resistant to corrosion, moisture, and insects. It is clearly revealed that the wisdom of the ancient people enabled them to express spiritual aspiration and sense of beauty.

6   INHERITING AND PROTECTION OF GREY SCULPTURES

A recognized good architecture must be complete in three elements: functionality, stability, and aesthetic. Grey sculptures and color painting, as a unique form of architectural art, is hugely related to the location, weather, living habits, architectural culture, aesthetics, religious belief, materials, and techniques of the place where it sprouted and developed. To inherit and protect this traditional architectural technique, we need to invest in the preparation of natural and green materials and techniques and skills so as to inherit the essence of the past wisdom. During the restoration of the damaged sculptures, we need to respect the original materials and design and preserve the original appearance, and well inherit its cultural value.

7   CONCLUSION

Hidden amid the mountains in the center of Fujian province, Anzhen walled-village is a unique and distinct large-scale rammed earth residential architecture that was constructed under the then special social environment and the purpose for a stable and safe life. The grey sculptures and color paintings of the village is a sheer manifestation of the unique techniques used and the people's wisdom and artistry. It beautifies the architecture and express the wish of people for a good life. It is the ancient folk architecture best preserved in the province, a prominent example of walled-village culture.

REFERENCES

D. Zhang, 2001. *Chinese Folk Art Dictionary*. Jiangsu Art Press.
J. Feng, 2004. *Architectural Art of Ancient China - Ancient Vibe*. People's Fine Arts Publishing House.
S. Hu, 2007. *Architectural Decoration of Ancient China. Carving*. Jiangsu: Jiangsu Art Press, 2007.
Z. Liu, 2010. *Guangzhou Grey Sculptures*. Guangzhou: Guangzhou Press.

Figure 4. The walls of the second floor.

Industrial design & design theory

Smart Science, Design & Technology — Lam et al. (eds)

Research on the one-piece washbasin of public toilets

Zhi-Xing Dai*, Siu-Tsen Shen & Huei-Min Lee
Department of Multimedia Design, National Formosa University, Yunlin County, Taiwan

ABSTRACT: With the rapid development of society and the continuous improvement of people's living standards, people often use weekend holidays to travel, and public places such as airports, stations, parks, amusement parks and so on become places where people gather. At present, there are many problems in supporting facilities in many public places, such as outdated equipment, low utilization rate, insufficient humanization, etc. However, as an essential equipment in public places, the sales volume and development potential of washbasins are huge [1]. However, in public places, we often find that we spend a lot of time waiting in line to wash our hands in public places with heavy traffic, or some public places have hand washing stations but lack functions such as hand sanitizer and hand dryer, and solving these problems has become an imperative at present.

1 RESEARCH BACKGROUND

In the 21st century, the overall social economy is in a process of rapid development, and the population growth rate is increasing rapidly, too. Correspondingly, people's demands for public facilities are also increasing. However, the current public facilities only stay in diversified style thinking, but they really lack design connotation as well as care for users. For instance, in public places, the washbasin is an indispensable product in people's daily life and the waste of a large amount of water resources is not uncommon, and most washbasins in public places can only satisfy the user's function of washing hands. Many places lack the equipment, such as hand dryer and hand sanitizer, thus ignoring the feelings of people when they use it [2]. Secondly, in accordance with my observations, during the peak period of rushing time, people often spend a lot of time waiting in line to wash their hands due to the large number of users using the rest room, which always reduces the efficiency of the use of the washbasin. And some users will make use of the washbasin to clean the fruits and plates that they will carry out, which causes the problem of blocking of the washbasin. The occurrence of this problem will result in problems for subsequent users and also increase equipment to maintain costs.

2 RESEARCH PROCESS

A. analysis of research status domestically and abroad

a. domestic situation

At present, there are very few professional works or papers on the design planning of washbasins in public places. Chen Zhenlong from Da-Yeh University takes the "Design Study of Improving the Splash of Water in the Washbasin" as a master's study. He mentioned that the design of the washbasin was relatively small, which would cause splashing of water during use and which would caused the phenomenon that the bathroom floor is wet and slippery. In order to deal with this problem, he used dynamic photography to observe the situation of splashing water from the washbasin. Through the collection of post-data and his own expertise, he put forward a new form of face-washing function to prevent splashing [3]. Yang Yalun, a student at National Taiwan University, combined software CAD/CAE as operating software in 2002 and made use of OLE technology (OBJECT LINKING AND EMBEDDING) to conduct system fetch component information. Thereby, an adjustable lifting washbasin is made [4]. In 2011, Chen Weiming's "Electronic Faucet Shape Design Study" in the Shih Chien University adopted questionnaires to understand consumers' preference for electronic faucet products, the research methods of variability data (ANOVA) and T-testing, providing a sense of style for future electronics faucet manufacturer [5].

b. present situation overseas

The frequency and duration of use of Japanese toilet are among the highest in the world. Since they like staying in the toilet to do their favorite things. In 1985, the Japan Public Toilet Association was established, which aimed at improving public toilets at famous attractions. The slogan "Creating a Toilet Culture" was proposed. The Association holds a toilet symposium every year and conducts the "Top Ten Public Toilets" selection event. The November 10th of

*Corresponding author: E-mail:10768115@gm.nfu.edu.tw

each year is also designated as "Japanese Toilet Day." In 2001, the first "World Toilet Summit" was held in Singapore to set up the "World Toilet Organization", which was set up to study and deal with a variety of problems related to toilets. In 2003, the World Health Organization held the "World Toilet Summit" in Beijing. The theme of the summit was to focus on toilet construction and management in developing and less developed areas. At the 67th session of the UN General Assembly in 2013, it was agreed that each year November 19th was designatedas "World Toilet Day", which advocated a clean, comfortable and hygienic toilet environment for everyone. 2015, the China Business Research Institute released the "2015-2020 China Washbasin Industry Market Forecast and Investment Strategy Analysis Report", which mainly analyzed the market size of the washbasin, the demands for washbasin products, the competition situation and the washbasin, the main business operations, the market share of major washbasins, and the scientific prediction of the future development of the washbasin [6].

2 RESEARCH METHODS

This study is mainly to explore the current situation of public washbasins, collect and summarize the problems of the existing washbasins through various research methods, and redesign a washbasin product that meets human's demand through professional knowledge.

A. Literature Analysis Method

Through consulting magazines and literature, analyze the existing relevant research results and understand the current product design situation. At the same time, collecting and sorting the product information of the hand washing facilities in the existing public space generally summarizes the products and makes solid preparations for early theoretical research and post-design practice.

B. On-site Investigation Method:

Conduct on-site investigations in subways, cinemas, pedestrian streets, universities, etc., and find the same points in the analysis process.

C. User Observation Method:

Observe the user's use process, feel the user experience, Analysis of existing product problems.

D. User Interview Method

Good design needs to meet the actual demands of users. Face-to-face interviews are good for the survey, which can really understand the users' experience and the demands for more products [6]. We conducted a field survey by using Huwei Tongxin Park in Yunlin County, Taiwan. This interview method: A total of 30 typical users were selected, including students, park staff, tourists and nearby neighbors., Their ages range from 13 to 67. Through the one-to-one interview with users on-site, we understood users' real demands in a more comprehensive and targeted way, as well as the problems encountered by users in actual using of public toilets faucets. The results of user's interview will be guided for the subsequent design of public toilet faucets

Interview Topic:

a. How long do you need to wait in the public places to wait for hand washing?

b. In which public place do you usually choose to wash your hands?

c. What do you think are the disadvantages of the washbasin now?

d. Where do you think the future washbasins need to be improved?

How long do you wait in the public toilet for hand washing?

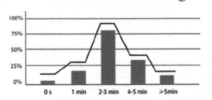

Where do you often go to the public toilet?

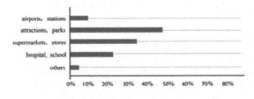

What do you think are the disadvantages of the sink now?

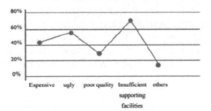

In the future, where do you want public washrooms to improve?

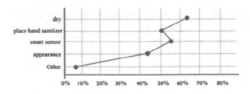

From the above table, it can be seen that most people in outdoor places choose to use public toilets in scenic spots or parks. At this stage, they take 2-3 minutes or so to wait in line to wash their hands. They are not satisfied with the existing hand washing facilities. The author hopes that in the future, the liquid soap and hand dryer will be concentrated on the washbasin.

3 DESIGN PRESENTATION

A. Sketch

According to research process and method, we have the following design concept: 1. Integrate the washbasin and hand dryer. 2. Change the traditional water droplet into flake water. 3. Use the infrared sensor to sense the palm size and automatically calculate how much water should be discharged according to the palm size. Therefore, the effect of saving water can be achieved. 4. In order to provide users with a better hand-washing experience, the terminal control system is used to make the washbasin produce hot water in winter and cold water in summer [7].

Based on the above conclusions and findings, the following concept sketches of washstand were drawn.

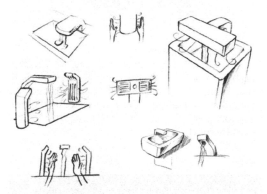

Innovation proposal:

a. The whole washbasin adopts the inductive operation mode.
b. Concentrating the whole hand washing equipment into the washbasin to form an integrated effect, such as liquid soap and hand dryer.

For this reason, the following sketch of the integrated annular washbasin was designed:

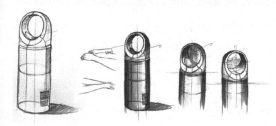

B. Model drawing

Some ideas based on hand drawing, Rhino software was used to make 3D modeling of circle hand washing and drying all-in-one machine. The product safety and functional integrity were inspected by ourselves in the process of modeling to achieve the final effect. The picture below shows the product model diagram.

C. Product use analysis and detail display

Step 1: In order to meet the demands of the washbasin, there is a liquid soap function to be prepared. For this reason, a liquid soap is provided directly above the washbasin. As long as the hand is close to the sensor window, the pump nozzle will discharge 0.5 ml of the liquid soap through the infrared sensor. The amount of liquid soap can be replenished at any time on the top of the washbasin.

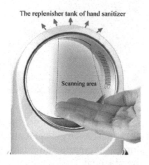

Step 2: Put your hands in the washbasin, and the faucet will automatically open after the infrared sensor senses. There is warm water in winter and cool water in summer. This design hopes to have the effect of warm winter and cool summer. Inductive water curtain faucets are more dispersed than the traditional faucets. When using an inductive water curtain faucet, people only need to shake their hands back and forth. The water outlet can automatically select the number of water curtain outlets so as to open in according with the size of the hand so as to achieve water saving.

Step 3: After the water curtain faucet is discharged for 10 seconds, the faucet will be closed. The hand dryers on both sides will open and the hands can be rotated to dry.

Step 4: The circular drain design of the traditional pool was abolished and changed to a linear drainage method, which aims at providing the most simple hand washing function for the public, thereby reducing the blocking of the washbasin.

Step 5: An exhaust function is provided under the unit to prevent unnecessary accidents because of high temperature of the parts caused by the internal operation of the machine.

D. Actual scene simulation:

The washbasin can be placed in a variety of scenic spots, parks and other public places. Considering the problem of power saving, the upper part adopts a circular shape design, which can enclose the wind discharged through the drying in the ring to form a circulation effect, thereby realizing the power saving effect. This design allows users to enjoy cleaning services the first time without the need to queue up.

E. Product rendering:

Using keyshot for product rendering, I chose dark and light colors to match, giving users a visual impact. The white on the top better reflects the streamlined appearance, the lower part of the uniform was used of frosted materials, Improve the use time of products.

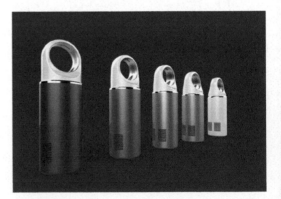

REFERENCES

[1] Kang Zongshe. Preliminary Study on the Research of the Design, Research and Methods of Washbasin [J]. Commerce, 2016(07):141–142.
[2] Tang Gensheng; Comprehensive Improvement of Product Design Level[J]; Journal of Anhui Electronic Information Vocational and Technical College; 2010,04.

[3] Chen Zhenlong. Research on the Design of Water Splashing When Using the Washbasin[J]. Da Ye University,2008.

[4] Yang Yalun. Adjustable Lifting Washbasin [J]. National Taiwan University, 2002.

[5] Chen Weiming. Research on the Shape Design of Electronic Faucet[J].Shih Chien University, 2011.

[6] China Business Research Institute; 2015-2020 China Washbasin Industry Market Forecast and Investment Strategy Analysis Report [J]; BeiJing, 2015.

[7] Lin Chonghong. The Principle of Shape Design [M]. Shichuan Culture Enterprise Co., Ltd., 2005:156.

Smart Science, Design & Technology — Lam et al. (eds)
© *2020 Taylor & Francis Group, London, ISBN 978-0-367-17867-3*

The needs of smartphones to provide medical services for the elderly

Lanling Huang*, Kaiqi Zhang & Shingsheng Guan
School of Design, Fujian University of Technology, Fuzhou, Fujian, PR China

Jianbo Dong
Department of Industrial Design, ARTOP DESIGN GROUP CO., LTD, Shenzhen, China

ABSTRACT: The purpose of this study was to survey the current state and needs of using smartphones about medical services for elderly people. A total of 40 participants were surveyed by questionnaire. The results of this study can be summarized as follows: 1) Applying medias (e.g. TV programs and telephone) to disseminate health care information that is often used by older people. 2) Design medical service applications with universal design concepts for the elderly, making the operation interface easy to operate. 3) Improving the quality of hospital services. For example: early training and evaluation mechanism for medical staff service attitude; medical bus service for the elderly or disability people; providing appointment services, 4) Combining artificial intelligence technology with home care products, track chronic diseases, conduct long-distance consultation services, deliver drugs by express delivery, and reduce the inconvenience of elderly people going to and from hospitals.

1 INTRODUCTION

The digital lifestyle has gradually become a new lifestyle for people. Medical service is one of the important basic requirements in our lives. According to the United Nations World Health Organization, by 2025, China's elderly over 65 will account for 13.2%, and the United States will be 18.5%. By 2050, there will be almost this many (120 million) living in China alone, and 434 million people in this age group worldwide. By 2050, 80% of all older people will live in low- and middle-income countries (World Health Organization 2018). There are more elderly people, and the demand for health care has increased.

Garçon et al (2016) studied to identify policy gaps in the delivery and availability of assistive health technology (AHT) and medical devices (MD) for aging populations. The results showed that practical, life-enhancing support for older people through AHT, MD, and related health and social services is a neglected issue. Health technologies, especially medical and assistive health technology, are essential to ensure older people's dignity and autonomy, but their current and potential benefits have received little recognition in low- and middle-income countries. Many countries need much greater official awareness of older adults' needs and preferences.

As an individual ages, there is an increased likelihood of having a multiple health problems or comorbidities (Anderson et al. 2004), which leads to an increasing need for health and/or disease management interventions. These demographic trends, combined with the growth of mobile phone telephony among the older adult population, suggest that using the mobile phone as a platform for interventions in health may be a viable way forward (Joe et al. 2013).

With increasing access to mobile technology, such as smartphones and smartwatches, the development and use of mobile health applications is rapidly growing. To meet the societal challenge of changing demography, mobile health solutions are warranted that support older adults to stay healthy and active and that can prevent or delay functional decline (Helbostad et al. 2017). How to improve the quality of life of the elderly has always been a design issue that researchers and designers should pay attention to in recent years. The rise of internet applications and the expectation of internet service design, we should pay attention to the experience and feedback of Internet users. At present, most Internet applications are the main design targets for young and middle-aged people, and neglected for the elderly.

The purpose of this study was to survey the current state and needs of using smartphones about medical services for elderly people.

2 METHODS

The questionnaire was used to investigate the needs of older people using their smartphones for medical services. The questionnaire consists of three parts: (1)

* Corresponding author. E-mail: anilhuang@163.com

basic data; (2) the status and problems of medical services for the elderly; and (3) the needs of medical services.

Basic information includes as following: age, living status, education level, occupation, the time it takes to use the phone every day and healthy status, etc.

The questions associated with the status and problems of medical services. The questions are as following: (1) how to obtain medical information sources (such as life, health knowledge, etc.); (2) what kind of smart health equipment are you using? (3) what problems have you encountered while visiting a doctor?

The questions associated with the needs of medical services. The questions are as following: (4) would you like to learn your smartphone to manage your health care data? (5) what kinds of smart health equipment do you expect to use? (6) what medical services do you expect to provide from smartphone app? (7) what method do you prefer to get medical care services?

3 RESULTS AND DISCUSSION

A total of 40 participants were surveyed, 27 males and 13 females. Most elderly people are 70-79 years old (50%) and 65-69 years old (32.5%). For living status, participants who live with the spouse (50%), followed by the spouse and children (30%), and live alone (13%).

For educational background, 35% subjects were graduated from high school (35%), followed by university graduates (20%) and college graduates (17.5%). The monthly employment income is 3000-4000 yuan (35%), followed by 4000-5000 yuan (25%). Thirty percent of the subjects were professional and technical, followed by workers from national enterprises (23%). Thirty percent of the subjects were professional and technical experts (e.g. doctor, teacher, etc.), followed by workers from national enterprises (23%). Smartphones use one hour (60%) per day, followed by 2 hours per day (20%), and 30 minutes per day (20%).

3.1 A. The status and problems of medical services for the elderly.

The results are as following:

(1) How to obtain medical information sources. The results showed that 93% older people received information from TV programs, followed by family members (50%) and friends from neighborhoods (33%).

(2) What kinds of smart health equipment are you using? The results showed that 53% older people never use smart health equipment. 40% older people using smartphones with health apps.

(3) What problems have you encountered while visiting a doctor? The results showed that 90% of the elderly thought that the waiting time for medical treatment was too long (n = 36), 88% of the doctors had a bad attitude, the service quality was poor (n = 35), 80% went to the hospital far away, and it was inconvenient (n = 32).

3.2 B. The needs of medical services for the elderly.

The results are as following:

(4) Would you like to learn your smartphone to manage your health care data? The results show that 50% of the elderly expressed they want to learn the functions they need. Another 45% said that smart medical services do not affect normal life, they will not actively learn.

(5) What kinds of smart health equipment do you expect to use? The results show that 85% of the elderly need an intelligent sphygmomanometer (n = 34), 83% smart electronic scale (n = 33), 50% smart bracelet or watch (n = 20).

(6) What medical services do you expect to provide from smartphone app? The results showed that 83% of the elderly wanted to book hospital services via mobile phone (n = 33), 58% of the elderly wanted the hospital to provide hospital transportation services (n = 23) and 43% to evaluate the quality of service for medical staff (n = 17).

(7) What method do you prefer to get medical care services? The results show that 90% of the elderly want government departments to inform aged care services by mobile phone (n = 36), and 63% of older people want the country's mobile phone operators to provide smart aged care services (n = 25).

From the results of the above survey, the results of problems in medical service are as following: (1) Half of the subjects have never used smart medical equipment. (2) In medical services, the problems often encountered are that the waiting time for medical treatment is too long, followed by the poor quality of medical services.

Needs of medical service for older are as following: (1) most elderly people in the medical service information are obtained from the TV programs, and relatives or friends. (2) They (50%) are willing to use smartphones to handle medical services. (3) they expect that the hospital to provide services (e.g. transportation and booking, etc.)

4 CONCLUSION

This research is a preliminary study to investigate the current state and needs of using smartphones about medical services for elderly people. To summarize the above results, the study proposes the following recommendations: (1) Applying medias (e.g. TV programs and telephone) to disseminate health care information that is often used by older people. (2) Design medical service applications with universal design concepts for the elderly, making the operation interface easy to operate. (3) Improving the quality of hospital services. For example: early training and evaluation mechanism for medical staff service attitude; medical bus service for the elderly or disability people; providing appointment services (4) Combining artificial intelligence (AI) technology with home care products, track chronic diseases, conduct long-distance consultation services, deliver drugs by express delivery, and reduce the inconvenience of elderly people going to and from hospitals.

ACKNOWLEDGMENT

This study was supported by Fujian Federation Of Social Science Circles with grant project No. FJ2018B150. This study was part supported by the Fujian University of Technology with grant project No. GY-S18093, and the Design-led Innovation Research Center, Fujian University of Technology.

REFERENCES

World Health Organization, 2018. *Ageing and life-course.* retrieved on https://www.who.int/ageing/events/world-report-2015-launch/en/

L. Garçon, C. Khasnabis, L. Walker, Y. Nakatani, J. Lapitan, J. Borg, A. Ross and A.V. Berumen, 2016. *Gerontologist 56* S293–S302.

G. Anderson and J. Horvath, 2004. *Public health reports 119*(3) 263–270.

J. Joe and G. Demiris, 2013. *Journal of Biomedical Informatics 46* 947–954.

J.L. Helbostad, B. Vereijken, C. Becker, C. Todd, K. Taraldsen, M. Pijnappels, K. Aminian and S. Mellone, 2017. *Sensors 17* 622.

Smart Science, Design & Technology — Lam et al. (eds)
© 2020 Taylor & Francis Group, London, ISBN 978-0-367-17867-3

Exploration of the teaching method for the course of "Bridge Engineering" under "the Belt and the Road" strategy

Lin Li*, Huaifeng Wang, Haitao Wang & Dandan Xia
School of Civil Engineering and Architecture, Xiamen University of Technology, Xiamen, Fujian, PR China

ABSTRACT: Based on the higher requirements of "the Belt and the Road" development strategy for engineers, combined with the development prospects of internationalization of bridge engineering, the teaching process of the bilingual course "Bridge Engineering" is reformed. Taking language as the medium, and professional knowledge as the main line, the basic knowledge fragments are comprehensively sorted out. By comparing the differences between Chinese and foreign projects, integrating aesthetics and cultural education, students' "global view" and "future view" are cultivated, and students' enthusiasm for exploration in professional fields is inspired to meet the requirements for engineers under the development strategy of "the Belt and the Road."

Keywords: the Belt and the Road, bridge engineering, bilingual course, teaching method

1   INTRODUCTION

As one of the traditional civil engineering disciplines, to meet the rapid development of China's infrastructure construction in the past 30 years, the course of bridge engineering needs improvement with the rapid development of modern science and technology. In recent years, new technologies such as assembly-type construction, BIM technology, and parametric design have occurred (Hongxue et al., 2012). As a traditional discipline, fundamental changes in bridge engineering have taken place in construction methods, design methods, and management models, etc. The cross-integration needs of civil engineering, mechanics, computer, management, and even art have reached an unprecedented level, so it also puts forward higher and more comprehensive training requirements for the professionals. With the successful construction of large and complex bridges such as Runyang Yangtze River Bridge, Hangzhou Bay Bridge, and Hong Kong-Zhuhai-Macao Bridge, bridge construction level in China is already ranked among the first in the world (Xiang, 2000; Fang, 1999).

In response to the "the Belt and the Road" plan put forward by the Chinese government, China's bridge construction team is stepping out of the country and gradually becoming internationalized. Project cooperation with other countries and regions has become the future export direction of the Chinese bridge engineering industry, which will bring a higher requirement and need for bridge engineering students in universities. The international classes of civil engineering, facing the needs of international students, which call for the teaching methods popular in Europe and the United States, and carrying out professional international training method, is a special educational need in the context of internationalization of civil engineering. Among them, "Bridge Engineering" is one of the most important professional compulsory courses with a theoretical class time of 64 class hours, which reflects the importance attached to the professional direction of bridge engineering in the professional teaching design.

The international background of the course and the relatively abundant course hours provide the necessary course carrier and implementation space to meet the needs of the "the Belt and the Road" development strategy.

2   INTRODUCTION OF THE BILINGUAL COURSE – BRIDGE ENGINEERING

The international class will take "Bridge Engineering" as a professional compulsory course. Based on the study of professional basic courses such as "theoretical mechanics," "material mechanics," "structural mechanics," and "structural design principles," we will expand the knowledge of students with practical aspects such as curriculum design and production internship, so that students can understand the basics of bridges, design principles, and procedures; understand the role of

---
*Corresponding author. E-mail: 2011110904@xmut.edu.cn

bridges and structural arrangements; know the structural system, structural characteristics, design methods, calculation points, and main construction methods of common beam bridges and arch bridges; understand the structural systems, structural features, design basic theories and construction methods of cable-stayed bridges; suspension bridges and composite bridges; and also can understand the maintenance technology of bridges.

Due to the future demand for international engineers, the English–Chinese bilingual teaching is applied in this course. The teaching materials such as syllabus, teaching plan and teaching course outline are all written in English. European and American textbooks are used as specific reference English textbooks (Troitsky, 1994; Chen and Duan, 2003). In light of students' English level, an auxiliary Chinese textbook (Chen, 2017) also is provided. The teaching is mainly in English, and the second explanation is given in Chinese in the individual concept and the necessary complex calculation interpretation. The whole course is divided into 10 parts for teaching, including: introduction, deck and bear of bridges, components and construction of simply support beam bridge, pre-stressed concrete bridge and rigid bridge, curved skew bridge and overpass bridge, composition and construction of arch bridge, calculation and design of arch bridge, suspension bridge, cable-stayed bridge, and the final curriculum review.

In accordance with the training requirements of the international class, the language in the class is mainly in English. The course score consists of four parts: 10% for attendance, 10% for class participation, 10% for completion of assignments, and 70% for exams.

3 THE CHALLENGE OF THE BILINGUAL COURSE – BRIDGE ENGINEERING UNDER "THE BELT AND THE ROAD" DEVELOPMENT STRATEGY

Professor Qian (2008) once said in an article, "As far as function is concerned, bridges are a part of traffic lines; in terms of beauty, bridges are only a part of the surrounding environment. In the field of technical science, bridge engineering is a small part of it. The world is huge and complex, and the bridge is only a small part of it." It can be seen that how to effectively connect "bridge engineering" as an applied professional knowledge with the disciplines closely related to its development, including machinery, materials, computers, is the challenge to modern professional education under "the Belt and the Road" strategy, that is, to cultivate students' overall view.

In addition, language is an important cultural carrier and another "bridge" for communication. In recently years, China's bridge engineering technology has become an important technology industry for external export. Therefore, bridge engineers are required to have good language skills under "the Belt and the Road" development strategy.

4 INNOVATION OF THE BILINGUAL COURSE – BRIDGE ENGINEERING UNDER "THE BELT AND THE ROAD" DEVELOPMENT STRATEGY

The traditional bilingual course teaching focused only on explaining the course content in English, which is helpful for improving students' English professional communication ability. However, under the current "the Belt and the Road" strategy, bridge engineering, as one of the most important professional courses, cannot just satisfy the improvement of language. It should make full use of the teaching process of the course, to help students fully remember the knowledge they have learned in the past, and develop their independent learning innovation ability by finding and solving problems. On the other hand, it should combine open teaching methods and foreign language auxiliary materials in the classroom to expand the view of the students, and cultivate students' curiosity and exploration ability. Therefore, how to integrate the existing knowledge and cultivate a new type of professional bridge engineers with creative spirit is an important part of the reform for this course.

The reform teaching method of the bridge engineering course aims to continuously guide students to broaden their horizons, expand their thinking, enhance their abilities, and stimulate students' independent ability to learn, innovate, and develop:

(1) Integration of knowledge fragments and multidisciplinary knowledge: Bridge Engineering is a professional course. Its preliminary study foundation course includes more than ten basic courses such as classic mechanics, hydraulics, soil mechanics, engineering geology, engineering materials, and structural design principles. During the three-year course of study, due to the complexity of the knowledge system and the complex relationship between the courses, the general situation of knowledge fragmentation occurred when the bridge engineering course was applied. Therefore, the teaching process of this course pays special attention to "memory wake-up" in the application of each knowledge, and systematically organizes the "knowledge fragments" that are incomprehensible in the memory of students, to realize the secondary learning of knowledge. At the same time, bridge engineering involves multidisciplinary knowledge such as aesthetics, mathematics, mechanics, mechanics, management, economy, etc. Therefore, in the teaching process, it is important to expand the students' ideas and inspire students' subjective initiative in independent learning to adjust the requirements of "the Belt and the Road" strategy.

(2) Integration of bridge aesthetics and the history and culture of bridges between the East and the West: The bridge is an important testimony to the development of human civilization and historical

changes. It represents the military, political, cultural, economic, and artistic level of a certain historical period. Due to the limitation of course hours in the traditional training methods, more attention was paid to training in technical knowledge, and the inheritance of the knowledge of the literature and history was neglected. This course aims to stimulate students' desire of exploration, to guide the study of the history and aesthetics of Chinese traditional bridges with Chinese traditional culture. While learning the language, it also helps students understand the promotion and representative level of the bridge in the process of human history development, and establish a global outlook and enthusiasm for the exploration of knowledge. At the same time, it also introduces the concept of aesthetics from the traditional art field to the engineering field, and guides students to understand the leading role of bridge aesthetics in regional planning and urban landmark aesthetics.

(3) Advanced technology of modern bridges and future trends: The limitations of construction techniques and methods have been overcome for modern bridges. The emergence of new materials, new processes, and new equipment has greatly improved the span of bridges, reduced construction costs, and especially reduced a large number of labor work, reducing engineering error caused by humans. More breakthroughs are expected in the field of special materials, deep foundation, artificial intelligence design, etc. Therefore, this course introduces modern and recent technologies, and helps students to understand the latest development status and technological development of bridge engineering by hiring industry experts for reports.

(4) Comparison of engineering regulations in the East and West: In the process of explaining the contents in each chapter of the course, it is necessary to involve various types of engineering regulations, and it is also necessary to guide students to think about the reasons of the relevant regulations in China. Regulations are the basic legal guarantee for the construction of bridges, but due to differences between the basic laws of the East and the West, there are also obvious manifestations in engineering regulations. Therefore, in the instruction of this course, students need to be further instructed in the similarities and differences between the two regulations. Students are also required to understand their origins, connections, and differences. This course will help students to establish the difference between the two types of regulation in terms of usage, rights, and responsibilities.

5 THE ACHIEVEMENTS OF THE CURRICULUM REFORM

"Bridge Engineering" is one of the most important professional directions of civil engineering. Therefore, in order to meet the requirements of bridge engineers under "the Belt and the Road" strategy, the course systemizes the basic knowledge and guides students in thinking about history, culture, aesthetics, economics, and law. This course takes the development of human civilization as the background, uses English–Chinese as the teaching language, develops students' vision, cultivates their vision of the world, and realizes the comprehensive knowledge of bridge engineering, leading students to learn modern bridge design and construction.

FUNDING

The research project "Research on the Reconstruction of the Professional Basic Course System to Adapt the Development of 'New Engineering' in Civil Engineering" (XGK201711) from Xiamen University of Technology is greatly appreciated.

REFERENCES

B. Chen, 2017. *Bridge Engineering*. Beijing: China Communications Press.
D. Qian, 2008. *Bridge Construction 08*(4), 82–88.
H.Xiang, 2000. *China Civil Engineering Journal 33*(3), 1–6.
L. I. Hongxue, G. Hongling, G. Yan, et al. 2012. *Journal of Engineering Management 12*(6), 48–52.
M. Fang, 1999. *Bridge Construction 99*(1), 58–60.
M.S. Troitsky, 1994. *Planning and Design of Bridge*. New York: John Wiley & Sons, INC.
W. Chen and L. Duan, 2003. *Bridge Engineering*. New York: CRC Press LLC.

Smart Science, Design & Technology — Lam et al. (eds)
© 2020 Taylor & Francis Group, London, ISBN 978-0-367-17867-3

Analysis of the application of lacquer thread sculpture in silk figure craft

Yangsheng Lin*
Qingyuan Colored Silk Store of Zhao'an County, Zhangzhou, Fujian Province, PR China

ABSTRACT: This paper talks about the application of lacquer thread sculpture in colored silk figure, provides a brief analysis on the combination of lacquer thread sculpture in silk figure and the production of various silk figures by "Qingyuan Colored Silk Store," and highlights the uniqueness of silk figures made by "Qingyuan Colored Silk Store" in Zhao'an County. They have clarified the relationship between different applications of lacquer threads in various festive lanterns, helmets, kermis, and other silk figure crafts; learned the strengths of other arts and crafts on the basis of inheritance of the one-thousand-year-old silk figure craft to enrich its content and fill gaps; and explored new techniques to manifest the characteristics of local folk culture and endow it with new culture value and connotation as well as create new aesthetic orientation.

Keywords: lacquer thread sculpture, silk figure, lacquer culture, lacquer material, innovation and development

1 INTRODUCTION

1.1 *Historical and cultural situations of lacquer thread sculpture*

Chinese lacquer is a kind of natural material uniquely possessed in China. Lacquering technology has developed in China for more than seven thousand years, starting from the red lacquer bowl with a wood body in Hemudu Culture to the carved lacquerware and gold-plated lacquerware of the Qing Dynasty; all of these have developed on the cultural basis of Chinese people's living utensils. After the gradual improvement of lacquering technology for several thousands of years, our ancestors have studied many special technological forms and created a splendid history in lacquer culture for the Chinese nation. Lacquer thread sculpture is an art treasure among Chinese lacquering culture, originating in Quanzhou, and belongs to a traditional craft of the Han people in Southern Fujian. Since the flourishing of painted sculptures in the Tang Dynasty, lacquer thread sculpture has been widely used in the decoration of Buddha statues. Lacquer thread sculpture features fine and elegant workmanship, vivid images, simple and solemn styles, true-to-life pictures, and can be reputed as an artistic miracle and a unique wonder in China (Xia and Run, 2006; Tao, 2017).

1.2 *Origin of Zhao'an silk figure and lacquer thread sculpture*

Lacquer thread sculpture derived from the Buddha statue and sculpture arts in the old time was inspired by thread carving techniques in the Song and Yuan Dynasties, especially embossed painting and mud thread carving. It formed in the late Ming Dynasty and early Qing Dynasty, and matured in the late Qing Dynasty. The lacquer thread sculpture crafts of Zhao'an "Qingyuan Colored Silk Store" also developed in that time. The flourishing of local opera culture and religious beliefs in Zhao'an makes the lacquer thread sculpture craft become popular and indispensable. The earliest lacquer thread sculpture is widely used in the carving of Buddha statues. It involves complicated and fine techniques: the Chinese lacquer, lime, and tong oil are mixed and made into mud; the original simple thread wrapping, carving, and engraving are developed into coiling, winding, piling, carving, and hallowing; after drying, it is painted with clear solid lacquer; after the lacquer dries, the gold foil is mounted. A total of more than ten pure manual working procedures are involved, which greatly increases its artistic expression patterns. The unique production techniques of lacquer thread sculpture contribute to its perfect combination of emotions and reason, thus it is widely recognized in the society and forms its own aesthetic pattern. As a folk handicraft of Han nationality, silk figure also has a long history in development and inheritance. According to records, it was ZhiGongshi who firstly advocated the production of silk figures before the Tang Dynasty, and later it gradually became prevalent. Artists from past dynasties integrated exquisite conception into it and simplified its techniques by cutting out superfluous procedures, and could make

*Corresponding author. E-mail: 1033329276@qq.com

silk figures in the shapes of various birds and animals as well as for the representation of dramas and stories. In the Song Dynasty, it developed into its period of great prosperity with more mature techniques. Since the Ming and Qing Dynasties, this craft had been developed rapidly. After the establishment of the People's Republic of China, the creative themes have become richer and the silk figure craft has manifested more vivid expressive force. Today, on the basis of inherited tradition, the silk figure does not confine itself to fixed stylization, but advocates the equal competition of all schools, shows inclusiveness and openness, learns from others' strong points to offset its weakness, and strives for a more perfect status in the new era.

Figure 1. The Mazu helmet.

2 APPLICATION OF LACQUER THREAD SCULPTURE ART IN THE SILK FIGURE CRAFT BY THE ZHAO'AN "QINGYUAN COLORED SILK STORE"

The Zhao'an "Qingyuan Colored Silk Store" was originally established in the Qing Dynasty, and was well-known in Zhangnan for its diverse workmanship and the precise work ethic of its founders. After the reform and opening up, the literature and arts have begun to thrive, and so has the silk figure craft. In the spirit of fine workmanship and the principle of selecting the essence and discarding the gross, Qingyuan store has integrated a wide range of traditional crafts and techniques into the production of silk figures. The application of lacquer thread sculpture art is one example. The successors to "Qingyuan Colored Silk Store" insist on the spirit of great country craftsman, practice rigorous requirements, keep improving, and have carried out a series of explorations on the basis of inheriting the excellent crafts of predecessors. The procedure of silk figure under application of lacquer thread sculpture craft is generally described as follows:

2.1 Application of lacquer thread sculpture in the helmets of various silk figures of immortals

For example, for the production of the helmet of Mazu silk figure, it is necessary firstly to select the paper board, plastic weaving, silk, and rattan paper as raw materials, and boil pastern for their mixing into the adhesive paper board, which can be used after being plaited for several days. Design the form of the helmet, then cut and hollow it. Use the rattan paper and iron wire well prepared to make the edge framework, and then apply the lacquer thread on the original line-drawing pattern under more than ten working procedures including carving, piling, wrapping, coiling, and so on. Then, finish the work with painting and the posting of gold foil. Assemble other components, such as the pompons, beads, and gold and silver threads, to complete the helmet and make it elegant, dignified, and magnificent (Figure 1).

2.2 Application of lacquer thread sculpture in traditional festive lanterns

In the production of a festive lantern, firstly, select and process the bamboo strip, bundle the rattan paper with aluminum wire, and complete the shape of the festive lantern. Then, stick silk fabrics and damask silk, etc., draw the pattern, and stick gold and silver threads. Later, carry out the line shaping with lacquer thread sculpture techniques, decorate it with calligraphy and traditional Chinese painting, and combine it with paper cutting, embroidery, beads, tassels, and lighting to enhance the stereoscopic impression of the work, highlight its effects, and increase its artistic appeal (Figure 2).

2.3 Comprehensive application of lacquer thread sculpture in various silk figures of kermis

The kermis has been kept relatively intact in Southern Fujian. Various kermises need the application of a large quantity of silk figures to enrich festival activities and increase their value in appreciation and artistry as well. Jiazhou Village of Qiaodong Town in Zhao'an holds one grand safety-praying ceremony every nine years, and the century-old brand Qingyuan store prepares various silk figures for the ceremony.

Figure 2. Production of traditional silk figure of festive lantern.

Figure 3. Avalokitesvara (silk figure as bottom and lacquer thread sculpture as decoration).

For example, consider the production of the four-meter-high silk figure of Avalokitesvara. It requires the application of lacquer thread sculpture technique; after the model is painted, lacquer thread production can be carried out on its clothes. It also needs silk figures of various characters in ancient costume dramas, such as Wangzai Peng and God of Doors, as well as silk figures of water lamp bowl, dragon, tiger, lion, and elephant, etc., by which the craftsmen can impress people with their extraordinary skills. The application of new techniques on the basis of inheriting tradition can increase the sense of beauty of works and make them resplendent, magnificent, and dazzling, so as to evoke people's reflection on the past and imagination of the future (Figure 3).

3 VALUE ORIENTATION OF SILK FIGURE CULTURE

In our current society of highly developed information, the silk figure craft does not withdraw from the historical arena as time goes by. The reason lies in the fact that artists in the past and nowadays have been persistent in inheritance and have kept making improvements in their creations. The constant supplying of new forces in the development process will keep an art everlasting. It is necessary for silk figure culture to recover its extraordinary splendor, not only for festivals, but also as a kind of inheritance and development of traditional folk culture. There is persuasive evidence to certify the integration of

lacquer thread sculpture craft into silk figure. On the principle of inheriting and carrying out traditional crafts, "Qingyuan Colored Silk Store" will continue to explore new creative modes and their inherent artistic value on deeper levels.

4 CONCLUSIONS

Talking about traditional national culture often evokes my native complex and makes me have a lot to say. As the inheritors and practitioners of traditional national culture, we are more obliged to treat each work with the spirit of a great craftsman. Even though careful and precise carving takes a lot of energy and time, we should insist on our pure faith. Silk figure art holds a special charm in our current society, for it can innovate and develop traditional things more systematically. The integration of lacquer thread craft injects new vital force for the traditional silk figure. Nowadays, the two supplement each other and give a concentrated reflection of the new aesthetic values. As Chinese people, we need to keep this complex and make a better inheritance of it so as to avoid its vanishing from history. This might be the idea advocated by Qingyuan predecessors in past times who innovated based on tradition and integrated development with inheritance! In a society featuring a bustling commercial economy, it provides a harbor for people to rest their spirits, which indicates the functions and significance of traditional culture in daily life, and also can inspire people of later generations to stay true to their mission and never forget where their happiness comes from. It is a requirement of the era to make efforts on innovation, inheritance, and development, to pursue truth, goodness, and beauty, and to manifest national characteristics. Young successors are expected to continue to make efforts and greatly contribute to the industrial development of silk figure integrated with lacquer thread sculpture.

REFERENCES

B. Xia and J. Run, 2006. *Agricultural Archaeology 06*(6), 206–208.
Y. Tao, 2017. *Identification and Appreciation to Cultural Relics 17*(3), 79–85.

Smart Science, Design & Technology — Lam et al. (eds)
© 2020 Taylor & Francis Group, London, ISBN 978-0-367-17867-3

A wearable 3D printed elbow exoskeleton to improve upper limb rehabilitation in stroke patients

Weite Tsai*
Department of Industrial Design, National Cheng Kung University, Tainan, Taiwan

Yusheng Yang
Department of Occupational Therapy, Kaohsiung Medical University

Chien-Hsu Chen
Department of Industrial Design, National Cheng Kung University, Tainan, Taiwan

ABSTRACT: Upper limb functional decline is a common result of neurological injury following stroke. Active recovery includes exercises to improve motor function. Machine-assisted systems have been widely used in stroke rehabilitation, and provide consistent, intensive repetitive upper limb motion exercises. However, the current exoskeletons are expensive and unsuitable for home use. We propose the use of 3D printing technology to develop a light-weight (~5 kg) wearable elbow exoskeleton designed with upper limb support, multi-axis hinge joints, and miniature electric push rods.

Keywords: 3D printing, robotic arm, rehabilitation and exoskeleton.

1   INTRODUCTION

Modern changes in the dietary habits of individuals contribute to a higher chance of acute stroke. However, medical advancements have greatly improved survival rates. A stroke refers to various clinical manifestations caused by abnormalities in the cerebral blood vessels or brain circulation, resulting in damage to brain tissue. Cerebrovascular diseases leading to stroke include occlusion and rupture. Both cause brain damage, which presents clinically as a sudden onset of neurological deficit. Hemiplegia is the most common form of stroke, causing the patient to become comatose, confused, and unable to move their limbs during the acute phase. After a few weeks of recovery, one side of the body may be non-responsive or, at least, weak. This is mainly due to the partial necrosis of brain cells caused by stroke, which abnormally inactivates or excites particular motor neurons, leading to muscle tension, paralysis or weakness. The symptoms of hemiplegia can be broadly classified into two groups: (1) negative symptoms, including flaccid limbs, which the patient cannot control during movement [4]; and (2) positive symptoms, including hypertonia, which is also called spasticity, and makes bending difficult [7]. Regardless of the presenting symptoms, hemiplegia will cause difficulty in limb movement execution and an inability to

complete desired cognitive activities, which affect daily routine and social functions of the individual. Among the problems associated with hemiplegia, motor dysfunction of the upper limbs and hands has a substantial effect on the daily life of patients [14].

In daily life, it is often necessary to rely on the dexterous operation of the hand in cooperation with the wrist and elbow to achieve a coordinated and smooth movement. Therefore, disability of the upper limbs may have a greater influence on functions of daily life than that of the lower limbs; unfortunately, the recovery of upper extremity motor function is often slower and the degree of recovery is slightly worse. Brunnstrom observed the recovery process of stroke patients and classified their movement into six stages [3]. In the first stage, no action occurs, muscle tension is very low, and the body is flaccid. In the second stage, spasticity begins to appear and the muscles show synergistic movement. In the third stage, voluntary movement is achieved, but follows a strong synergistic pattern with high spasticity of the limbs. In the fourth stage, the patient is able to break the pattern of synergy and perform isolated movements, and muscle tension begins to decline. In the fifth stage, muscle tension gradually approaches normal, and isolated movements are more mature. In the sixth stage, movement is fast or the height is alternately coordinated [3]. These six

*Corresponding author: E-mail: p38021067@ncku.edu.tw

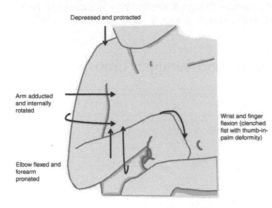

Figure 1. Elbow flexion coordination posture [19].

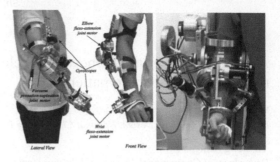

Figure 2. Wearable exoskeleton robotic arm drive system configuration located on the outside of the upper arm (left) [16] or on both sides of the upper arm (right) [18].

classifications are commonly used in clinical practice. Brunnstrom suggested that several muscles around a joint form a unit in hemiplegia patients; when a muscle contracts, other muscles of the same joint unit will contract simultaneously. These movements are very primitive and rigid, and cannot be controlled consciously. The basic limb synergies of hemiplegia patients have fixed modules, which can be divided as the upper and lower limbs, and as extensor and flexor synergy. Flexion is more common in the upper limbs (e.g., "Figure 1"), accounting for about 80%–90% of total synergy. Elbow flexion is the most powerful action, and leads to tall shoulder blades, external rotation of the shoulder, and forearm rotation [3]. It is used clinically to induce the appearance of other components of this action. Accordingly, elbow movement is important for the recovery of upper limb function in stroke patients. Joint movement can be controlled by bending and straightening movements of the elbows, thereby improving the upper limb function.

2 RELATED WORK

Machine-assisted arms are expensive and unaffordable for the average person. They are currently only used in large-scale teaching hospitals and are limited in number. Only a small number of patients receive such ma-chine-assisted treatment at specific times. Recent studies have indicated that home-based machine-assisted treatment is viable if a significant difference is found compared to standard treatment [5]. Allowing patients to carry out rehabilitation activities anytime and anywhere in the home increases treatment intensity, and may improve the treatment effect and their motivation. Accordingly, a wearable machine-assisted exoskeleton arm has more therapeutic significance and potential than current machine-assisted arms; so many research teams have invested significant resources to develop such wearable devices. A review of publications reporting wearable exoskeleton robots identified broadly two types, defined as those in which (1) both the support skeleton and the power-drive system are

placed on the outer side of the upper arm [11] (Figure 2, left panel) or (2) the support skeleton and power-drive system are placed on both sides of the upper arm [16] (Figure 2, right panel). Both designs have the disadvantage of using a single shaft to set the axis of motion of the elbow joint. According to Perry et al., the concept of exoskeleton design is defined as "an external support mechanism with joint functions and conforms to the corresponding human movements" [17]. Therefore, in terms of the elbow anatomy, an exoskeleton consists of the distal humerus and the proximal humerus and ulna, including the ulnar, ankle, and proximal joints, which are posterior and can be flexed. Stretching exercises also contribute to the pronation and supination of the forearm. The distal humerus and proximal ulna form the ulnar hinge joint, the main part of the elbow [1]. However, human arthrokinematics are not simply comprised of a single action; the elbow joint includes rolling, gliding, and spinning motions [15]. When the elbow is bent and stretched, the axis of rotation in the elbow joint changes with the angle of the limb. Therefore, the simplified single hinge joint design cannot be adapted to fit the human elbow joint movement pattern. These exoskeleton robotic arms need to be redesigned to incorporate a polycentric hinge joint (Figure 3), enabling an exoskeleton pivot

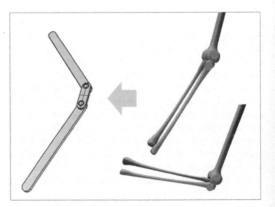

Figure 3. Elbow multiaxial flexion joint design.

hub that is more consistent with the normal elbow joint trajectory.

3 DESIGN SYSTEM

The wearable 3D printed elbow exoskeleton described in this study requires two main functions: (1) it can assist a stroke patient with upper limb weakness to bend the arm; and (2) it can help straighten the arm to reduce muscle tone when the upper extremity is spastic. Therefore, the elbow exoskeleton bracket needs a power-drive system that can pull the arm to bend and straighten the joint. This goal was achieved by a smooth move. The following is a description of each design element:

A. Hinge joint design

A double-rotor design was used at the elbow joint. The upper arm and the forearm each have a pivot axis (Figure 4). When the two segments are straightened or bent, they use one another's limbs. The axis is positioned so that there is no pulling or pressing during movement (Figure 5). Additionally, there is a combined axis at the junction of the upper arm and the forearm with the electric push rod. We presuppose that the shaft comprises part of the rotating space, allowing the upper arm and forearm support brackets to be partially raised and lowered. When the microelectric push rod is driven to bend and straighten, the whole elbow is affected, the movement is smoother and the normal elbow joint trajectory is better, consistent with the double-shaft design of the elbow.

B. 3D Printing

This study used the UP Box+ 3D Printer to produce the exoskeleton parts (Figure 6). In addition to its convenient installation, simple operation and high printing precision, key features of this model include an enlarged printing size to 255 × 205 × 205 mm and a reduced width of each layer to 0.1 mm; these sizes are aligned with the specifications required for this study. Acrylonitrile butadiene styrene (ABS) will be used as the printing material because it is stronger than polylactic acid. When subjected to high pressure, ABS maintains its bending elasticity, is resistant to immediate breakage and demonstrates good mechanical properties such as toughness, hardness, and rigid phase balance. According to the proposed size of the exoskeleton described in this study, printing requires approximately 16–20 h to complete. Table 1 summarizes the design considerations described above.

Table 1. Design specifications of the proposed exoskeleton arm.

Item	Specification	Item	Specification
Upper arm support bracket length	24 cm	Electric push rod stroke	20 mm
Upper arm support bracket width	10 cm	Maximum bending angle	140°
Forearm support bracket length	16 cm	Maximum straightening angle	0°
Forearm support bracket width	10 cm	Total net weight	5 kg

Figure 4. Wearable 3D printed elbow exoskeleton design.

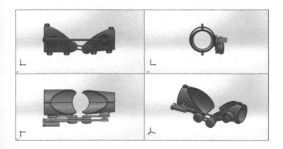

Figure 5. 3D model diagram of four perspectives.

Figure 6. 3D printing machine. Source: https://www.idea-diy.com/products/3d-printer/.

4 CONCLUSIONS AND FUTURE WORK

3D printing has played a key role in the next-wave industrial revolution, known as Industry 4.0; however, the research and development process, software integration and application industry strategy cannot be completed simultaneously. The use of 3D printing technology in this project is aligned with the government's industrial policy of promoting high value and employment. If the research results are successful, the finished product could replace the more expensive auxiliary machine-assisted exoskeleton arms. Through low-cost 3D printing technology and wearable design, the elbow exoskeleton could be manufactured, repaired or modified to suit the anatomy of the patient for their use at home. The stent can be used to perform upper limb rehabilitation activities at home, effectively improving the function of the upper limb on the affected side and reducing the inconvenience on daily life. Additionally, if the exoskeleton is found to be efficacious, it could also be provided to patients with upper limb dysfunction resulting from spinal cord injury. The controllable component can be used to manipulate elbow flexion and extension, enabling patients to gain cooperation of the affected limb using their other hand. This application enables autonomous performance of upper extremity movements, which could provide substantial assistance to patients with spinal cord injury who have complete loss of their upper limb function.

REFERENCES

J.G. Alcid, C.S. Ahmad and T.Q. Lee, 2004. *Clin Sports Med 23* 503–517.

L. Bishop and J. Stein, 2013. *Neuro Rehabilitation 33* 3–11.

S. Brunnstrom, 1970. *Movement therapy in hemiplegia: a neurophysiological approach*, 1st ed. New York: Medical Dept.

J.G. Colebatch and S.C. Gandevia, 1989. *Brain 112* 749–763.

A.V. Dowling et al., 2014. *IEEE J Transl Eng Health Med. 27*, 2:2100310.

J.J. Gerhardt et al., The practical guide to range of motion assessment, 1st ed., Chicago, Ill American Medical Association.

J.M. Gracies, 2005. *Muscle Nerve 31* 552–571.

S. Hesse, et al., 2003. *Arch Phys Med Rehabil 84* 915–920.

L.E Kahn, et al., 2006. *J Neuroeng Rehabil 21* 3–12.

G. Kwakkel, B.L. Kollen and H.I. Krebs, 2008. *Neurorehabil Neural Repair 22* 111–121.

R. Looned, et al., J Neuroeng Rehabil, 7, 11:51, 2014.

C.J. Mottram, et al., 2009. *J Neurophysiol 102* 2026–2038.

I.A. Murray and G.R. Johnson, 2004. *Clin. Biomech 19* 586–594.

H. Nakayama, et al., 1994. *Arch Phys Med Rehabil 75* 394–398.

C.A. Oatis, 2009. *Kinesiology: the mechanics and pathomechanics of human movement*, 2nd ed. Baltimore: Lippincott Williams & Wilkins.

A.U Pehlivan, O. Celik and M.K. O'Malley, 2011. *IEEE Int. Conf. Rehabil. Robot* 597–602.

J.C. Perry, J. Rosen and S. Burns, 2007. *IEEE/ASME Trans-actions on Mechatronics 12* 408–417.

J.L. Pons, et al., 25 Upper-Limb Robotic Rehabilitation Exoskeleton: Tremor Suppression.

R. Teasell and R. Viana, 2014. *Textbook of Neural Repair and Rehabilitation*, M. Selzer, S. Clarke, L. Cohen, G. Kwakkel, & R. Miller, Eds., Cambridge: Cambridge University Press 601–614.

Innovation design & creative design

Smart Science, Design & Technology — Lam et al. (eds)

An acupuncture VR game

Chienshun Lo*

Department of Multimedia Design, National Formosa University, YunLin, Taiwan

ABSTRACT: In this study, an acupuncture virtual reality game with VIVE was introduced to train and test trainee acupuncturists. Eight videos were made to introduce the techniques of meridian massage for the relief of common symptoms. Trainees watch these videos to learn the appropriate meridian massage to relieve different symptoms. The VIVE VR game is then played by the trainees to practice and test what they have learned. The experimental results show that the proposed system can improve memory of the map of acupoints of diseases and the positions of acupoints using a method that is fun and encourages learning.

1 INTRODUCTION

Recent studies have shown that acupuncture and meridian massage are popular preventative therapies and useful for the provision of pain relief (Lee, et al. 2018). During the process of training the mapping of acupoints to their corresponding symptoms, however, a significant challenge for trainees is to memorize which acupoints are related to which symptoms and the location of those acupoints on the body. In this study, an acupuncture virtual reality game with VIVE was introduced to train and test trainee acupuncturists. There are two phases in the running of this game. The first phase involves watching videos to learn how to massage to relieve the symptoms. The second phase uses the VIVE helmet to run the VR game. The VIVE system was made by HTC. In the first phase, eight videos were made to introduce how to carry out meridian massage to relieve common symptoms referenced in (Deadman, et al. 2007). These symptoms include headache, eye fatigue, nasal congestion, tinnitus, toothache, cough, frozen shoulder, and stiff neck. Trainees watch these videos to learn the appropriate meridian massage to relieve the symptoms. The video frame configure of a 3D model shows the name and position of the acupoints, an animation illustrates the symptom, a paragraph explains how to find the position of the acupoints, and a film demonstrates how to find and perform the massage for this acupoint. The configuration is shown in Figure 1. In Phase 2, the VIVE VR game is played by the trainees to practice and test the information learned in Phase 1. The eight symptoms were shown as occurring on the virtual human body in the game. The player was given 20 seconds to touch the relevant acupoint to relieve the symptoms. The player was only able to make up to 10 incorrect choices of acupoints. Eight symptoms involved 29 different acupoints. There were 8 grade levels awarded to measure the achievement of the player. The top grade indicated that the player chose the correct 29 acupoints in time and made incorrect choices less than 10 times. The experimental results showed that the proposed system was able to improve memory of the map of acupoints of diseases and the positions of acupoints using a method that is fun and encourages learning.

Figure 1. The configurations of the video frame.

2 STORYBOARD OF THE VR GAME

The second phase involved playing the VR game for the purpose of practice and testing whether players have understood the relationship between the acupoints and the symptoms. The storyboard of the game is illustrated in the following steps.

---

*Corresponding author: E-mail: cslo@nfu.edu.tw

Figure 2. The first frame of the game to start or exit.

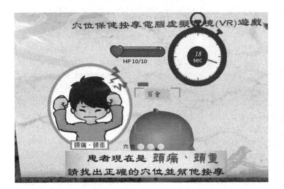

Figure 3. The second frame shows the first symptom, a headache, and allows the user to find the relevant acupoint for relief.

Storyboard 1. Start the game
As shown in Figure 2 there are two buttons: one is to start the game, and the other is to exit the game. The choice is made by looking at the button for 3 seconds.

Storyboard 2. Relief of the symptom
When the game starts, the first symptom is shown in the animation. It appears as shown in Figure 3. The user may make up to 10 incorrect choices of acupoints. The number of chances remaining is shown in the top middle of the screen. The user has 20 seconds to choose an acupoint. The center of the screen is the aim box. The user can look at the acupoint for 3 seconds to choose the acupoint. When the aim box appears under the acupoint, it shows the name of the acupoint to let the user judge whether it is the correct choice.

Storyboard 3. Choose the correct acupoint
In this sample case of a headache, there are three acupoints that are related to relief of the headache. They are Baihui, Jiaosun, and Quchi. It is not necessary to choose these three acupoints in sequence. Figure 4 shows that if the user chooses Baihui correctly, the first red circle of three radian circle buttons will be enabled. The timer will also reset to 20 seconds.

Figure 4. The third frame shows the user choosing the correct acupoint to obtain one point.

Figure 5. The fourth frame shows the animation of the headache symptom being reduced when a correction rate of over 50% is obtained (two correct acupoints of three achieved).

Storyboard 4. Relief of the symptom animation
When the user chooses over 50% of the correct acupoints, the animation changes to illustrate easing of the symptom. Figure 5 illustrates that the headache has been relieved to show less severe symptoms.

Storyboard 5. Health Point reduction
The user can only make up to 10 incorrect answers. Each time the decision must be made within 20 seconds. So, the user's life blood represents 10 full Health Points (HP). In Figure 6, the user chooses the wrong acupoint, so that user's remaining HPs change from 10 to 9, shown in the top of game screen as a green line. It should be noted that HPs are reduced when the time runs out. In other words, if the user does not give a choice within 20 seconds, it also results in HP reduction.

Storyboard 6. A completed 3D animation
In the case of a headache, when three correct acupoints have been choosen, a 3D animation is

238

Figure 6. The sixth frame illustrates when the user chooses the wrong acupoint.

Figure 7. The fifth frame shows how, when finished, all the correct acupoints for this symptom are lit up on a 3D model animation.

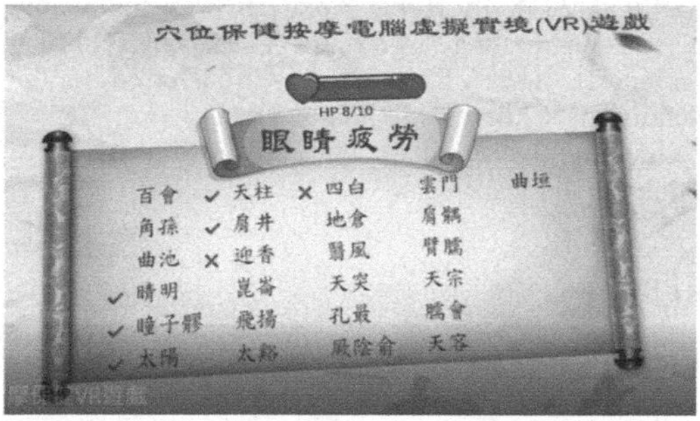

Figure 8. The sixth frame shows the correct acupoints and wrong acupoints that the user chose.

used to light the correct acupoints and help the user to understand where on the body they are. This is shown in Figure 7.

Storyboard 7. The massage record

For the relief of eye fatigue, it is necessary to massage 5 acupoints. Figure 8 shows that the user massaged 5 correct acupoints and 2 incorrect

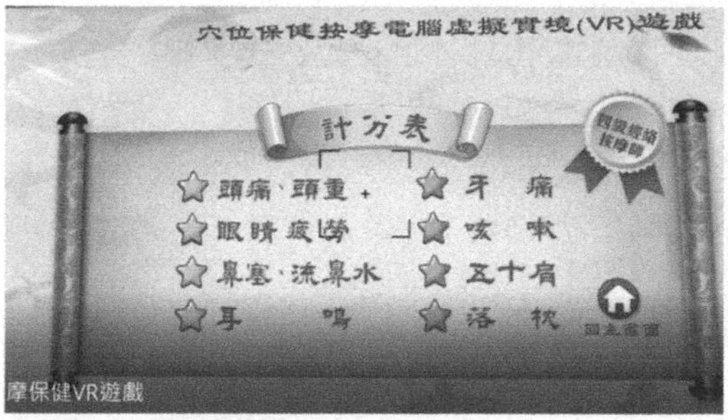

Figure 9. The seventh frame shows the achievement of this round to pass the symptoms and the grade achieved.

acupoints. When a symptom test is complete, the massage record of this symptom is shown before the next symptom is started.

Storyboard 8. The achievement

At the conclusion of a round, the achievement shows the passed symptoms and the still-to-be-attempted symptoms. There are 8 grades corresponding to how many symptoms have been passed. In Figure 9, the list of 8 symptoms is shown in the center of the screen. The yellow star sign indicates the symptom has been passed. In this case, the user passed 4 symptoms and achieves a Grade 4 medal as shown in the top-right of the screen. In the bottom-right of the screen is a button to return to the home screen shown in Figure 2.

3 KEY TECHNIQUES

A. Look at to choose

VIVE provides two controllers to improve the user interaction with the VR system shown in Figure 10. In this VR game, however, using controllers to choose the acupoint is more difficult, especially with time limited to only 20 seconds, so the design of this

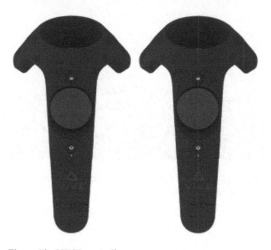

Figure 10. VIVE controllers.

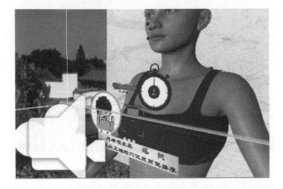

Figure 11. The game UI attached on the front of the camera.

VR game uses only the VIVE helmet to choose the acupoint. An aim box is shown in the center of the screen. This aim box is attached to the front of the virtual camera of the VR system shown in Figure 11. The system produces a virtual laser line from this aim box along the direction the user is looking. When the laser line touches any 3D object in the scene, it will bring up the name of the touched 3D object as shown in Figure 3. The aim box has a red top bar. This is a progress bar which counts 3 seconds when the aim box is located on an acupoint. If the aim box is located on an acupoint for over 3 seconds, the user confirms the decision and then the system judges whether the selection is correct. If the answer is correct, then the "correct" message is shown as in Figure 4. Otherwise, the "wrong" message is shown as in Figure 6.

B. Spherical panorama by aerial photography

The VR system environment is applied using a new technique named spherical panorama. It is shot using aerial photography. Several pictures are taken to build a spherical panorama. The picture shown in Figure 12 was taken at Chiayi Performing Arts Center in Taiwan. For fidelity of

Figure 12. The spherical panorama captured from Chiayi performing arts center in Taiwan.

Figure 13. A panorama is applied in a cylinder.

vision, a cylinder is applied to form the whole picture as shown in Figure 13. A cylinder has more fidelity than a sphere for control of the depth of vision.

4 CONCLUSION

In this study, an acupuncture VR game is proposed to learn, practice, and test use of meridian massage points. It helps the user to learn how to relieve various symptoms by massage in their daily life. This system was demonstrated at an exhibition in Inno-Carnival 2018 in Hong Kong. It was favourably received, attracting large numbers of people to play the game. The VR game was transferred to the School of Chinese Medicine, Hong Kong Baptist University, where it is currently being used for teaching and learning.

REFERENCES

T.L. Lee and B.L. Marx, 2018. Noninvasive Multimodality Approach to Treating Plantar Fasciitis: A Case Study, *Journal of Acupuncture and Meridian Studies*. 11 (4),162–164.

P. Deadman, M. Al-Khafaji, K. Baker, 2007. A Manual of Acupuncture, 2nd ed., Hove, East Sussex, England; Vista, CA, USA: *Journal of Chinese Medicine Publications*.

Smart Science, Design & Technology — Lam et al. (eds)
© 2020 Taylor & Francis Group, London, ISBN 978-0-367-17867-3

Analysis of mutation trend of influenza A viruses using a computational evolution simulator

Xinting Chung & Pujen Su

Institute of Biomedical Engineering, National Chiao Tung University, Hsinchu, Taiwan

Yuhjyh Hu*

Institute of Biomedical Engineering, National Chiao Tung University, Hsinchu, Taiwan
College of Computer Science, National Chiao Tung University, Hsinchu, Taiwan

ABSTRACT: According to Darwinism, all species of organisms arise and develop through natural selection of inherited variations that are more able to compete, survive, and reproduce so as to adapt better to the change of environments. We propose an evolution simulator to analyze the evolution of influenza A viruses. From the simulated trajectories, we are able to observe and analyze every single hypothetical mutational step in evolution, which cannot be achieved easily from a phylogenetic tree. We expect the evolution simulator to provide insights into the adaptive variability of viral proteins, and present the potential of predicting prospective adaptation trends that warrant further investigation by biologists and doctors.

1 INTRODUCTION

Most phenotypic evolution within species as well as most morphological, physiological, and behavioral differences between species can be explained by adaptation due to natural selection [1–2]. Given a change in environment, the task of adaption for a species is to move the population from its current state toward the new phenotypical optimum state. This study analyzes the evolution of influenza A viruses (IAV) by applying an evolution simulator. The simulator generates a hypothetical evolutionary trajectory from an initial IAV to a set of prespecified target IAVs. From the trajectory, we can observe and analyze every possible single mutational step in evolution, which cannot be done easily based on a phylogenetic tree. Because the task of adaptation is complicated by random mutations, pleiotropy, and population genetics, we adapted the sequence-based models [3] to develop a simulator that considers adaptation in the discrete state space of amino acid sequences to avoid a long extended adaptive walk toward the optimum state when the model space is continuous as in Fisher's [1].

2 METHODS

Given the initial wild-type IAV, and a set of prespecified target IAVs, the goal of the evolution simulator is to simulate a hypothetical evolutionary path from the starting IAV to a target IAV. We consider the finding of evolutionary paths as a stochastic search process that is defined by its state space, initial state, target state, operators, and goal test as follows.

2.1 State space

The simulation proceeds in a discrete state space represented by amino acid sequences. Given a wildtype IAV sequence of L amino acids, the simulator generates a possible evolutionary path that records the continual changes in amino acids from the wild type to a prespecified target IAV sequence. The entire state space is the set of all possible amino acid sequences of length L. It has a size of 20^L, and is denoted by $\{m_{ab}^k\}$, where a is the initial wild type, b is the target, and k marks a particular time point in evolution.

2.2 Initial state and target state

The initial state, denoted by m_{ab}^0 such that $m_{ab}^0 = a \in \{m_{ab}^k\}$, is the amino acid sequence of a prespecified wild-type IAV at time 0. It is the starting point for the stochastic search process. The target state $m_{ab}^{final} = b \in \{m_{ab}^k\}$ is the amino acid sequence of a target IAV. It is a final state of the stochastic search process. The states $\{m_{ab}^0, m_{ab}^1, \cdots, m_{ab}^{t-1}, m_{ab}^t, \cdots, m_{ab}^{final}\}$ on a path from the initial wild type, m_{ab}^0, to a target IAV, m_{ab}^{final}, i.e.,

---

*Corresponding author. E-mail: yhu@cs.nctu.edu.tw

$$m_{ab}^0 \to m_{ab}^1 \to \cdots \to m_{ab}^{t-1} \to m_{ab}^t \to \cdots \to m_{ab}^{final},$$

constitute an evolutionary path. We can observe the mutations on the amino acid sequences by traversing the evolutionary trajectory to analyze the dynamics in evolution.

2.3 Operator

When the fitness of a current wild type decreases and becomes unfit due to a change of environment, it may adapt to the environmental change by going through a series of random mutations until a fitter mutant is available to replace the current unfit wild type. For a search problem, the purpose of an operator is to modify the current state to a new state that is closer to the goal. In our simulator, the operator modifies the current wild type to a fitter mutant that is closer to a target IAV in terms of the sequence similarity.

Assuming mutation is very rare [4–5], the operator applies a one-point mutation to the current wild-type sequence to create a pool of candidate mutants. The pool size is $19L$, where L is the length of the wild-type sequence, because a one-point mutant is different from the wild type by only one amino acid. To simulate natural selection, from the pool of candidates the operator selects a mutant that is fitter than the current wild type for substitution. The substitution of a fitter mutant for the current wild type constitutes a single operational step in the stochastic search process that corresponds to a one-step mutation in evolution. The substitution operation continues until the fitness of the current fixed sequence can no longer be improved, at which point the adaptation of the initial wild type has arrived at an optimum state, that is, the initial wild type has evolved to a stabilized mutant.

If this final mutant is a target IAV, the simulator has reached the global optimum state, and the simulation of evolution is complete; otherwise, to leave the local optimum state, the simulator initiates another stochastic search from this final mutant that is treated as a new initial wild type, and then repeats the same process until it reaches another optimum state. The mutants recorded in a stochastic search process starting from an initial wild type to an optimum state pave an adaptive walk. We call an adaptive walk a partial evolutionary path if it ends at a local optimum state. By connecting all the partial evolutionary paths between the first initial wild type and the target IAV, we construct a complete evolutionary path, as illustrated by Figure 1.

The operator selects the next mutant that is fitter than the current fixed sequence for substitution subject to two aspects: (a) the attraction from the targets, and (b) the influences of different constraints on the mutants. The target attraction and the constraint influences direct the operator to search different portions of the state space for the next fitter state. By confining the

Figure 1. A sample complete evolutionary path consisting of three partial paths. The evolution starts with an initial wild type sequence w_0, and goes through a series of one-point mutations until it reaches a local optimum state (e.g., w_1 or w_2) in which the fitness of the current mutant is higher than all its one-point mutants. If the current mutant is not a target sequence, it becomes a new wild type, i.e., w_1 and w_2, and the evolutionary process continues until the final converged mutant is one of the target sequences, e.g., t ($t \in T$, where T is a prespecified target sequence set).

search space, not only can we reduce the search time before we reach a target, but we can also increase the stability of the simulation results. We considered two types of constraints in this study: temporal constraint and geographic constraint. Simulation regulated by various factors can shed some light on the study of evolution from different perspectives.

2.3.1 Target attraction

Because our goal of simulation is to produce an evolutionary path from a wild type to a target IAV through a series of amino acid mutations, we measure the target attraction for a mutant by the sequence similarity between the target and the mutant. We score the similarity Seq_{sim} between two sequences by the sum of the substitution scores based on a substitution matrix, e.g., BLOSUM 62. A higher sequence similarity between a candidate mutant and a target suggests the candidate mutant is more likely to become the target IAV after subsequent mutations. We define the attraction from a set of IAV targets, T, for a mutant m_{ab}^t as

$$Attraction(T, m_{ab}^t) = argmax_{s \in T} Seq_{sim}(s, r_{ab}^t),$$

where Seq_{sim} is the sequence similarity, and r_{ab}^t is the real IAV that best matches m_{ab}^t in sequence similarity such that

$$r_{ab}^t = argmax_{r_{ab}^k \in Real\ IAV} Seq_{sim}(m_{ab}^t, r_{ab}^k),$$

When the target attraction is considered in the stochastic simulation process, the operator selects a candidate that is more similar to a target in sequence to drive the mutations toward the targets more rapidly.

2.3.2 Temporal constraint

The temporal constraint enforces a temporal order of the IAV mutants on an evolutionary path such that the real IAV sequences best matched by the mutants follow an actual temporal order in evolution. With the temporal constraint, we mitigate the deviation of the simulated evolution from the correct time line. For a mutant m_{ab}^t at time t, we define

$$r_{ab}^t = argmax_{r_{ab}^k \in Real\ IAV} Seq_{sim}\left(m_{ab}^t, r_{ab}^k\right).$$

We denote the year of the discovery of r_{ab}^t by $Yr\left(r_{ab}^t\right)$. Given a prespecified time threshold θ, a mutant m_{ab}^{t+1} at time $t+1$ satisfies the temporal constraint if $Yr\left(r_{ab}^0\right) < Yr\left(r_{ab}^1\right) < \cdots < Yr\left(r_{ab}^{t-1}\right) < Yr\left(r_{ab}^t\right)$ and $Yr\left(r_{ab}^{t+1}\right) - Yr\left(r_{ab}^t\right) \leq \theta$, where r_{ab}^0, $r_{ab}^1, \cdots, r_{ab}^{t-1}, r_{ab}^t$ are the real IAV sequences best matched by the mutants $m_{ab}^0, m_{ab}^1, \cdots, m_{ab}^{t-1}, m_{ab}^t$ on the current simulated evolutionary path from time 0 to time t.

2.3.3 Geographic constraint
The geographic constraint is employed to contain the mutation within a geographic area such that the most similar real IAVs to the current wild type, and to its mutant, respectively, both have the same city name in their sequence identifiers, or they were discovered in the cities within a geographic proximity. Let m_{ab}^{t+1} be a mutant of m_{ab}^t, and let $GeoProx\left(m_{ab}^i, m_{ab}^j\right)$ denote a geographic proximity measure between r_{ab}^i and r_{ab}^j, where

$$r_{ab}^i = argmax_{r_{ab}^k \in Real\ IAV} Seq_{sim}\left(m_{ab}^i, r_{ab}^k\right),\ and$$

$$r_{ab}^j = argmax_{r_{ab}^k \in Real\ IAV} Seq_{sim}\left(m_{ab}^j, r_{ab}^k\right).$$

If $GeoProx\left(m_{ab}^t, m_{ab}^{t+1}\right) \leq \rho$ (ρ is a geographic proximity threshold), m_{ab}^{t+1} is claimed to satisfy the geographic constraint. We designed the geographic constraint-based operator to reflect the real-world evolution in which a new mutant of a virus is usually similar to those viruses found in the same geographic area, or in the proximity. We exert the geographic constraint on simulation not only to characterize the geographic property in real-world evolution by preventing the mutants from jumping between geographically remote regions, but also to produce a feasible geographic trajectory of the evolutionary path.

2.3.4 Fitness assignment in evolution
Adaptation of DNA due to environmental changes proceeds in two steps: (1) alleles with different fitness arise by mutation, and (2) alleles fitter than the current wild type increase in frequency by natural selection [5]. Several studies of the distribution of fitness effects among the beneficial mutations, using experimental microbial populations, have been conducted. However, the problems caused by stochastic loss, clonal interference, or low sensitivity of detecting beneficial mutations of small effects

have greatly limited the applicability of this type of empirical method [6–7]. By contrast, the population genetic theory has provided an alternative approach to characterizing the distribution of beneficial effects among new mutations. Particularly, the pioneering works of Gillespie [5,8] has attracted significant attention from various research communities to the study of evolution and genetics. Several works and theories have been extended and generalized [9,10].

For a one-point mutant m_{ab}^{t+1} of the current fixed m_{ab}^t, we followed [5,8] and defined its selection score s_{ab}^{t+1} by the influences of the target set, and m_{ab}^t's k-nearest neighbors. We denote the k-nearest neighbors of m_{ab}^t by $N_{k-nearest}\left(m_{ab}^t\right)$, and define it as

$$N_{k-nearest}\left(m_{ab}^t\right)$$
$$= \left\{ m_{ab}^u \middle| \begin{array}{l} Yr\left(r_{ab}^u\right) - Yr\left(r_{ab}^t\right) \leq \theta\ and\ GeoProx\left(m_{ab}^t, m_{ab}^u\right) \leq \rho \\ and\ Seq_{sim}\left(m_{ab}^u, m_{ab}^t\right) \leq Seq_{sim}\left(m_{kth-nearest}, m_{ab}^t\right) \end{array} \right\},$$

where $m_{kth-nearest}$ is the k-th nearest neighbor of m_{ab}^t according to the sequence similarity. A k-nearest neighbor of m_{ab}^t must not only meet the sequence similarity criterion, but also satisfy the prespecified temporal and geographic constraints. This neighborhood reflects the properties of the present environment for m_{ab}^t, and consequently affects the selection scores of its mutants. Similar to the target attraction, we measure the influence of the k-nearest neighbors of m_{ab}^t by the average sequence similarity, and define it as

$$Influence\left(N_{k-nearest}\left(m_{ab}^t\right)\right) = \frac{1}{k} \cdot$$
$$\sum_{n \in N_{k-nearest}\left(m_{ab}^t\right)} Seq_{sim}\left(r_{ab}^t, n\right).$$

The selection score s_{ab}^{t+1} of a mutant m_{ab}^{t+1} from the current fixed m_{ab}^t is defined as

$$S_{ab}^{t+1} = \lambda \cdot Attraction\left(T, m_{ab}^{t+1}\right) \cdot$$
$$+ (1 - \lambda) \cdot Influence\left(N_{k-nearest}\left(m_{ab}^t\right)\right),$$

where T is the target sequence set, and $0 \leq \lambda \leq 1.0$, which specifies the weight of the target attraction in selection scores. We sort the $19L$ one-point mutants in the descending order of their selection scores. We assign the fitness to the $19L$ mutants of the current fixed m_{ab}^t based on their selection scores. Following [5,8], the fitness of a mutant follows a probability distribution that belongs to the Gumbel domain [11]. In the current study we draw the fitness from a normal distribution. We randomly generate $19L$ numbers according to a normal distribution and sort them in descending order. We then assign these sorted numbers as the fitness values f_i to the $19L$ previously sorted mutants in descending order, respectively.

2.3.5 *One-point mutation and goal test*

Based on the fitness values, we define the selection coefficient of a new mutant m_{ab}^{t+1} from m_{ab}^{t} by $s_{t,t+1} = \frac{f_{t+1}-f_t}{f_t}$, where f_{t+1} and f_t are the fitness values of m_{ab}^{t+1} and m_{ab}^{t}, respectively. Subsequently, the probability of one-point mutation from m_{ab}^{t} to m_{ab}^{t+1} is defined as

$$p_{t,t+1} = \frac{s_{t+1}}{\sum_{s_i < s_t} s_i}.$$

When a one-point mutation occurs, a new mutant m_{ab}^{t+1} is selected according to the probability $p_{t,t+1}$. By repeating one-point mutations, we can simulate an evolutionary path until it converges. When a mutant converges, we identify its best matching real viral sequence. As a goal test, we compare it with the prespecified target sequences. If the best matching sequence is one of the target sequences, we terminate the simulation. The partial evolutionary paths to be connected form a complete evolutionary path, as illustrated in Figure 1.

3 RESULTS

We tested the proposed evolution simulator on the internal protein segment, PB2, of IAVs. We selected 16 avian PB2 proteins as the initial wild types from North America, Europe, and Oceania. We prepared two sets of target IAV sequences from different regions, one from six regions and the other from only one region. We show the initial wild types and the two sets of PB2 targets in Table 1.

Table 1. Summary of initial wild types and targets.

PB2 Wild-Type Set			Target Set (All Regin)			Target set (Only NA)		
NA			NA			NA		
	A/duck/Alberta/35/1976 (H1N1)	ACZ36512		A/RhodeIsland08/ 2009(H1N1)	ACV67255		A/RhodeIsland08/ 2009(H1N1)	ACV67255
	A/mallard/Wisconsin/ 142/1976(H1N1)	AHN01123		A/California/ VRDl225/2009 (H1N1)	ADN89280		A/California/ VRDl225/2009 (H1N1	ADN89280
	A/blue-wingedteal/Wis-consin/336/1976(H1N1)	AHN01210		A/California/ WR1320P/2009 (H1N1)	ADMI4634		A/California/ WR1320P/2009 (H1N1	ADMI4634
	A/green-wingedteal/Wis-consin/535/1976(H1N1	AHN04036		A/NorthCarolina/37/ 2009(H1N1)	AGI54749		A/NorthCarolina/37/ 2009(H1N1)	AGI54749
	A/mallard/Wisconsin/ 683/1976(H1N1)	AHN04048		A/MexicoCity/ WR1706T/2009 (H1N1)	ADMI4892		A/MexicoCity/ WR1706T/2009 (H1N1)	ADMI4892
	A/mallard/Alberta/46/ 1977(H1N1)	ABB19528		A/MexicoCity/ WR1683T/2009 (H1N1)	ADMI4819		A/MexicoCity/ WR1683T/2009 (H1N1)	ADMI4819
	A/pintailduck/ALB/219/ 1977(H1N1)	ABB19539		A/California/ VRDL92/2009 (H1N1)	ADG42682		A/California/ VRDL92/2009 (H1N1)	ADG42682
	A/widgeon/Wisconsin/ 505/1979(H1N1)	AHN01810		A/Texas/JMS386/ 2009(H1N1)	ADF27698		A/Texas/JMS386/ 2009(H1N1)	ADF27698
	A/mallard/Wisconsin/73/ 1979(H1N1)	AHN02659		A/NewYork/6943/ 2009(H1N1)	ADF83822		A/NewYork/6943/ 2009(H1N1)	ADF83822
	A/mallard/Wisconsin/ 216/1979(H1N1)	AHN02755		A/Boston/617/2009 (H1N1)	ADO30416			
	A/mallard/Wisconsin/ 524/1979(H1N1)	AHN02791						
	A/mallard/Alberta/965/ 1979(H1N1)	ABB19561		A/Nertherlands/602/ 2009(H1N1)	AGU92710		A/Boston/617/2009 (H1N1)	ADO30416
			EU	A/Nertherlands/ 947b/2009(H1N1)	AFH97149			
EU	A/duck/Bavaria/2/1977 (H1N1)	AID48461		A/Moscow/ WRAIRI627T/2009 (H1N1)	ADX96544			
		AID48449	OC					

(Continued)

Table 1.   (Cont.)

PB2 Wild-Type Set				Target Set (All Regin)		Target set (Only NA)
OC	A/duck/Schleswig/21/ 1979(H1N1)					
	A/duck/Australia/749/ 1980(H1N1)	ABB20386	SA	A/NewZealand/ 1212d/2009(H1N1)	AHI97775	
	A/duck/Victoria/23/1981 (H1N1)	ADT64867		A/Brazil/AVS05/ 2009(H1N1)	AFK78196	
			NEAsia	A/Amagasaki/1/2009 (H1N1)	ACR49296	
				A/Hubei/99/2009 (H1N1)	AEW25703	
				A/Hubei/76/2009 (H1N1)	AEW25489	
				A/Jiangsu/S62/2009 (H1N1)	AEW32080	
			SEAsia	A/Jiangsu/S62/2009 (H1N1)	ACR01011	
				A/Mum/NIV398/ 2009(H1N1)	AEM62833	
				A/Ngp/NIV14725/ 2009(H1N1)	AEM62806	

*NA: North America, EU: Europe, OC: Oceania, SA: South America, NEAsia: North East Asia, SEAsia: South East Asia.

Table 2.   Percentage of expected natural evolution.

Wild-Type	Target Set (only NA)	Target Set (All Region)
ABB19528	21.90%	15.80%
ABB19539	22.50%	13.20%
ABB19561	36.40%	35.20%
ACZ36512	14.40%	15.10%
AHN01123	12.90%	11.60%
AHN01210	18.10%	15.10%
AHN01810	24.70%	28.00%
AHN02659	24.00%	21.20%
AHN02755	44.10%	55.00%
AHN02791	36.60%	43.20%
AHN04036	21.40%	19.40%
AHN04048	22.40%	18.60%
AID48449	45.00%	34.50%
AID48461	33.00%	23.60%
ABB420386	14.00%	21.60%
ADT64867	47.30%	42.80%
total	**27.42%**	**25.87%**

By varying the temporal constraint threshold θ and the geographic proximity threshold ρ, we ran 1000 simulations for each parameter setting. We summarize in Table 2 the percentages of the simulated evolutionary paths that concurred with the expected natural evolution trends, namely, Avian→-Triple Reassortment→2009 Human Pandemic. From Table 2, we noted that the percentages of the correct simulated paths could vary among different initial wild types, while they were generally higher for the

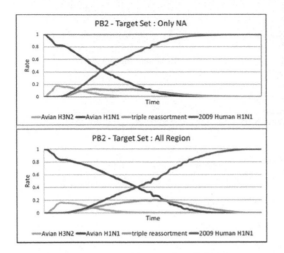

Figure 2. Changes in proportion of mutants in different species along with time.

targets selected from North America alone than from multiple regions.

We normalized the time required of a complete evolution for each simulation because the simulated evolutionary paths from an initial wild type to a target could be different. We show the changes in proportion of the mutants in different species along with the normalized evolutionary time in Figure 2. It suggests that as the avian PB2 mutants started to decrease, the proportions of triple reassortment and human pandemic mutants commenced rising. In addition, we identified the four most frequent triple reassortment PB2 sequences on the simulated evolutionary paths. They were AEK70338(A/swine/Texas/4199-2/1998(H3N2)), AHB21002(A/swine/Minnesota/00938/2005(H1N1)), AHB51159(A/swine/Oklahoma/00153/2003(H3N2)), and AFR76801(A/swine/Wisconsin/H02AS8/2002(H3N2)). Among the four, AEK70338 was the most frequent one. Notably, it is the most widely studied triple reassortment sequence [12], and our simulated results agree with the previous reports.

4 CONCLUSION

We developed a simulator to extend previous adaptive models and generate hypothetical evolutionary trajectories from an initial IAV to a set of prespecified target IAVs. From the simulated trajectories, we are able to analyze every single hypothetical mutational step in evolution, which cannot be observed from a phylogenetic tree. Because the task of adaptation is complicated by random mutations, pleiotropy, and population genetics, we adapted the sequence-based models to design a simulator that considers adaptation in the discrete state space of amino acid sequences instead of a continuous space to avoid a long extended adaptive walk toward the optimum state. To demonstrate the feasibility of the simulator, we tested it on the internal protein PB2 on IAVs, and the results correlated well with previous findings.

REFERENCES

[1] R.A. Fisher, The genetical theory of natural selection. Oxford Univ. Press, Oxford, U.K., 1930.
[2] P. Gerrish, "The rhythm of microbial adaptation," *Nature*, vol. 413, pp. 299–302, 2001.
[3] H. A. Orr, "The population genetics of adaptation: the adaption of DNA sequences," *Evolution*, vol. 56, pp. 1317–1330, 2002.
[4] J. H. Gillespie, *The causes of molecular evolution*. Oxford Univ.Press, New York, 1991.
[5] H.A. Orr, "The distribution of fitness effects among beneficial mutations," *Genetics*, vol. 163, pp. 1519–1526, 2003.
[6] M. Imhof, C. Schlotterer, "Fitness effects of advantageous mutations in evolving *Escherichia coli* populations," *Proc. Natl. Acad. Sci.USA*, vol. 98, pp. 1113–1117, 2001.
[7] D. E. Rozen, J.A.G.M. de Visser, P. J. Gerrish, "Fitness effects of adaptations in microbial populations," *Curr. Biol.*, vol. 12, pp. 1040–1045, 2002.
[8] J. H. Gillespie, "A simple stochastic gene substitution model," *Theor. Popul. Biol.*, vol. 23, pp. 202–215, 1983.
[9] L.M. Wahl, D.C. Krakauer, "Models of experimental evolution: the role of genetic chance and selective necessity," *Genetics*, 156, pp. 1437–1448, 2000.
[10] H. A. Orr, "The population genetics of adaptation on correlated fitness landscapes: the block model," *Evolution*, vol. 60, pp. 1113–1124, 2006.
[11] E.J. Gumbel, *Statistics of Extremes*. Columbia University Press, New York, 1958.
[12] Q.Liu, C.Qiao, H.Marjuki, B.Bawa, J.Ma, S.Guillossop, R. J.Webby, J. A.Richr and W.Ma, "Combination of PB2 271A and SR Polymorphism at Positions 590/591 Is Critical for Viral Replication and Virulence of Swine Influenza Virus in Cultured Cells and In Vivo," *Journal of Virology*, vol. 86, pp. 1233–1237, 2011.

Smart Science, Design & Technology — Lam et al. (eds)
© 2020 Taylor & Francis Group, London, ISBN 978-0-367-17867-3

Visual design evaluation of consumers' responses to bottled shower gel packaging

Hungyuan Chen, Kuoli Huang* & Cheni Huang
Department of Visual Communication Design, Southern Taiwan University of Science and Technology, Tainan, Taiwan

ABSTRACT: An effective package design enables a product to stand out from its competitors and create an initial impression which generates a favorable consumer psychological reaction. This phenomenon is conspicuous in the visual design of bottled products, such as shower gels, shampoos, cosmetics, and so forth. To create a satisfactory bottled shower gel packaging before the product is launched onto the market, it is essential to comprehend the consumers' responses regarding the visual design of bottled shower gel packaging and to explore the corresponding visual design features which influence the consumers' responses so as to provide designers with useful insights. The present study conducts a series of visual evaluation to explore consumers' responses to the visual design of bottled shower gel packaging. The evaluation results are analyzed using conjoint analysis and TOPSIS algorithm to determine the critical features regarding the visual design of bottled shower gel packaging for satisfying the consumers' responses. The results of present study could provide package designers with useful insights and are thus more likely to achieve commercial success when the bottled shower gel products are brought to the market.

1 INTRODUCTION

In an increasingly global marketplace, every manufacturer continually seeks to gain a competitive advantage in order to strengthen their market position. A consumer's decision to purchase a particular product is motivated not only by its physical requirements, but also by their psychological response induced by its physical appearance. Hence, satisfying consumers' psychological responses has become a major concern of almost every product manufacturer (Yan et al., 2008). The physical appearance of a product consists in its visual design including the visual design of product and its packaging. The importance of product visual design is crucial to the success of a product by ensuring the product attracts consumers attention, and communicates information to the consumer (Bloch, 1995). Similarity, product packaging is also one of important aspects of product strategy, and it plays the role of a silent salesman in enhancing the initial impression of the product in the eyes of its potential consumers' and influencing their purchase decisions (Vazquez, 2003).

Several researchers indicated that the label and the form of a package are the major design attributes of typical package and also play an important role in attracting consumer attention. For instance, the labeling of bottled product package could evoke consumers' perceptions (Alcx ct al., 2010; Giese, et al.,

2014). In addition, Becker et al. (2011) indicated that the container form of yoghurt packaging may inspire consumers' intense taste sensations and potency-related associations or perceptions, and Luo and Fu (2012) argued that conspicuous and distinctive form design for bottled product package could attract consumers' attention and increase the brand identification. Therefore, a bottled product packaging with well-done form design and label/graphic design is beneficial to a product with a powerful advantage by setting it apart from its competitors and generating a favorable consumers' psychological response.

In this study, the bottled shower gel product is specifically chosen here since it is one of the representations of mature consumer bath products and hence the decision making of purchase or the consumers' responses (CR) may be determined intuitively by the packaging details of bottled shower gel rather than merely its product attributes. In order to evaluate the task of consumers' responses to bottled shower gel packaging, it is essential to comprehend the consumers' responses regarding the visual design of bottled shower gel packaging and to explore the corresponding visual design attributes/features which influence the consumers' responses so as to provide designers with useful insights. This study conducted a series of visual evaluation of the bottled shower gel packaging to explore and analyze the relationship between the consumers' responses and the bottled shower gel packaging. The remainder of

---

*Corresponding author. E-mail: hungyuan@stust.edu.tw

this study is organized as follows: Section 2 illustrates the definition and evaluation samples of bottled shower gel packaging. Section 3 presents the selection of appropriate consumers' responses (CR) to bottled shower gel packaging. Section 4 describes the visual evaluation of bottled shower gel packaging. Section 5 presents the analysis of visual evaluation result. Section 6 offers the conclusions.

2 DEFINITION AND EVALUATION SAMPLES OF BOTTLED SHOWER GEL PACKAGING

To explore the CRs to the visual design of bottled shower gel packaging, this study commences by collecting a large number of bottled shower gel packaging pictures with similar view-angles, and approximate 80 shower gel packaging pictures are collected. A focus group consisting of 5 designers in the field of package/graphic design, is conducted to discuss a morphological analysis of the shower gel packaging pictures in order to identify the attributes and associated features which collectively define the generic shower gel packaging variables most likely to influence the CRs.

As shown in Table 1. 8 attributes are identified, with each attribute having 2 or 3 features. These attributes and the corresponding features are determined by the discussion of a focus group and the following principles are applied:

- The attributes must be immediately cognizable as components of a bottled shower gel packaging.
- Any two attributes must be distinguishable from one another and collectively all of the attributes must be able to capture all the aspects of diversity associated with the bottled shower gel packaging.
- Each attribute must contain multiple features, where the relations between these features must be distinct and independent.

Furthermore, in order to accommodate the 8 attributes of shower gel packaging and their corresponding features, the orthogonal design shown in Table 2 is applied to design the conditions of shower gel packaging samples. Finally, the 16 shower gel packaging samples shown in Figure 1 are created in accord with the conditions of orthogonal design table.

3 SELECTION OF APPROPRIATE CONSUMERS' RESPONSE TO BOTTLED SHOWER GEL PACKAGING

The adjectives provide an explicit representation of psychological perceptions and personal experiences, so that consumers can use simple adjectives to express their affective response to a shower gel packaging. In this study, approximately 60 adjectives relating to shower gel packaging are collected from commercial advertisement wording. Subsequently, 3 designers and

Table 1. Definition results of shower gel packaging.

Attributes	Features 1	2	3
Type of pump head and nozzle (X1)	Circle form with round tube (X11)	Circle form with flat tube (X12)	Plane-curve form with flat tube (X13)
Proportion of pump head width and nozzle length (X2)	1:1 (X21)	1:2 (X22)	
Width-to-height proportion of bottleneck (X3)	1:0.5 (X31)	1:0.8 (X32)	1:1 (X33)
Shape of bottle body (X4)	Rectangle body with round corner type (X41)	Bottle body with convex curve type (X42)	Bottle body with concave curve type (X43)
Width proportion of upper bottom-to-lower bottom for bottle body (X5)	1:1 (X51)	1:2 (X52)	
Width-to-height proportion of bottle body (X6)	1:3 (X61)	1:1.5 (X62)	
Height proportion of bottle cap-to-bottle body (X7)	1:3 (X71)	1:2 (X72)	
Graphic design on bottle body (X8)	Graphic design (X81)	No Graphic design (X82)	

Table 2. Orthogonal design of shower gel packaging samples.

No.	X1	X2	X3	X4	X5	X6	X7	X8
1	1	2	3	1	1	2	1	2
2	3	2	2	2	2	2	1	1
3	1	1	2	2	1	2	1	2
4	2	1	3	3	2	2	1	1
5	3	1	1	1	2	1	1	2
6	2	2	2	3	1	1	2	1
7	3	1	3	2	1	1	2	1
8	3	2	1	3	1	2	2	2
9	1	2	1	2	2	2	2	1
10	1	1	2	3	2	1	2	2
11	1	1	1	1	2	2	2	1
12	1	1	3	1	1	1	1	1
13	2	1	2	3	1	2	2	2
14	2	2	1	2	2	1	1	2
15	3	2	3	1	2	1	2	2
16	2	2	1	3	1	1	1	1

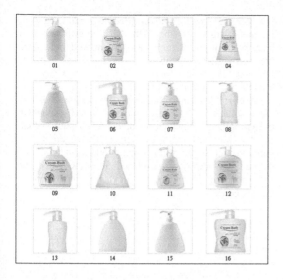

Figure 1. 16 bottled shower gel packaging samples.

2 individuals (non-design background) are recruited to join in a discussion for identifying the suitable adjectives. These adjectives are induced from the participants in accordance with the following three steps: (1) The Focus Group method is used to select 23 adequate adjectives if they relate entirely to the shower gel packaging. (2) The Kawakida Jirou (K.J.) sorting procedure is introduced to categorize the 23 adjectives according to the semantic similarities. (3) Finally, three basic adjectives groups are identified and the representative CR descriptions are chosen to stand for the entire adjective grouping meaning. The results in the formation of three groups with the following titles: "Cosy and Refreshing", "Aesthetic and Attractive" and "Easy to use", respectively.

4 VISUAL EVALUATION OF BOTTLED SHOWER GEL PACKAGING

In the visual evaluation procedure, the CRs induced by each shower gel packaging are quantified using 7-point Likert scales, with anchors ranging from "not at all (1)" to "intensely (7)". The 16 shower gel packaging samples and the three representative CR descriptions with 7-point Likert scales are integrated into a questionnaire-type document of the form shown in Figure 2.

The evaluation is implemented via 53 subjects including 21 designers with 2 years experience in packaging or graphic design, and 32 individuals from non-design backgrounds. Having acquired evaluation data from all 53 subjects, the evaluation data are processed using Conjoint Analysis and TOPSIS techniques in order to explore the relationships between the 3 representative CR descriptions and the shower gel packaging features.

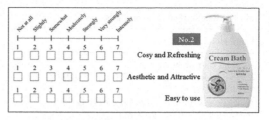

Figure 2. Questionnaire for evaluating consumers' response to bottled shower gel packaging.

5 ANALYSIS OF VISUAL EVALUATION RESULT

5.1 Conjoint analysis models

The Conjoint analysis (CA) is a popular research technique with the discrete choice analysis (Rohae, 2003) and it is often used to build the correlation between product attributes and the consumers' evaluations. In the present study, the independent variables in the CA model correspond to the attributes/features of the bottled shower gel packaging, while the dependent variables correspond to the 3 CR descriptions shown in used to characterize the consumer response to the shower gel packaging. Referring to the morphological definitions of the 16 shower gel packaging samples shown in Table 2 and the 3 CR descriptions used to evaluate the consumers' responses. Table 3 indicates the three CA models that are the relationships between the shower gel packaging and the corresponding 3 CR descriptions. These models also show that the Adjusted R2 values is 0.792 (Cosy and Refreshing), 0.874(Aesthetic and Attractive) and 0.836(Easy to use), respectively. As a consequence, the overall fits of these models are good. The 3 CA models provide designers with the means to obtain the predictive value of likely single CR description to shower gel packaging in terms of its "Cosy and Refreshing", "Aesthetic and Attractive" and "Easy to use" characteristics if only offer the number coding of corresponding feature for each of its eight attributes.

5.2 Integration of conjoint analysis models with TOPSIS algorithm

The CA models described in Table 3 enable the designer to predict the likely CR description induced by a particular bottled shower gel packaging. In this section, the 3 CA models are integrated with the TOPSIS (Techniques for Order Preference by Similarity of Ideal Solution) algorithm (Yoon and Hwang, 1995) to determine the optimal package design which achieves a certain set of CR description targets simultaneously. To acquire the weights of the 3 CR descriptions to bottled shower gel packaging, this study adopted the Analytic Hierarchy Process (AHP) approach (Satty, 1980) to identify the weights of the 3 CR descriptions. 56 representative

Table 3. Result of conjoint analysis.

CA models	R^2
"Cosy and Refreshing" model $= 4.236 - 0.156X_{11} + 0.312X_{12} - 0.156X_{13} + 0.221X_{21}$ $- 0.221X_{22} - 0.116X_{31}$ $+ 0.224X_{32} - 0.108X_{33} - 0.158X_{41} - 0.157X_{42}$ $+ 0.315X_{43} + 0.161X_{51} - 0.161X_{52}$ $+ 0.112X_{61} - 0.112X_{62} + 0.091X_{71} - 0.091X_{72}$ $+ 0.392X_{81} - 0.392X_{81}$	0.792
"Aesthetic and Attractive" model $= 3.947 - 0.191X_{11} - 0.191X_{12} + 0.382X_{13}$ $+ 0.108X_{21} - 0.108X_{22} + 0.284X_{31}$ $- 0.142X_{32} - 0.142X_{33} - 0.206X_{41} + 0.412X_{42}$ $- 0.206X_{43} - 0.082X_{51} + 0.082X_{52}$ $+ 0.108X_{61} - 0.108X_{62} + 0.086X_{71} - 0.086X_{72}$ $+ 0.476X_{81} - 0.476X_{81}$	0.874
"Easy to use" model $= 4.134 + 0.173X_{11} + 0.173X_{12} - 0.346X_{13}$ $+ 0.216X_{21} - 0.216X_{22} - 0.158X_{31}$ $- 0.079X_{32} + 0.079X_{33} + 0.193X_{41} - 0.386X_{42}$ $+ 0.193X_{43} - 0.064X_{51} + 0.064X_{52}$ $+ 0.101X_{61} - 0.101X_{62} - 0.086X_{71} + 0.086X_{72}$ $+ 0.162X_{81} - 0.162X_{81}$	0.836

consumers are recruit ed to perform pair-comparisons in order to identify the relative importance of each CR description. The results reveal the following weights: "Aesthetic and Attractive"= 0.438, "Easy to use"=0.313, and "Cosy and Refreshing"=0.250. Then, the integrated calculation of conjoint analysis models with TOPSIS algorithm comprises eight steps.

Step 1: Calculate the TOPSIS decision matrix. The TOPSIS decision matrix, D, i.e., the predictive values of the 3 CA models, is calculated as

$$D = \begin{bmatrix} P_{11} & P_{1j} & \cdots & P_{1n} \\ P_{21} & P_{22} & & P_{2n} \\ \vdots & \vdots & & \vdots \\ P_{m1} & P_{m1} & \cdots & P_{mn} \end{bmatrix}, \quad (1)$$

where P_{ij} is the predictive value of the i-th shower gel packaging sample ($i = 1, 2...m$, where $m = 16$) for the j-th CR description ($j = 1, 2,...n$, where $n = 3$).

Step 2: Calculate the normalized predictive value of each sample. The normalized predictive value of each sample over all the CR descriptions, r_{ij}, is calculated as

$$r_{ij} = \frac{S_{ij}}{\sqrt{\sum_{i=1}^{m} S_{ij}^2}}, \quad i = 1, \ldots, 16, \; j = 1, \ldots, 3 \quad (2)$$

Step 3: Calculate the weighted normalized decision matrix. The weighted normalized predictive values, v_{ij}, are calculated as

$$v_{ij} = w_j r_{ij}, \quad \sum_{j=1}^{3} w_j = 1, \; i = 1, \ldots, 16,$$
$$j = 1, \ldots, 3, \quad (3)$$

where r_{ij} are the normalized predictive values and w_j is the weight of the j-th CR description. For example, for the design case considered here (i.e., the normalized weights are "Aesthetic and Attractive"=0.438, "Easy to use"=0313, and "Cosy and Refreshing"=0.250, respectively.

Step 4: Determine the ideal solution and negative ideal solution. The ideal solution is a hypothetical design solution in which all of the CR values correspond to their best levels. Conversely, the negative ideal solution is a hypothetical design solution in which all of the CR values correspond to their worst levels. The two solutions, denoted as A^+ and A^-, respectively, are expressed as

$$A^+ = \{v1^+, \; v2^+, \; v3^+\}, \quad (4)$$

and

$$A^- = \{v1^-, \; v2^-, \; v3^-\}. \quad (5)$$

In this study, the solutions are obtained as $A^+ = (0.0466, 0.0792, 0.0526)$ and $A^- = (0.0262, 0.0439, 0.0337)$, respectively.

Step 5: Calculate the distances from each shower gel packaging sample to the ideal solution, D^+, and negative ideal solution, D^-, respectively, i.e.,

$$D_i^+ = \sqrt{\sum_{j=1}^{n} (v_{ij} - v_j^+)^2}, \quad i = 1, \ldots, 16, \quad (6)$$

and

$$D_i^- = \sqrt{\sum_{j=1}^{n} (v_{ij} - v_j^-)^2}, \quad i = 1, \ldots, 16, \quad (7)$$

Step 6: Calculate the preference index (PI) of each shower gel packaging sample. For each sample, the PI is calculated as

$$PI_i = \frac{D_i^-}{D_i^- + D_i^+}, \quad i = 1, \ldots, 16, \quad (8)$$

Table 4 shows the calculation results for the distance D_i^+ between each shower gel packaging sample and the ideal solution, the distance D_i^- between each shower gel packaging sample and the negative ideal solution, and the PIs of each shower gel packaging sample.

Step 7: Evaluate the utility of each feature of each attribute upon the PI. The utility of the i-th feature of the j-th attribute is calculated by summing the PIs of all the conditions involving the i-th feature of the j-th attribute and then dividing the result by the number of PIs.

Step 8: Select the optimal features. The feature of each attribute which has the greatest utility upon the PI is selected and added to the optimal set of features for the shower gel packaging sample.

Table 4. Three CR descriptions predictive values and corresponding values of D+, D− and PI.

	Predicted values of CR descriptions					
No.	Cosy and Refreshing	Aesthetic and Attractive	Easy to use	D^+	D−	PI
1	3.341	3.266	3.950	0.037	0.127	0.776
2	4.136	5.027	3.146	0.022	0.202	0.900
3	4.116	4.100	3.645	0.025	0.179	0.877
4	5.186	4.052	4.834	0.019	0.230	0.925
5	3.677	4.315	3.956	0.023	0.186	0.889
6	5.440	3.880	4.490	0.022	0.221	0.911
7	4.610	5.287	3.982	0.012	0.240	0.952
8	3.624	3.711	3.366	0.033	0.137	0.808
9	3.614	5.090	3.758	0.020	0.209	0.912
10	4.308	3.526	4.726	0.028	0.191	0.872
11	4.055	4.688	4.769	0.015	0.229	0.939
12	4.791	4.650	4.908	0.011	0.244	0.957
13	4.874	2.928	4.396	0.036	0.170	0.825
14	3.704	4.144	3.464	0.027	0.165	0.857
15	3.061	3.501	3.933	0.035	0.130	0.787
16	5.282	4.478	4.239	0.014	0.232	0.943

Table 5. Feature utility and attribute importance.

Attributes	Feature utility			Importance (%)
	1	2	3	
X1	0.762	0.744	0.867*	**34.99%**
X2	0.904*	0.862		**12.02%**
X3	0.891*	0.877	0.872	5.33%
X4	0.869	0.900*	0.881	8.55%
X5	0.881	0.885*		1.20%
X6	0.896*	0.870		7.26%
X7	0.890*	0.876		4.18%
X8	0.930*	0.836		**26.46%**

The asterisk indicates the biggest utility features.

Table 5 shows the utility of each feature of each attribute and the overall importance of each attribute upon the preference index (PI). It is seen that the "Type of pump head and nozzle" (34.99%), "Graphic design on bottle body" (26.46%), "Proportion of pump head width and nozzle length" (12.02%) have the significant influence on the multiple CR descriptions to shower gel packaging since these attributes account for 70% of the correlation between the shower gel packaging and the multiple CR descriptions. In other words, the designer should focus particularly on these attributes when attempting to create a new bottled shower gel packaging. Furthermore, an inspection of the utility values (i.e. higher is better) in Table 5, it is inferred that the optimal shower gel packaging design should comprise a plane-curve form with flat tube

(X13), ratio 1:1 proportion of pump head width and nozzle length (X21), ratio 1:0.5 width-to-height proportion of bottleneck (X31), bottle body with convex curve type (X42), ratio 1:2 width proportion of upper bottom-to-lower bottom for bottle body (X52), ratio 1:3 width-to-height proportion of bottle body (X61), ratio 1:3 height proportion of bottle cap-to-bottle body (X71) and Graphic design on bottle body(X81).

6 CONCLUSIONS

In this study, a series of visual evaluation of the bottled shower gel packaging has conducted to explore the relationship between the consumers' responses and the bottled shower gel packaging. The major attributes/features of bottled shower gel packaging have been identified, and the relationships between the descriptions of consumers' responses and the attributes/features of bottled shower gel packaging have been constructed such that the efficacy of the design process in developing a new shower gel package design for meeting consumers' psychological expectations can be accelerated. However, consumers' responses of a bottled shower gel packaging are both changeable and intricate over time. Different consumers are almost certain to experience a different psychological response when presented with the same shower gel packaging. The consumers' responses to a shower gel product are multi-dimensional in the sense organ. The consumers' responses are governed not only by the package graphic/bottled form of the visual sense, but also by its material, roughness or hardness of the product surface (tactile sense), fragrance (smell sense), and so on. As a consequence, there is no certainty that the design result obtained using the results of proposed procedure will satisfy all customers, either now or in the future. Finally, it is noted that although this study has focused on the relationship between the attributes/features of the bottled shower gel packaging and the designated multiple CR descriptions, the notions embodied in the proposed procedure are equally applicable to other package designs such that the product packaging satisfies all aspects of the consumers' psychological expectations.

REFERENCES

Yan, H.B., Huynh, V.N., Murai, T., Nakamori, Y. 2008. Kansei evaluation based on prioritized multi-attribute fuzzy target-oriented decision analysis, Information Science, 178, pp. 4080–4093.

Bloch, P.H. 1995. Seeking the ideal form: product design and consumer response, Journal of Marketing, 59, pp. 16–29.

Vazquez, D., Bruce, M. and Studd, R. 2003. A case study exploring the packaging design management process within a UK food retailer, British Food Journal, 105(9), pp. 20–31.

Alex G. and Howard R. Moskowitz. 2010. Accelerating structured consumer-driven package design, Journal of Consumer Marketing, 27(2), pp. 157–168.

Giese, J.L., Malkewitz, K., Orth, U.R., Henderson, P.W. 2014. Advancing the aesthetic middle principle: Trade-offs in design attractiveness and strength, Journal of Business Research, 67, pp. 1154–1161.

Becker, L., van Rompay, T.J.L., Schifferstein, H.N.J., Galetzka, M. 2011. Tough package, strong taste: The influence of packaging design on taste impressions and product evaluations, Food Quality and Preference, 22, pp.17–23.

Luo, S.J., Fu, Y.T., Korvenma, P. 2012. A preliminary study of perceptual matching for the evaluation of beverage bottle design, International Journal of Industrial Ergonomics, 42, pp. 219–232.

Rohae M. 2003. Conjoint analysis as a new methodology for Korean typography guideline in Web environment, International Journal of Industrial ergonomic, 32, pp. 341–348.

Yoon, K.P., and Hwang, C.-L. 1995. Multiple Attribute Decision Making: An Introduction (Sage Univ. Paper series on Quantitative Appl. in the Social Sciences 07-104). Thousand Oaks, CA: Sage.

Satty, T. 1980. The Analytical Hierarchy Process. McGraw Hill, New York.

Smart Science, Design & Technology — Lam et al. (eds)
© 2020 Taylor & Francis Group, London, ISBN 978-0-367-17867-3

Discussion on how human characters in different types of animated MVs transformed to animated characters affect the young generation's perceptions

Chianfan Liou
Department of Visual Communication Design, Southern Taiwan University of Science and Technology, Tainan, Taiwan

Chaochih Huang*
Department of Popular Music Industry, Southern Taiwan University of Science and Technology, Tainan, Taiwan

Hungyuan Chen
Department of Visual Communication Design, Southern Taiwan University of Science and Technology, Tainan, Taiwan

Yangkun Ou
Department of Creative Product Design, Southern Taiwan University of Science and Technology, Tainan, Taiwan

ABSTRACT: In recent decades, the technology of wireless internet and smart phones has been growing fast, and social media information is transmitted instantly. Music videos (MVs) especially have exerted a great influence on the young generation. Animations have been used as a means to attract customers' attention with a commercial purpose. In most cases, some human characters in videos are transformed to animated characters in the animated MV. The design of an animated character determines the success of an animation video. This research has adopted the focus group methodology, questionnaire, One-Way ANOVA, to analyze the design of different characters and discuss whether they will affect the young generation's perceptions. The results indicate that people are more attracted to puppet animations than to 2D animations and 3D animations, and puppet animations also impress the audience more.

Keywords: animation types, music video, character design, human character transformation

1 INTRODUCTION

As the wireless internet is used more widely and smart phones have become more advanced in recent years, the demand for digital media is growing in a speedy manner. In internet music videos, the demand for all types of animations is also increasing each day. The convenience of music video platforms on the internet also makes it easier to watch music videos. Ward et al. indicated, "In existence since 1981, music videos represent an important area of study ... because of their popularity with younger viewers" (Ward et al., 2005). Younger viewers are also the main users of smart phones and wireless internet.

Past research also indicated that the target audience for music shows on TV, like MTV, is mostly the ages between 12 and 34, thus the median age is 23 (Englis et al., 1994, Tiggemann and Slater, 2004).

The audience spends one-half hour to two hours watching MTV (Sun&Lull, 1986). The young generation is not only the main audience but also the people who would watch music videos regularly. This shows that MVs have a greater influence on the younger viewers.

In the commercial market of popular culture with massive and fast production of MVs, the animations are often used as presentations to attract audience's attentions, as well as to seek new changes in promotion styles. Back in 1967, The Beatles' song "Yellow Submarine" was presented using OP art images in a rebellious rock style. Thus this song has attracted a lot of attention in England (Wang, 2005). This MV was not only the first animated MV product in the commercial market but also a successful example of the cooperation between animation and music creators to promote and market the artist's records.

*Corresponding author. E-mail: z7m@stust.edu.tw

253

To unify the style in the entire video, transforming the performer (the human singer or musician) into an animated character is a common approach for character design. In animated videos, the character design is a very important section. Most animations need the characters to present the plots to connect to the audience's feelings. The well-known animated characters are usually artistic, stylish, and commercial. A well-designed animated character also has more chances to develop into an independent brand and obtain the best commercial value and market. A successful animated character could be more popular than a singer or a famous star (Shiu and Chung, 2010).

The success of an animated video is determined by whether it can attract the audience's attention and impress them. Therefore, the design of characters and the plots in the contexts are very important (Lin and Yang, 2014). Power also indicated, "Appropriate triggering of agency and mentalizing is fundamental both in terms of character design and narrative believability (and at a deeper level underpins the whole narrative experience)" (Power, 2008). Characters and plots are two fundamental elements in the animations. In the past, the research indicated that only a good design of the characters is able to express the story or the plots in the video properly. If a good story isn't represented by good characters, the film won't be very appealing. For viewers, the design of characters will decide the popularity and impression of a video presented to the audience (Huang and Chen, 2016).

The design of characters has been the main factor that determines a success of an animated video. Many researchers in the past have discussed how the appearance and the personality of a character are accepted by the audience. However, the character was usually discussed and analyzed in one type of animation. Only a few discussions tried to compare the same character in different animation types and whether they caused different audience perceptions.

This article uses the three most common types of the animated movies, 2D Animation, 3D Animation, and Puppet Animation, to discuss different presentations of animations. As for whether the design of an animation character after it is transformed from a human character will affect the target audience's perceptions, we selected the MV "Which one you like" nominated in the Ottawa International Animation Festival. From this MV, five human characters of the music band "We Save Strawberry" were transformed to animated characters through three different presentations in the illustration cards to discuss how they affect younger viewers' perceptions.

The three most common types of animation presentations from the human characters in the MVs will be discussed, as well as whether different animation presentations will affect the younger viewers' preference. The intent is twofold:

1. To provide a reference to selecting a type for presenting animation characters in MVs in future production.
2. To provide a reference to the style design of an animation character in the future when transforming a human character into an animated character.

2 EXPERIMENTAL

Our research used the focus group methodology, inviting six commercial movie producers with at least three years of experience to prepare and create the questions on the survey after understanding the main issue discussed in this research—the design of animated characters transformed from human characters and how they affect the audience's perceptions.

After the focus group methodology was used to ensure the questions on the questionnaires, this research created the illustration cards for the animated characters transformed from the human performers in the music band. The five characters consisted of a vocalist, a keyboard player, a bass player, a guitarist, and a drummer. The presentation of the illustration cards was 2D Animation, 3D Animation, and Puppet Animation. Their gestures, picture size, and background color in the cards were mostly consistent to avoid the extra difference for perceptions. Those illustration cards were listed at the top of the questionnaire. After the subjects viewed the cards, they answered the questions.

A questionnaire is designed for the three different presentations of animations in the illustration cards. Except for the difference in the presentations of animations, all the questions are the same. The questionnaire was divided into three groups: 2D Animation, 3D Animation, and Puppet Animation, where their tested subjects were not repeated. The ages of the 30 different subjects are between 18 and 28, totaling 90 subjects being tested. The survey collected scores using the Seven-Level Likert scale for the subject's perceptions.

This research mainly discussed the difference in the three different types of animations; so in order to avoid the perception inaccuracy caused by one single character, the data of the same presentation of the five characters were collected and averaged for a median value. After the survey was done, the score was averaged by the Seven-Level Likert Scale and then run by One-Way Analysis of Variance (ANOVA). The purpose was to discuss the difference between different types of animations but not between different characters.

Three animation types were shown in the illustration cards at the top of the questionnaire first, and

Chart 1. Professionals.

Professionals	1	2	3	4	5	6
Gender	M	M	F	M	F	F
Years of Experience	8	6	6	3	3	4

Figure 1. 2D animations (vocalist, keyboard player, bass player, guitarist, drummer).

Figure 2. 3D animations (vocalist, keyboard player, bass player, guitarist, drummer).

Figure 3. Puppet animations (vocalist, keyboard player, bass player, guitarist, drummer).

Chart 2. Questions.

1 I like the animated style transformed from the human character.

2 The animated character transformed from the human character looks attractive.

3 The animated character transformed from the human character impresses me.

4 The animated character transformed from the human character makes me curious about the actual performer.

Seven-Level Likert scale scoring: 1. Totally disagree, 7: Totally agree.

each card was placed together with the photo of a human character in comparison. Below the illustration cards were the questions. The picture size was 800×600 pixels, 72 dpi. All the subjects answered the questions on the internet. The selected characters were the vocalist, the keyboard player, the bass player, the guitarist, and the drummer in that order.

3 RESULTS AND DISCUSSION

As the One-Way ANOVA shows, there is an obvious difference in a preference for the animated characters transformed from human characters, $F(2,87) = 6.12$, $p < 0.01$. Scheffe post hoc tests indicated that Puppet Animation ($M = 5.54$, $SD = 0.9$) is higher than 2D Animation ($M = 4.74$, $SD = 1.31$) and 3D Animation ($M = 4.63$, $SD = 1.02$).

In terms of the attraction, there is also an obvious difference, $F(2,87) = 6.8$, $p < 0.01$. Scheffe post hoc tests indicated that the Puppet Animation ($M = 5.62$,

$SD = 1.0$) is higher than 2D Animation ($M = 4.78$, $SD = 1.33$) and 3D Animation ($M = 4.62$, $SD = 0.99$).

In terms of how much an animated character impresses a viewer, there is an obvious difference, $F(2,87) = 5.12$, $p < 0.01$. Scheffe post hoc tests indicated that Puppet Animation ($M = 5.58$, $SD = 1.01$) is higher than 2D Animation ($M = 4.63$, $SD = 1.38$).

There is not a big difference in degree in terms of whether the transformed animated character makes an audience curious about the real performer, $F(2, 87) = 2.94$, $p > 0.05$.

4 CONCLUSION

Through the survey data, we learned that the character design transformed from a human performer to an animated one in the most common animation presentations will cause a difference in terms of younger viewers' perceptions as follows:

1. In terms of the character design of an animated style from a human character, Puppet Animation is more favored by the viewer than 2D and 3D Animation.
2. In terms of the character design of an animated style from a human character, Puppet Animation is more appealing than 2D and 3D Animation.
3. In terms of the character design of an animated style from a human character, Puppet Animation is more impressive than 2D Animation. And 3D Animation didn't make a difference compared with other two.
4. In terms of the curiosity triggered by the animated character design transformed from the human character, there isn't any obvious difference among these three animation presentations.

This research reached its conclusion using focus group methodology and questionnaires to provide a reference for producers to choose an animation presentation when they want to create animated MVs, as well as when they want to create an animated character transformed from a human character. The analysis result tells them which animation design seems to attract and impress the viewers more.

Lastly, this research discussed how the animated characters transformed from human characters affect the young generation's perceptions using the three common presentations of animations to ask the viewers of the MVs four different aspects of questions. Looking at the long-term benefit in the future, this research might help researchers add more presentations for animations or increase the number of subjects and aspects of questions for a questionnaire so as to have more in-depth and solid discussion on the attraction power of animated characters transformed from human characters in different types of animations.

REFERENCES

Ward, L.M., Hansbrough, E., and Walker, E., 2005, "Contributions of Music Video Exposure to Black Adolescents' Gender and Sexual Schemas," *Journal of Adolescent Research*, Vol. 20 No. 2, pp. 143–166.

Englis, B.G., Solomon, M.R., and Ashmore, R.D., 1994, "Beauty before the eyes of the beholders: The cultural encoding of beauty types in magazine advertising and music television", *Journal of Advertising*, Vol. 23 No. 2, pp. 49–64.

Tiggemann, M. and Slater, A., 2004, "Thin ideals in music television: A source of social comparison and body dissatisfaction," *International Journal of Eating Disorders*, Vol. 35 No. 1, pp. 48–58.

Sun, S.E., and Lull, J., 1986, "The adolescent audience for music videos and why they watch," *Journal of Communication*, Vol. 36 No. 1, pp. 115–125.

Wang, I.W., 2005, "Creation Descriptions of 'Sing no sad songs for me' and Investigations musical animation (Master's thesis)," available from https://hdl.handle.net/11296/vpp26f

Shiu, Y.S. and Chung, S.K., 2010, "An Application of Computer Animation Arts to the Taiwanese Folklore-Hero 'Pat-kaChiòng' Characters: A Case Study," *Journal of Humanities and Social Sciences*, Vol. 6 No. 1, pp. 31–47.

Lin, Y.C. and Yang, M.C., 2014, "The Design Evaluation for Characters in the 3D Computer Animation Movies," *Taiwan Journal of Arts*, Vol. 94, pp. 103–131.

Power, P., 2008, "Character Animation and the Embodied Mind–Brain," *Animation - An Interdisciplinary Journal*, Vol. 3 No. 1, pp. 25–48.

Huang, J.S. and Chen, T.Y., 2016, "The Research of the Kansei Image's Analysis About Animation Character's Vision - Take 'Rise of the Guardians' For Example," *Journal of Performing and Visual Arts Studies*, Vol. 9 No. 2, pp. 35–72.

Smart Science, Design & Technology — Lam et al. (eds)
© 2020 Taylor & Francis Group, London, ISBN 978-0-367-17867-3

The influence of musical fitness to design theme on designer's creative idea generation behavior

Chimeng Liao*, Lanling Huang, Chihwei Lin & Lifen Ke
School of Design, Fujian University of Technology, Fuzhou City, PR China

ABSTRACT: There have been many studies showing the effects of music listening in many domains, but research regarding the impact of music listening on design is rare. The experimental task adopted in this study uses the within-subject design method to investigate how music affects idea generating behavior. The subjects were invited to draw lamp sketches by freehand while music related or unrelated to the design theme was playing. The subjects' retrospective reports, which they generated in the design process, were decoded and statistical methods were applied to compare the quantities of lamp sketches, as well as with previous relevant studies. The results show that, when listening to music related to the design theme, the subjects spent more time in drawing sketches and generated more lamp sketches than when listening to music unrelated to the design theme. In the case that music unrelated to the design theme was playing, the subjects spent a longer time in arbitrary graffiti and thinking. Writing text, however, was not significantly different under these two music situations. This study demonstrates that designers will have a more fluent idea manipulating procedure when music suited to the design theme is playing. We speculate that the reasons might be attributed to the attention theory of cognitive psychology as well as episodic memory which was induced by music fit to the design theme.

1 INTRODUCTION

Music has been closely related to humans' lives since ancient times. There have been numerous studies on the influence of listening to music. For example, playing music in the workplace can improve efficiency and mood, and enhance perceptive ability (Lesiuk, 2005). The music that fits in consumption places can improve customer purchasing rate (North et al. 2003), or enhance corporate brand recognition rate (Zander, 2006). Previous studies have shown that most designers often listen to music while working, and reached the consensus that listening to music has positive effects (Liao, 2003). Music permeating our life has almost become part of the design. The multiple benefits of music, as presented by previous studies, make it worthwhile to explore the impact on design when listening to music.

2 LITERATURE REVIEW

The diversity of music enriches the human experience and makes people intoxicated with the information characteristics presented by the melody. A previous study shows that music speed is linked with and impacts human behavior. Under fast music situations, the subjects draw faster, and the fluency of creative ideas increases. In the case of slow music, however, the subjects are more concentrated, showing better novelty of creative ideas (Chang et al. 2016). Music that fits in the environment can convey specific ideas or influence people's consumption behavior (Jacob et al. 2009). Moreover, music can evoke autobiographical memories of past experiences and events that are peculiar to the listener, thus presenting itself in memorial scenarios and emotions (Sternberg, 2005). Since music can evoke the concepts and memories of various objects in our brain, the priming effect probably exists when people are listening to music (Koelsch et al. 2004). When the priming effect is activated, it will stimulate the human memory network nodes, enabling people to extract these related events more smoothly (Tsai, 2015). In other words, if the site where people recall information is exactly where they acquired such information, the context-dependent memory facilitates the recollection of this kind of information (Santrock, 2004). The above research shows that appropriate music is like a cue role that activates cognitive behavior related to memory, such as contextual memory and autobiographical memory, promoting human association connections.

Designers usually collect information related to the design theme, such as pictures or samples, in the design environment to create a design context conducive to inspiration, particularly in the stage of

---

*Corresponding author: E-mail: jameslgm88@sina.com.cn

design concept development. In addition, designers normally apply memory to retrieve information conducive to design tasks. When information appears, people can use attention to eliminate inappropriate information, and accept information that is useful or fitting in tune with the current situation, which enables extraction through long-term memory (Best, 1999). Clues and records are interlinked and exist in the environment. Appropriate clues help make the records highly activated, easily extracted, and more effective than free recalling (Sternberg, 2005). In the early concept development stage, designers often draw conceptual sketches free-hand to quickly capture flashes of ideas in their minds (Purcell et al. 1998). Conceptual sketches can be regarded as the unique design behavior of designers in the concept development stage. Therefore, in the research of design conceptual behavior, the application of sketches for exploration is closer to the actual design behavior of designers.

3 RESEARCH METHOD

In the experiment, the subjects were asked to generate lamp ideas with Japanese style by freehand sketch or text annotation while listening to music. In the concept development stage, however, designers sometimes produce messy sketches or ideas, the meanings of which are difficult for others to interpret. Therefore, this study applied retrospective report by asking the subjects to inform their ways of thinking as an effective way to understand their thinking (Ericsson et al. 1993). Quantity of conceptual elements generated by the subjects was analyzed by statistics, expecting to present the experimental results objectively. In addition, due to the inevitable individual differences in the design ability of the designers recruited for this experiment, this experiment adopted the within-subject design; each subject performed two experiment tasks in two different musical situations to reduce the error variance.

3.1 Experimental tasks and equipment

In this study, 14 graduate students of industrial design were invited to participate in the experiment. The subjects were asked to record ideas associated with Japanese-style standing lamps/pendant lamps in a free-hand manner while music was played. For exploring design behavior, two digital cameras were used, one of which was placed above the subjects to capture how the subjects generated design concepts, while the other recorded the interaction between the subjects and their surroundings. During the task, the subjects were asked to put on earphones to listen to music during the experiment. Given that the experiment adopted the within-subject design, each subject designed a lamp with Japanese style while listening to Japanese music and non-Japanese music, each experiment interval one week.

3.2 Music used in experiment

The music played in this experiment includes Japanese and non-Japanese music, each of which has six pieces. The principle for selecting Japanese style music was the methods for singing and transliteration with Japanese features such as Enka; or the lyrics may obviously of Japanese style; or the musical instruments exhibited Japanese characteristics, such as Shakuhachi, Taisho Koto, and drum. Thus, Japanese pop music that presents more like Western pop music was excluded from this experiment. In terms of non-Japanese music, the results of the author's previous questionnaires show that Western pop music and Chinese pop music are the most popular among designers (Liao et al. 2015), thus, three pieces of Western pop music and three of Chinese pop music were played in this experiment.

4 EXPERIMENT RESULTS AND ANALYSIS

4.1 Decoding retrospective reports and design behaviors

The subjects generated numerous ideas for Japanese-style lamps, such as the partial conceptual sketch of subject No. 5, as shown in Figure 1. To decode the design behavior, researchers compared the video footage of how the subjects drew and their retrospective report data, then coded into verbatim draft. Based on the characteristics of this experimental task, the creative design conception behavior of this experiment is divided into four categories: drawing, written instructions, design thinking, and random graffiti. In addition, the time is noted on each drawing paragraph to inspect time of each design's conceptual generation behavior.

4.2 Music characteristics and design conception behavior

After reviewing the 40-minute video footage of the experiment and the subjects' retrospective reports,

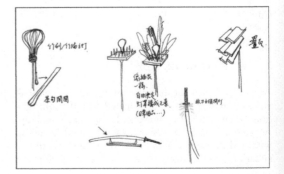

Figure 1. Partial sketches of subject no. 5.

Table 1. Paired sample t-test of design activity.

Design Activity	J-Music		NJ-Music				
	Mean	SD	Mean	SD	t	fd	p
Drawing	1243.1	318.8	1076.5	322.8	3.01	13	0.01*
Writing	523.4	263.1	570.7	314.8	−1.26	13	2.31
Thinking	590.4	206.5	687.5	165.9	−2.99	13	0.01*
Graffiti	44.5	30.1	65.3	47.6	3.03	13	0.01*

$\alpha = 0.05$, *$p < 0.05$

Table 2. Paired sample t-test of design output.

Design Output	J-Music		NJ-Music				
	Mean	SD	Mean	SD	t	fd	p
Lamp sketch	14.07	5.33	10.93	4.25	3.15	13	0.01*
Graphic element	13.36	4.70	11.21	4.92	1.72	13	0.11
Word element	25.50	9.23	21.86	9.69	2.63	13	0.02*

$\alpha = 0.05$, *$p < 0.05$

a paired sample t-test was used to test whether the drawing, writing, design thinking, and graffiti of subjects were significantly different for Japanese and non-Japanese music, respectively, and whether the time used was significantly different. The results of t-testing are shown in Table 1. The p value of the subjects' drawing behavior (t = 3.01, p = 0.01) is < 0.05, reaching the significantly different level, which indicates that the subjects invested more time in sketching lamps when Japanese-style music was played. The p values of design thinking (t = −2.99, p = 0.01) and graffiti (t = −3.03, p = 0.01) are < 0.05, reaching the significantly different level, which clearly indicates that the subjects invested more time in design thinking and random graffiti when non-Japanese music was played. However, the p value of writing words (t = −1.26, p = 2.31) is >0.05, which fails to reach a significantly different level.

4.3 *Music characteristics and quantity of design conceptions*

The classification results show that the subjects produced a large number of sketches, graphics, and word elements with Japanese characteristics. In the case of Japanese music, 197 conceptual sketches of lamps, 187 graphic elements, and 382 word elements were produced, while in the case of non-Japanese music, 152 conceptual sketches of lamps, 157 graphic elements, and 306 word elements were produced. This study adopts paired sample t-test to verify whether the quantity of conceptions varies significantly from Japanese music to non-Japanese music. As shown in Table 2, the p values of both the sketches (t = 3.15, p = 0.01) and the conceptual elements (t = 2.63, p = 0.02) of the lamps are both < 0.05, reaching the significantly different level, which indicates that the subjects produced more sketches and conceptual elements of lamps while listening to Japanese music. However, the figure elements (t = 1.72, P = 0.11) have a p value of > 0.05, as there was no significant difference between the figure elements produced in the case of Japanese and non-Japanese music being played.

5 DISCUSSION

5.1 *Hint effect generated by music fitting in design theme*

People will be more willing to perform a task continuously without pausing to think or looking around if they perform it smoothly; by contrast, they stop to think and review what they have done if the task does not go well. By deconstructing the video data, retrospective, and paired t-test results (as shown in Table 1) of subjects' drawing processes, this study finds that most of the subjects obviously devoted themselves to generating lamp sketches while listening to Japanese music, but less time on thinking and graffiti, demonstrating that the design creative behavior was smoother. However, in the case of listening to non-Japanese music, the subjects spent more time on random graffiti, thinking, and browsing previous drawings. The reasons for these phenomena may be attributable to interference or distraction from listening to music not corresponding to the design theme, as the partial retrospective data of subject No. 6 show.

> *08:13_S6: The music is pretty and similar to what I often listen to, but it doesn't seem to be very helpful for design. Instead, it will take my thoughts elsewhere and it's not easy to associate things related to Japan.*

The greater amount of concepts of Japanese-style produced by the subjects when Japanese music is being played may be attributed to their awareness of the fact that Japanese music fits with the theme of the experimental tasks, and their consequent willingness to be stimulated by such music. This can be explained by the attention and memory extraction of human cognitive behavior. When external stimuli appear (music played), they will arouse people's attention to choose to accept the stimulation in line with the current context (designing Japanese lamps), which enters the long-term memory area to extract related information (Best, 1999). In other words, music related to a design theme played as a clue for

259

design tasks that activate Japanese-related cognition and association (Byrens, 2001). Therefore, listening to Japanese music will stimulate more conceptual sketches and elements of Japanese-style lamps, as is shown in the partial oral data of the subject No. 2.

> *15:07_ S2: This song reminds me of the Japanese rites in which there are carp flags in my impression. The carp flag is a strong Japanese image to me because it is so unique that it can only be found in Japan. I tried my best to paint the carp flag like a carp, thus, this lamp is quite Japanese.*

This study shows that music in line with the design task not only attracts the attention of designers, it also provides design clues that designers can explore and create appropriate design elements.

5.2 Episodic memory triggered by music fitting the design Theme

Previous studies have shown that appropriate music can shape the environment and hence influence human behavior (Jacob et al. 2009). Such music also triggers episodic memory related to events, places, and moments in life, especially autobiographical memory related to personal life experiences (Sternberg, 2005). Therefore, playing Japanese music can trigger subjects to evoke their past exposure to Japanese music and then to associate it with certain memories of that time, including specific events, places, objects, or persons. Such memory is indicated by the oral data of subject No. 5.

> *10:25_S5: This song is soft, so I see falling sakura. I see the image once I listen to the music. I am just thinking about how to put what I envisage on a lamp that fits in the context, because sakura is more specific and the lamp post is like branches.*

According to the subject's retrospective data, listening to music fit to the design theme triggers episodic memory, helping them to devote more time to design tasks and associate more concepts related to the design theme.

6 CONCLUSION AND SUGGESTIONS

Listening to music related to the design theme activates the information and memories relevant to the design theme, which can create another way for designers as they generate design concepts. This study shows that playing music related to the design theme can create a theme-specific context that makes designers feel as if they are in the real situation. Music that matches the theme exerts a cue effect, such as providing auditory clues, which allows designers to follow this clue and explore the concepts related to the design task. In addition, it enhances generation of design concepts fluently and efficiently.

This study demonstrates the impact of listening to music on design from the perspective of auditory perception. The influencing factors of design are highly complex, however, and this paper does not intend to propose the suggestion of replacing visual perception with auditory perception in the process of design. On the contrary, we believe that visual stimulation accompanied by appropriate music should stimulate more design conceptions. This study merely marks the beginning of research related to music and design, and there should be more topics worthy of discussion.

ACKNOWLEDGMENT

This study was partly supported by the Design-led Innovation Research Center of Fujian University of Technology, and partial funding from the Fujian University of Technology with grant No. E1600144.

REFERENCES

T. Lesiuk, 2005. *Psychology of Music.* 33(2), 173–191.
A.C. North, A. Shilcock, D.J. Hargreaves, 2003. *Environment and Behavior.* 35(5), 712–718.
M.F. Zander, 2006. *Psychology of Music.* 34(4), 465–480.
C.M. Liao, W.C. Chang, 2015. *JSSD.* 61(5), 47–56.
W.C. Chang, C.M. Liao, 2016. *LNCS* 9735, 223–234.
C. Jacob, N. Gue´guen, G. Boulbry, S. Sami, 2009. *The International Review of Retail Distribution and Consumer Research.* 19(1), 75–79.
R.J. Sternberg, 2005. *Belmont, CA: Thomson Wadsworth.* 206–207.
S. Koelsch, E. Kasper, D. Sammler, K. Schulze, T. Gunter, A.D. Friederici, 2004. *Nature Neuroscience* 7, 302–307.
C.G. Tsai, 2015. *Taipei: National Taiwan University.* 89–90.
J.W. Santrock, 2004. *New York: McGraw-Hill.* 249–251.
J.B. Best, 1999. *Belmont, CA: Wadsworth.* 50–51.
A.T. Purcell, J.S. Gero, 1998. *Design Studies.* 19(4) 389–430.
K.A. Ericsson, H.A. Simon, 1993. *Cambridge, MA: The MIT Press.* 19–20.
J.P. Byrens, 2001. *New York: Guilford Press.* 53–54.

Smart Science, Design & Technology — Lam et al. (eds)
© 2020 Taylor & Francis Group, London, ISBN 978-0-367-17867-3

The natural beauty in carving – ingenious carving of wooden fossil

Tingfang Lin*
Opposite to the Telecom Company, Congwu Town, Hui'an County, Fujian Province, PR China

ABSTRACT: The artistic expression of wooden fossil carving focuses on nature and natural delight. This paper drew from the author's experiences in carving wooden fossils and explored the way to highly integrate mankind with nature in the work.

Keywords: wooden fossil, expression of carving, nature and natural delight

1 OVERVIEW OF WOODEN FOSSIL CARVING

The rarely seen wooden fossil seems to be near and distant from us at the same time. Under the fine work by the craftsmen, wooden fossil carvings appear to so live in our hearts that you can even sense their breath and spirit. Coming through the hundreds of millions of memory of the wooden fossil, you listen to it talking to you about the time and the waves and tides on the other end of the time stream. For over 150 million years, it has not vanished nor even ceased to transform. If the wooden fossil is hailed as the wonder of nature, then wooden fossil carvings should be called the divine work of artisans.

Wooden fossils, also known as silicified wood, formed hundreds of millions of years ago. When trees became buried deep in the ground, they exchanged and replaced molecules under great pressure with surrounding colloid solutions containing compounds of various elements. Chemical substances like silicon dioxide, ferric sulfide, calcium carbonate, and others that enveloped the stalks, entered the secondary xylem cells, and replaced the original substance while maintaining the shape of the wood. Thus the wood, under petrifaction, became a wooden fossil that is wondrous as fossil and beauteous as jade stone. Because the silicified wood is non-renewable, resource departments in many countries introduced injunctions prohibiting the unrestricted extraction of silicified wood, making wooden fossils highly valuable and popular among collectors. Hence, there is a saying among the collectors: "ancient coins, gold, jade, and precious stones bear increasing value over the last one, and the most supreme one is wooden fossils."

In recent years, because collectors keep searching for wooden fossils, appreciating and wearing jade, and treasuring and collecting gems, the price of wooden fossils has skyrocketed. The wooden fossil, under its innate beauty and polishing by artists, appears more noble and magnificent without any "intended decoration." Hence, the wooden fossil evolved into a recognized treasure among collectors that brims with artistic and aesthetic tastes (Lu, 2014).

By inheriting regional styles and developing the styles of different schools, Huian carving art stands as an integration of traditional culture and classic artistry. Over the past one thousand years and more, Huian carving has maintained the fusion of its history and status quo, filiations, spirits of masters, modern thinking and art, and new conceptions, inheriting and surpassing past achievements and reaching a whole new level in terms of art and thinking. The encounter of Huian carving masters with wooden fossils and their soul interaction seems surprisingly coincidental and harmonious, given the pursuit for precious materials that derive from the regional history background and cultural heritage.

2 WOODEN FOSSIL CARVING ART

Wooden fossils, as combined wood and stone, feature a hard and dense texture with gem-like quality that is much superior to that of ordinary stones. It also demonstrates the beauty of nature, science, and life from the fusion of wood and stones that is unparalleled by other materials. Hence, this supreme material became the optimal choice for artistic creation among the new-generation folk artists and masters in Huian. Nature's making and fine artistry has transformed the

---

*Corresponding author. E-mail: momingstone@163.com

unique patterns, colors, and shapes of the material into a dazzling star of its kind in the cultural world.

Wooden fossil carving art takes advantage of and conducts the carving on the natural form and texture of the materials, with a focus on exhibiting its natural beauty and giving rise to the uniqueness of the carving. The most exclusive and ingenious part is to present the miraculous work done by nature. Years of stone carving have led to more experience in aesthetic forms of Chinese traditional culture, and I came to love wooden fossils during my interaction with jade and stones. I would like to share some of my work and experience in relation to wooden fossil carving from my continued exploration and learning.

3 CREATION OF WORKS

While creating the carving *Fishing Alone in the Snow*, I utilized the natural shape of silicified wood and its pattern, which is white on the outside and black in the inside, worked under a meticulous conception, and presenting an image as follows: in a world covered by snow where icicles hang like waterfalls, an old man in a straw cape fishes in complete ease. A setup of such a landscape and character inscribed by an ancient poem serves to accentuate the underlying implication and give the viewers sudden enlightenment.

Figure 1. The carving work *Fishing Alone in the Snow*.

Figure 2. The carving work *Shang Shan Ruo Shui*.

Figure 3. The carving work *Totem*.

This work is created on a piece of a natural wooden fossil and followed the image of an ancient Chinese philosopher and sage "Laozi." By adhering to his philosophy "The greatest truths are the simplest," only the top of the material was moderately carved. The whole piece of work takes into account the natural flowing patterns of the material and delivers an idea that is expressive, inclusive, natural, and ultimately kind and tender. The work reaches a unity of its soul and appearance.

The raw wooden fossil looks quite "unattractive," the surface of which is enveloped by a layer of coarse and white powder-like bark. The silicified bark needs to be polished before the tender, bright, and porcelain-like inside can be shown.

The work "Totem" strikes one as a stately dragon carved based on the shape, outthrusts, and different colors in and out of the original material, from which some unnecessary parts were removed. The totem embodies the sacred might and unyielding soul of the vast country and represents the divinity, evil-counteracting, and auspice. This is undoubtedly an emblem of the Chinese nation and the foundation of culture and history of Chinese descendants.

Created by nature, the wooden fossil is a witness of prehistoric times and the vicissitudes of the world. Hence, there is the saying, "Ancient wood is immortal for its soul; the spiritual stones relate the history without making sounds."

The ancient people held such a view: "The wise find pleasure in water, and the virtuous find pleasure in hills." My view, however, is that "the wise and virtuous find pleasure in stones." The wooden fossil is a symbol of longevity, auspice, evil-counteracting, and abundance, and, as a work of art, bears more significance and value in being appreciated and collected. We have achieved an artful combination of nature and senses by making use of its nature-given ingenuity. In the course of creating each piece of work, I endeavor to extract the natural qualities and make sure to integrate nature's work with man's carving to enhance the philosophy and conception. By being imbued with inherited history and traditional culture; an understanding of life, humanity, and human affections; and enlightenment in the truth, kindness, and beauty, my works are endowed with a never-ending life philosophy.

REFERENCES

Q. Lu, 2014. *The definition of implication and the extension of shape – study on the application of petrified wood materials in the decoration design arts*. M.S. Thesis, Shenyang Ligong University.

Smart Science, Design & Technology — Lam et al. (eds)
© 2020 Taylor & Francis Group, London, ISBN 978-0-367-17867-3

Exploration of artistic aesthetics of Field stone and carving

Youlong Pan*

Xiamen Academy of Arts and Design, Fuzhou University, Xiamen, Fujian Province, PR China

ABSTRACT: Two kinds of beauty are recognized in Chinese arts history, namely, the beauty of nature and the beauty of carving. The beauty of Shoushan stone carving art is the natural blending and perfect unity of these two kinds of aesthetic feelings. The beauty of superficial carved Field-yellow stone as mentioned by the author is representative of reaching the proverbial realm—"the extreme of splendor relies on plainness."

Keywords: the beauty of Field stone, the beauty of implication, the beauty of carving

1 INTRODUCTION

Excellent works of Field stone carving in Shoushan stone make full use of the beauty of materials. As the saying goes, "Jade cannot become a utensil without carving," which shows that natural stone must be carefully carved artificially and given a certain shape and decoration to show the beauty of stone and shape. It can be seen that carving is particularly important, but what kind of carving conforms to the natural attributes of stone? What kind of creative graphics can be fully reflected in stone? That's the question of exploring the beauty of a skilled craftsman's implication (Chen, 2012).

First of all, the beauty of materials is the natural beauty of Field stone; secondly, the beauty of Field stone's identification is the beauty of implication; thirdly, it is the beauty of carving. These three characteristics are a natural and reasonable embodiment, and the key to these characteristics is to rely on human carving to achieve them. *Rites of Zhou. Book of Diverse Crafts* mentions "fair weather, favorable climate, excellent materials and exquisite workmanship." Only by combining these four conditions can we make excellent utensils (Wang, 2017).

It can be seen that the ancient people paid great attention to the natural beauty of materials. Material has its own aesthetic value, and its aesthetic attributes include its physical and chemical properties, natural color, texture, and so on.

2 THE NATURAL BEAUTY OF FIELD STONE

Shoushan stone, a kind of stone comparable to jade, is produced in the mountains of Shoushan Village in the northern suburb of Fuzhou City. Shoushan stone presents a riot of colors, the source of its natural beauty. Field stone is produced in the paddy field irrigated by streams of Shoushan Village. This field has a hole called "Kengtou Hole," which is several miles long. The range of the paddy fields irrigated by streams usually is the limit of the producing area of the Field stone. The producing area includes the upper slope, the middle slope, the lower slope, the cultivated field and, so on. The colors of Field stone include red, yellow, white, and black. The red one is called "red Field stone," the yellow one is called "yellow Field stone," the white one is called "white Field stone," and the black one is called "black Field stone." In addition, the external white and internal yellow one is called "gold wrapped in silver," the external yellow and internal white one is called "silver wrapped in gold," and the one with thin black skin is called "raven-skinned Field stone" (Yu, 2017; Lin, 2017).

"Field stone" is divided into upper class, middle class, and lower class, with more than ten categories, such as orange peel yellow, gold yellow, loquat yellow, fragrant yellow, lucid Ganoderma field-yellow, cooked millet yellow, etc. Several varieties such as white Field stone, black Field stone, black-skinned Field stone, gold wrapped in silver, the lower Field stone, the cultivated Field stone, Jiushou Field stone, and hard Field stone also belong to Field-yellow stone. Their values, appeal, characteristics, and textures are varied. Among these stones the black-skinned Field stone and gold wrapped in silver are both upper class stones and are not easy to obtain.

---

*Corresponding author. E-mail: 5102452@163.com

The best Field-yellow stone is brown-skinned fine Field stone, most of which is red and yellow, especially dark red, resembling orange peel and red amber. The texture contains extremely detailed reddish veins, which is very rare. Calcined red Field stone comes from slope field, and its red layer resembles painting, often with black color on the surface. The stone body is rather small, which is slightly inferior to the orange peel Field stone. Its appearance is slightly transparent, and it is slightly hard and rough in texture, usually without obvious reddish veins. Field-yellow stone refers to Field stone of yellow color, which comes from paddy fields irrigated by streams at the upper slope, the middle slope, and the lower slope. It's the best and most productive variety. It's well-known in the world, so there is a saying: "one teal of Field-yellow stone equals to three teal of gold." Field stone is commonly referred to as Field-yellow stone by the folk.

Field-yellow stone is bright and lustrous, multicolored, crystal clear, warm, and moist as jade. The natural beauty of Field-yellow stone has made it a rare treasure for sculptors and collectors to obtain.

3 THE IMPLICATION BEAUTY OF DESIGN AND CONCEPTION OF FIELD STONE

The Field stone itself has an implicit inner beauty. The key is that good stone sculptures need skilled craftsmen to shine brilliantly without burying the spirit of the earth. Field-yellow stone is famous all over the world because of the superb skills of the carving artists who work with it. It is carving that gives stone new life. However, although Field-yellow stone is beautiful and pleasant, it's still an ordinary stone without the proper choice of subject matter and careful carving by skilled craftsmen. Carvers need to apply their craft skillfully, and the expressive subject matter also shall provide a strong breath of life. The beauty of the stone and the ingenuity of the carving are revealed by turning defects into advantages. Only in this way can the ornamental value, artistic value, and economic value of such works be multiplied.

About design conception, the predecessors of today's carvers once said, "One time of observation is worth nine times of carving." Observation means thinking, namely, how to "observe" varies with each individual, each style, and each artistic level. With a stone in the hand, the heart will be full of thoughts. When the thinking is narrow, the art space is little; when the thinking is broad, the art space is big. This space is comprehended from "observation" of stones. The beauty of nature is enhanced by carving. It's what we called the beauty of implication.

4 THE CARVING BEAUTY OF FIELD STONE

The beauty of Field-yellow stone carving includes the beauty of implication, as the two are inseparable, because the beauty of carving consists of the shape and the pattern. Most of the patterns come from the unique connotation and aesthetic taste of our nation. In order to convey the beautiful information of natural stones, the most concise, elegant, and vivid lines are used in carving. The natural and beautiful form and perfect pattern of stone, whether the expression of the line of shape or the line of pattern, flow out in nature. This is carved nature, a kind of deliberate nature and nature in business.

The beauty of carving is the beauty of cutting. For thousands of years, carvers have acquired a series of techniques for cutting stones. Numerous works have demonstrated the artist's ability to control stones and tools. They use the knife calmly and steadily, and with proper strength they make fine stones. They draw the lines with a magical skill, using fine and firm strokes; the way of cutting is diversified and extremely skillful. The rich and exquisite cutting language cause the formal beauty and implication beauty of Field-yellow stone to be fully displayed. Appreciating stones has the theory of "extreme gorgeous," which proves the wonderful carving beauty of Field-yellow stone.

To embody the natural beauty of carving, we shall also follow the principle of "faithful to texture of natural Field stone." Therefore, through past dynasties, a set of carving methods and artistic characteristics of "applying according to one's aptitude" have been formed, which also contains an important aesthetic idea, that is, "the true beauty lies in the light of the texture itself." After hundreds of millions of years of natural experience, Field-yellow stone contains unique natural beauty, and each stone has a different shape, texture, and color. Carvers shall not only have superb skills, but also have extraordinary wisdom and profound understanding of the internal relationship between the stone quality, intention, and skill; use them skillfully and learn them well; and seek a kind of form language that matches the shape, color, and gentle temperament of Field stone in carving. In modeling, we shall follow the essential characteristics of stone. Modeling, graphics, and craftsmanship shall be subordinated to the stone, not override it.

Carving with ingenuity shall be guided by circumstances, so that the artificial carving can be promoted to "nature's work" to achieve the perfect unity of natural beauty, implication beauty, and carving skill beauty of stone.

Skillfully utilizing stone color makes the stone work interesting naturally in the creative conception. Removing defects is regarded as a good opportunity to reshape the stone form and make the stone more vivid, and sometimes the glittering and beautiful stone seems to have a trace of shyness, not easily shown on the surface, but residing deep within it. The carver shall watch the stone quietly, observe it silently, and read it carefully; that is the key to carving stone. Only by seeing through the stone can the carver have a thorough understanding of the stone and be able to carve with ease, and can the natural beauty of Field-yellow stone be fully and perfectly reflected.

Lin Qingqing, the most influential artist in modern Shoushan stone carving, applied painting theory and pitch rhyme of Chinese painting to the superficial carving creation of Field-yellow stone, creating the highest realm of Field-yellow stone. He used the ancient stone carving, portrait brick, and bamboo flat carving for reference to enrich and develop the superficial carving art of Field-yellow stone. Lin Qingqing's first work of painting started from repeatedly observing a yellow-black double-skinned Field-yellow stone. At first, he created "Washing Feet with Canglang Water." Although he painted repeatedly on the Field-yellow stone with his brush, he was never satisfied. Later, he changed his original intention and drew a clump of bamboo on the yellow skin and two chickens scrambling for food on the black skin, carving carefully using only two layers of stone skins, a process that lasted more than half a month. This groundbreaking work of art has a strong charm and fresh composition. It can be said that after the carver has learned how to paint, superficial carved Field-yellow stone can have a different look.

Lin Qingqing's work "Evening Banquet in the Peach and Plum Garden" is a raven-skinned and orange peel red Field-yellow stone, which weighs up to 250 grams. It is the top grade of Field-yellow stone. Lin Qingqing devoted all his effort to it. He used more than two months to observe the implication of the stone. This work is extremely exquisite, excellent in both carving and stone quality. His works have the characteristics of a creative artistic style; whether observing stone, scribing lines, removing extra skin, or carving, all have their own unique features. Usually when he created, he finished and polished the stone, played with it at any time, and conceived and designed it. He said, "Before carving, we shall observe the stone. When the stone color is exquisite, it shall be carved into flowers. When the stone has cracks, it shall be carved into mountains and rivers to make full use of the stone potential. Before carving, we shall draw the painting in our minds to express the poetic illusion." Observing the stone is the beginning of creation. Whether the design is exquisite or not determines the success or failure of the work.

With regard to Lin Qingqing's superficial carved seal work "Journey to the Autumn Mountains," in the small picture, the sky is high and cloudy; the ancient trees are sparse; the trestle lies across the river; several wine shops are located in it; waiters are greeting businessmen outside the wine shops, and guests are drinking freely in the wine shops; four donkey riders and three porters are on their way. The characters'

demeanors echo each other, and the main scene and the matching scenery are dense and compact. Each detail is lifelike, and can be described as trying "to achieve a broad macro realm while penetrating into the fine and detailed micro-level."

Lin Qingqing's depiction of the surface of the superficial carved scenery is even more unique. He carved the trunk, scars, knots, petals, leaves, characters' structures, and clothing patterns with a small round knife. He used the knife smoothly and compactly with excellent skill. For smoke, clouds, rocks, etc., he used a round or semi-round knife. His carving method is smooth, simple, rigid, and flexible, highly proficient. He usually reserved the smooth surface of freehand bamboo leaves, reeds, etc. He intentionally blended the tail of bamboo leaves with the ground. His skills of carving the watermarks are very special and admirable. He applies just a few carvings, and the position, length, and density of carving are exactly right and cannot be changed. It is just like magic.

From the 19th-century carving master Lin Qingqing's Field-yellow stone works to the contemporary master Lin Wenju's works Grass Pond, master Liu Aizhu's works Samsung Cup Baduk Championship, and works of other carving masters, they all embody the beauty of nature, implication, and carving in aesthetic treatment of superficial carving, intentionally and unintentionally, in the aspect of skill and nature, and in imagination of beauty of artistic conception.

5 CONCLUSION

In carving, first of all, we shall conform to the law of natural beauty of stone. At the same time, carvers shall have consummate skills and proficient techniques of carving stone. From being highly skilled in technology, they shall gradually enter the flexible artistic realm, and they shall control freely and intelligently. From intentional skill pondering to unintentional skill revealing, push the carving skills to a new higher level and the beauty of nature can be better figured out.

Although carving stone is a skill, it's a unique way of design thinking. It perfectly integrates "natural taste" and "intention," delivers beautiful information, and nourishes people's hearts through its warm and moist texture, diverse shapes, and elegant rhythm, all of which are based on the skill of carving.

REFERENCES

X. Chen, 2012, *Literary Theory* (6), 27–29.
Y. Wang, 2017, *Literature Life (Trimonthly Publication)* (12), 195.
R. Yu, 2017. *Literature Life (Next Trimonthly Publication)* (1), 43.
X. Lin, 2017. *Masterpieces of Nature* (1), 48–49.

Smart Science, Design & Technology — Lam et al. (eds)
© *2020 Taylor & Francis Group, London, ISBN 978-0-367-17867-3*

Rethinking design: An innovative approach to sunglasses design and manufacture

Safa Tharib*
Department of Entertainment Management, I-Shou University, Kaohsiung, Taiwan

Tun Hsueh Chan & Yung Chun Lin
Department of Creative Product Design, I-Shou University, Kaohsiung, Taiwan

ABSTRACT: Approaches to design are changing. Glasses as well as other wearable fashion items need not be restricted to traditional approaches in their creation, wherein flat components are manufactured and faceted together to create the final item. 3D printing technology approaches allow for manufacture of said items in single flexible pieces, reducing manufacture time and allowing for new design approaches to be envisioned. Sunglasses, while being a practical item that a person would wear to protect their eyes from the sun, are also a fashion statement. Therefore, new design approaches need to be explored.

1 INTRODUCTION

3D printing offers new ways of implementing specific creative and manufacturing choices. 3D printing itself could be considered to be a form of rapid prototyping (Wang 2013). However, it is more accurately described as additive manufacturing. Within the manufacturing industry there are three different approaches to realising a design; these are additive, subtractive, and moulded. Traditionally, the most common approaches to manufacturing have been subtractive and moulding. Utilisation of technologies such as CNC milling to create moulds for injected plastics and/or metals have been, and still are prevalent in the manufacturing industry (Bikas et al. 2016). Aside from 3D printing there are traditional approaches to additive methods such as welding and to some extent ceramics.

The use of 3D printing technologies allows for the fabrication of complex shapes such as lattice structures and interlocking curves (Ishutov et al. 2017). The rapid fabrication in which a design is modelled and sent to a 3D printer as one would print a text document allows for corrections and adjustments to be made on-the-fly. In the utilisation of 3D design software, the designer can quickly modify a design to fit the specification of a consumer. Otto and Wood describe a configurable design method in which a product could use a fabricate-to-fit approach. This approach allows for some variables to be adjustable, which provides a customised product that more ideally fits the needs of the consumer (Otto & Wood 2000). A Belgian shoe retailer known as Runners Service Lab has taken this approach with the utilisation of 3D printing technology in the production of customised running shoes. Nike also offered customised sport shoe components in 2013. These components made use of 3D printing technology (Weller et al. 2015). Abdalla and Chan suggest that companies could offer customers the ability to customise their product before manufacturing via online web platforms in the future. In turn this gives the retailers and manufacturers insights into future treads as they begin to emerge (Abdalla & Chan 2011).

In this study we actuate an approach to creating sunglasses with the utilisation of 3D printing technology. In this study we hope to address some issues with traditionally manufactured sunglasses which 3D printed approaches can diminish or mitigate entirely. A problem with traditionally manufactured sunglasses is that at times they can present a health hazard, a particularly important consideration when creating an item that is to be worn not only on the face but also close to the eyes. With 3D printing technology antibacterial plastics can be incorporated into the design with little or no impact on the aesthetic value of the product or the production time. Antimicrobial composites have already been used in medical applications of 3D printing (Pattinson & Hart 2012). Using stereolithographic materials also provides the option of adding an FDA-approved coating to the finished product. Other considerations could be waste plastics and material that can be damaging to the environment. However, some filaments such as PLA are biodegradable.

In regards to production time, 3D printing offers a much more streamlined approach to manufacturing than tradition methods which incorporate CNC milling and injection moulding. 3D printing offers a tool-less approach to design and manufacture, as such production time is shortened and the need for

---

*Corresponding author: E-mail: safa@isu.edu.tw

expensive tooling is mitigated (Bak 2003). However, 3D printing itself presents challenges. Among these challenges are the fill, structural strength, and elasticity of the finished product. 3D printed wearable elastic materials have been explored with the application of embedded electronic devices. Research conducted by Muth et al. explored this application; however, the final output, while complex, was a flat object with complex circuity embedded inside the design (Muth et al. 2014). Complex and bendable 3D objects are still challenging in 3D printing methods; this pertains to the durability of the item. To address the issue affecting the viability of 3D printed glasses, a method for assessing the quality of the output of this study is required. The FAST method described by Otto and Wood (Otto & Wood 2000) is a practical approach which fits with the objectives of this study to explore and actuate unique design approaches with 3D printing technology that are problematic for traditional manufacturing methods. The FAST method, also known as The Function Analysis System Technique, involves breaking down a design and assigning different levels of importance to each component that makes up that particular product. This allows one to surmise which components are necessary for functionality and which are not. In eliminating nonessential components, one can fully concentrate on the remaining features and fully develop a design appropriate for the medium.

2 METHOD

We began our approach by breaking down the glasses into key component parts (see Figure 1). This allowed us to explore which parts were absolutely necessary for the basic design to function as sunglasses. We decided to merge the important components into a single piece. This meant that glasses could be printed in one go. In exploring how 3D printing could provide new design approaches we developed a pair of glasses that had a twist and dimpled design (see Figure 2).

The approach to fitting the glasses made use of 3D scanning technology. This allowed us to measure the important areas of a human face and adjust the model to perfectly fit (see Figure 3).

Figure 2. Initial design with perforations and twist.

Figure 3. 3D scanning for the application of fitting.

Figure 4. Result of SLA printer.

Once basic criteria were met, we moved on to 3D printing the design. For the first model we opted to print using a stereolithography 3D printer and cured the output with a UV machine (see Figure 4).

Upon reviewing the result, we found the small perforations had not materialised. Aside from this, the design had printed well. The twist was printed as expected and the glasses were wearable. However, due to the hingeless design and inflexibility of the cured resin, the final output was impractical. From this outcome we found two separate branches of exploration, one being the flexibility of the output and the other being the unique twist design. To address the first issue, we printed another single piece design using thermoplastic polyurethane (TPU) filament. This filament has elastic properties and an almost clothlike feel to it. The output, while untidy, allowed for an extreme amount of flexibility (see Figure 5). The glasses were

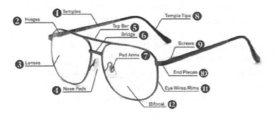

Figure 1. Component parts of a typical pair of glasses (Anon 2015).

Figure 5. Glasses printed with TPU filament.

strong enough to support a heavy amount of distortion and return to their original shape. However, due to this flexibility these glasses were deemed unable to effectively support any type of lens.

Following on from the first output, we further explored harder materials and once again used the SLA 3D printer to print an even more twisted design based on our original concept (see Figure 6).

This time, further twists were added and the glasses were thickened. The result was a much sturdier set of glasses with a more complex twist in the design (see Figure 7). However, some of the original problems still existed. At this stage we decided to include components that were deemed of lesser importance than initially thought.

To further explore the twists in our initial designs and provide a greater level of user practicality, a radically new design was envisioned (see Figure 8).

In the process of the fabrication of this design, we explored different printing materials. The various materials used in the realisation of this design were ABS, PLA, SLA, TPU, and PVA. When printed with SLA, the output was less than desirable. The legs of

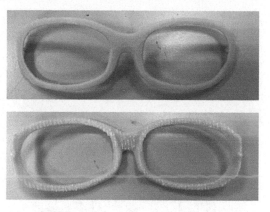

Figure 9. PVA result.

the glasses did not print successfully even after they were thickened. A similar problem occurred when we used TPU filament to print this design. Both TPU and SLA printing methods provided unusable outputs. This design proved challenging due to the delicate spirals. We did further tests using PVA and ABS filament. In this test a dual extrusion system was employed; again, due to the delicate nature of the design, this proved problematic. Slight errors in the printing process due to extra filament occasionally leaking from the extruder led to misalignment of the overall print. This led to poor results, and the application was deemed undesirable for intricate design at this stage. That said, we noted the PVA material alone provided high amount of detail. We printed the glasses again with only PVA filament (see Figure 9).

Unfortunately, due to the brittle nature of the PVA material the legs were particularly difficult to remove from the supports without causing damage. The best results were obtained with the use of ABS and PLA material. Both PLA (see Figure 10) and ABS (see Figure 11) provided comparable results with a lower, but still high failure rate (see Figure 12). Problems encountered were due to three details. One of these details was support filament becoming unstuck from the tray, which in turn causes a misaligned print. The second was slight fluctuations in print temperature, which caused burning of the small filament strands, once again leading to a misaligned print

Figure 6. Twisted glasses design.

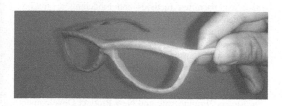

Figure 7. Second result of the SLA printer.

Figure 8. New twisted spiral design.

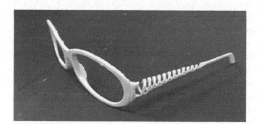

Figure 10. PLA result.

Figure 11. ABS result.

Figure 12. Failure rate.

result. Finally, as with other materials, the extra deposits from the extruder that would leak and solidify would on occasion impede the movement of the printing head which moved printed filament to an offset position. Positioning the model in a correct orientation avoided viscous material collapsing, which was of particular importance to the small screw holes on the hinges of the design. The best method was to print the legs at a flat angle so that the cavity was facing straight down. However, the spirals were better suited to a straight upright printed angle. Therefore, a compromise between the screw guide and spirals needed to be made.

3 DISCUSSION

Results showed that printing complex designs with 3D printing technology is possible and viable. However, there is a high failure rate involved. This is often due to the variable temperatures between which the extruder fluctuates during the printing process. Other problems pertain to leaking extruder heads and printed filament being detached from the heated platform. These problems lead to a misalignment during the printing process, which in turn compounds to further misalign the print. Different materials in different 3D printers provide dissimilar results, as one would expect with any mechanical device. Depending on the design of the model, different materials respond in different ways, thus some materials are more suitable and codependent on a particular design. Depending on the particular design, usable sunglasses were obtained from this study. This proves that unique designs are obtainable; however, as the complexity of the design increases, so does the failure rate.

4 CONCLUSION

One could conclude that results would not necessarily match when the same object is printed in a different 3D printer or when different materials are used. There is still a need for troubleshooting and the process of 3D printing a complex structure with fine details is not a simple automated task, but one which requires operator input at all stages. That said, the production pipeline is drastically reduced when compared to traditional methods. In turn, this reduces production time and cost. Adjustments and corrections can easily be made on the fly without the need for tooling. Therefore, this approach with further refinement can be a viable alternative to traditional methods of manufacturing and as demonstrated provide a larger choice of design and manufacturing options over traditional methods.

REFERENCES

Abdalla, H., & Chan, T.H., 2011. An integrated design framework for mass customisation in the consumer electronics industry. *International Journal of Computer Applications in Technology*, 40(1–2), 37–52.

Anon, 2015. Frame Diagram [online]. Readers. Available from: https://www.readers.com/blog/glasses/eyeglass-frame-diagram/ [Accessed 12 February 2019].

Bak, D., 2003. Rapid prototyping or rapid production? 3D printing processes move industry towards the latter. *Assembly Automation*, 23(4), 340–345.

Bikas, H., Stavropoulos, P., & Chryssolouris, G., 2016. Additive manufacturing methods and modelling approaches: a critical review. *The International Journal of Advanced Manufacturing Technology*, 83(1–4), 389–405.

Ishutov, S., Hasiuk, F.J., Jobe, D., & Agar, S., 2017, Using resin-based 3D printing to build geometrically accurate proxies of porous sedimentary rocks. *Groundwater*, doi:10.1111/gwat.12601

Otto, K., & Wood, K., 2000. Product Design: Techniques in Reverse Engineering and New Product Development, 1st ed, New Jersey: Pearson, 154–313I.

Jacobs, S., & Bean, C.P., 1963, *Magnetism*, 3, Rado, G.T., & Suhl, H., Eds., New York: Academic Press, 271–350.

Pattinson, S.W., & Hart, A.J, 2017. Additive manufacturing of cellulosic materials with robust mechanics and anti-microbial functionality. *Advanced Materials Technologies*, 2(4), 1600084.

Muth, J.T., Vogt, D.M., Truby, R.L., Mengüç, Y., Kolesky, D. B., Wood, R.J., & Lewis, J.A., 2014. Embedded 3D printing of strain sensors within highly stretchable elastomers. *Advanced Materials*, 26(36), 6307–6312.

Weller, C., Kleer, R., & Piller, F.T., 2015. Economic implications of 3D printing: Market structure models in light of additive manufacturing revisited. *International Journal of Production Economics*, 164, 43–56.

Smart Science, Design & Technology — Lam et al. (eds)
© 2020 Taylor & Francis Group, London, ISBN 978-0-367-17867-3

A study on app design of a smart happy farm

Shuhuei Wang*

Department of Digital Design, MingDao University, ChangHua, Taiwan

ABSTRACT: Due to agricultural transition and an aging agricultural population on average for the past few years, young people that are limited to a traditional stereotyped impression of agriculture neither know much about it nor are willing to dedicate themselves to it. As a result, it is the objective of this research that children of school age may develop knowledge and an interest in agriculture through interactive farm application games. Literature review was conducted for an understanding of children's cognition of agriculture. The method of case study was also used to locate and interpret the issues of the objects under study. It was found that the interactive app game of Little Angel's Farm could motivate childrens' interest in agriculture via incorporation of technology since they had fun playing the game, felt comfortable with agriculture, had an understanding of planting vegetables and fruit, and created little farms that belonged to them.

1 INTRODUCTION

A. *Research motives*

Chen Chong, the Director of Post Modern Agriculture of Ming Dao University, indicated that farmers in advanced countries like Germany and America were required to have a certificate. They were very professional, aware of acquiring new knowledge, able to use computers, and engaged with agriculture via technological machines. Besides, they had a sound agricultural organization to integrate agriculture and enhance efficiency. Farmers overseas have specialty certificates and know how to apply technological products to their work as well as to develop an all-around organization to integrate agriculture. Our farmers, however, tend to be older, and it takes a tremendous amount of time to teach them how to use technological products. If we develop our children's interest in agriculture at an early age, agricultural progress in Taiwan for the future will be possible.

B. *Research issues*

The objects under study are children of school age (under twelve years old), as we intend to motivate kids to have an interest in and knowledge of agriculture at an early age. Most current farm games give only a simple introduction of the seasons to plant a certain crop instead of fun and knowledge of planting in reality. Furthermore, no content on vegetables is contained in the games and no actual planting methods are provided either since fun is the orientation. Consequently, the issues to be studied are:

- What kinds of games can attract children?
- What is children's cognition about app games?
- Can interactive app games help children learn more about agriculture?

2 LITERATURE REVIEW AND EXPLORATION

A. *Development of interactive farm app games*

Farm games have become must-play games, starting with the Happy Farm on the Facebook page, which is a social game of running a farm, suitable for all ages and easy to manipulate with a relaxing atmosphere. Specific crops and animals are available, and the game is popular with more than twenty million players around the globe, causing game companies to research and launch farm games (e.g., the Brown Farm of Line). Plenty of farm games are played by running them on the interactive app game market for the moment. Unique crops are planted, animals are kept, delicate tableaus are owned, and buying and selling among friends is maintained, which is easy for all ages.

B. *Popularity of mobile devices*

According to the popularity data of mobile devices surveyed by Foreseeing Innovative New Digiservices (FIND), in 2014 the tendency of people to use mobile devices was increasing (see Figures 1 and 2).

*Corresponding author: E-mail: angelawang36@gmail.com

Figure 1. Popularity and growth rates of smart phones in Taiwan [FIND, 2014 (H2)]; [1].

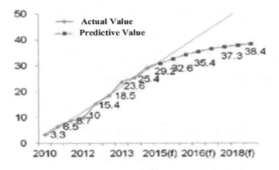

Figure 2. Popularity and growth rates of iPads in Taiwan [FIND, 2014 (H2)]; [1].

Table 1. Erikson's theory of psychosocial development (Peng Cuei-Hua, p. 5, 2011).

Stage	Age	Psychosocial Crisis (Key Development)	Good Development	Developmental Hindrances
1	Infancy	Basic trust vs. basic mistrust	Trust in others; have security	Anxious about interpersonal communication
2	Toddlerhood	Autonomy vs. shame and doubt	Self-control; act with confidence	Self-doubt; act without confidence
3	Early Childhood	Purpose, Initiative vs. Guilt	Objective oriented; act independently	Flinching and holding back without confidence
4	Middle Childhood	Competence, Industry vs. Inferiority	Able to study, do tasks and associate with people	Lack of basic life ability with frustration
5	Adolescence	Fidelity, Identity vs. Role Confusion	Precise concepts about self; direction of life confirmed	Lack of goals in life, feelings of lost and confusion
6	Early Adulthood	Love, Intimacy vs. Isolation	Successful love relationships and a good career	Loneliness; can't maintain intimate relations
7	Middle Adulthood	Care, Generativity vs. Stagnation	Passion for family; help protégé	Self-indulgence and disregard future
8	Late Adulthood	Wisdom, Ego Integrity vs. Despair	Acceptance of life in its fullness	A guilty conscience about the past leading to depression and hopelessness

C. Exploration of children's learning

Reference: Peng Cuei-Hua (May 10, 2011); Lesson 1: Ego Growth and Preparation of Becoming a Citizen. [source of data: Jhang Chun-Sing (1992), [2]

Erikson presented his theory of psychosocial development in 1950 and divided human life into eight stages with respective explanations. He proposed two unique perspectives. First, physical and mental developments at any stage are closely related to those in the previous stage (good development helps further development in the following stage). Second, every stage plays a critical part in crises and turning points for the better, which means when people encounter specific problems or difficulties at each stage, their mental crises will continue before problems are solved. Development will continue successfully after difficulties are overcome [2].

D. Erikson's eight stages of development

Erikson thought that ego awareness of a person could develop continually through a whole life and divided formation and development of ego identity into eight stages, which were determined inherently. Successful development at any stage, however, was decided by the environment, and no stage could be neglected. In addition, the bases and contents of education for each stage were important since any educational mistakes at any stage might result in developmental obstacles for an individual for life [2].

E. Case analysis of happy farm games

Games of happy farms include the following:
(1) Happy Farm
 More than ten million farmers are using their little fingers to

- plant fresh crops
- cook delicious food

272

- keep cute pets
- make friends with their neighbors
- trade with other farmers around the world
- complete challenging tasks
- design their most beautiful farms by selecting hundreds of ornaments (Google Play, [3])

(2) Brown Farm of Line

You may experience a colorful farm life with ease on the Brown Farm of Line by visiting your Line friends' farms, making money from the star figures of Line, and enjoying funny dialog with members of the brown bear family (Google Play, [3]).

(3) Innovation Drive Studio

You need to work and develop your farm day and night to become a leader in the commercial

world of farmers. You have to bake and make dairy products and clothes to be a happy farmer (Google Play, [4]).

(4) Farm School

You can express your own style at Farm School. Design and decorate your own farms to show their uniqueness. Till and pasture on your own land. Then, you can get more gold to expand and upgrade your farm through harvests of plants and animals. You don't have to wait. Enjoy a happy life on the farm right now (Google Play, [3]).

Aforesaid app games are summarized in the following table. As a matter of fact, there are a lot of similar games.

3 APP DESIGN OF SMART HAPPY FARM

Most of the current farm games available on the market only give a simple introduction to how to plant a certain crop instead of the fun of planting and knowledge about vegetables and fruit in reality. Little Angela's Farm, an App game produced in this research, was under testing and expected to help kids know the actual methods of planting crops, vegetables, and fruit during the process of game playing. In this way, children are not only playing games but also learning. Little Angela's Farm is a game on agricultural knowledge designed specifically for children. Simple operation helps kids learn about crops as well as planting methods. Besides, embedded little games motivate children to learn during playing.

Nothing is too special for popular App game players nowadays since they will get bored with highly repetitive games that are without innovation and significance quickly. In App design of the Smart Happy Farm, knowledge of vegetables and fruit will be incorporated for all age groups. Items designed for each age group are described as follows.

Table 2. Data analysis related to happy farms (summarized by this research).

Name	Characteristic	Method	No. of Players	Difference
Happy Farm	Top 5 family games in over 100 countries	Plant and sell vegetables and fruit	319,439	A family game in over 100 countries
Brown Farm of Line	Visit your Line friends' farms, make money from the star figures of Line, and enjoy funny dialog with members of the brown bear family	Create your own farm	1,109,480	Play with your Line friends by sharing your own farm
Innovation Drive Studio	1. The only offline farm game 2. Use buildings and decoration to create your dream farm 3. Increase cute animals and manage the pasture for cows 4. Plant fresh crops and cook delicious food 5. Upgrade barns/granaries for big families 6. Eye-catching character animation of high definition 7. Easy control facilitates farm management 8. Create a brand new world in a country setting 9. More harvest, more levels, and more experience	Vegetables, fruit, baking, dairy products, and clothes are included	102,162	The only offline farm game
Farm School	Players create their own farms	Players design and decorate by themselves	100,356	It is a place where you may express your own style

Note: the numbers of players are extracted from Google Play.

Table 3.   Age grouping of happy farm app players.

Age Group	Level	Content	Result
Below 7	Matching vegetables and fruit - 1	Learn about vegetables and fruit	Kids learn during playing
7–12	Lives of flowers, vegetables and fruit - 2	Learn about the growth process of every flower and vegetable	A further understanding of the growth of vegetables and fruit
Above 13	Knowledge of planting flowers, vegetables, and fruit	Teach planting	Application to real life is achieved via virtual games

References: App Store push children's game classification by age 3 - Database & Sql Blog Articles [5]

Table 4.   Little Angela's farm.

Interactive app design cover of Little Angela's Farm

Operation

Click seed and drag to the farm

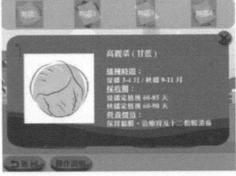

Introduction of vegetable

Select wrong places

Selecting three wrong places within a specified time can pass through three red lights

274

Apple's app store added a game app for kids recently and classified three age groups of below 5, 6–8, and 9–11 based on the difficulty of the games. Games corresponding to ages were placed together so that children and parents might choose suitable games [5].

4 CONCLUSION

The popularity of smart phones is increasing year by year in our modern society in which mobile devices are available everywhere. As a result, the research tool of this study was aimed at mobile devices. Kids at school age are curious about the world and desire knowledge. It will be a great opportunity to provide an experience on a virtual farm for children to enjoy and learn about agriculture and methods for planting crops and vegetables if they are encouraged to play farm app games. An app game, Little Angela's Farm, was constructed preliminarily and used to test children's cognition of and attraction toward agriculture. Defects and shortcomings of the game will be corrected after gathering statistics.

ACKNOWLEDGEMENTS

The Ministry of Education for a subsidy provided to the Innovative Alliance Program for Smart Living in 2017– Innovation Alliance for Smart Agriculture.

REFERENCES

[1] Popularity and Growth Rates of Smart Phones in Taiwan [FIND, 2014 (H2)]; Shen Ying-Yin, Yan Syuan, Lin Su-Jhu and Jheng Ren-Fu.
[2] Peng Cuei-Hua (May 10, 2011); Lesson 1: Ego Growth and Preparation of Becoming a Citizen. [source of data: Jhang Chun-Sing (1992), *Modern Psychology*, Taipei: Tung Hua Book Co., Ltd.; pp. 386–387].
[3] Google Play, Jan. 31, 2018.
[4] Google Play, Dec. 31, 2017.
[5] Mofang Inc., Jan. 14, 2015, App Store push children's game classification by age 3 - Database & Sql Blog Articles.

Smart Science, Design & Technology — Lam et al. (eds)
© 2020 Taylor & Francis Group, London, ISBN 978-0-367-17867-3

Research on the design of urban public environmental facilities based on regional image

Xiaoyan Zhang*
School of Design, Fujian University of Technology, Fuzhou, Fujian Province, PR China
School of Art and Design, Wuhan University of Technology, Wuhan, Hubei Province, PR China
Fujian University Humanities and Social Science Research Base Design Innovation Research Center, Fuzhou, Fujian Province, PR China

Yunshan Wang
Center of Modern Education Technology, Fujian University of Technology, Fuzhou, Fujian Province, PR China

Kun Wu
School of Art and Design, Hankou University, Wuhan, Hubei Province, PR China

Ming Hong
Department of Art and Design, Wuhan College, Wuhan, Hubei Province, PR China

ABSTRACT: This paper is aimed at the current problems of urban public environmental facilities, which are lack of personality and cultural characteristics and other issues. It puts forward that the regional image system theory should be adopted in the design of urban public environmental facilities. This paper first expounds the significance of the theory of the regional image system to the design of urban public environmental facilities. On the one hand, the regional image system can promote the inheritance and innovation of regional culture. On the other hand, it is conducive to the expression of urban characteristics and personality. Then the paper discusses the design of urban public environmental facilities, which should be combined with the natural resources and geographical characteristics of each city, as well as the adoption of systematic, people-oriented design principles. Finally, it highlights the local urban culture and other aspects. The conclusion is that the design of urban public environmental facilities should adopt the theory of the regional image system.

Keywords: regional image, public environmental facilities, design

1 INTRODUCTION

Urban public systems have always been a symbol of urban modernization level, and street public environmental facilities are an important part of urban public systems, so the image design of urban street public environmental facilities is particularly important. Urban public environmental facilities are the pillar to promote direct dialogue between mankind and nature. To some extent, they can help people to have a more direct and convenient dialogue with nature and play a role in coordinating the relationship between people and the urban environment. Reasonable layout and elaborate design of a series of public environment facilities let users feel the concern of the builder and managers. Nowadays, people's living habits and consumption concepts have changed significantly, and traditional public environmental facilities can no longer meet people's needs. Every designer needs to think about the new image design of urban public environmental facilities from a diversified perspective (Zhang & Dong, 2007; Zhou and He, 2016; Zhang, 2012; Xu and Zhang, 2013; Xue and Chen, 2012; Li, 2012; Sun, 2011; Qiu, 2009; Yang, 2014; Li, 2015).

Now, with the improvement of social and economic level, urban public environmental facilities have been greatly improved compared with the past. However,

*Corresponding author: E-mail: 610491762@qq.com

the difference between cities is still not obvious, and the regional image is not clear, which is inseparable from the cultural background of the city. Therefore, research on the design of urban public environmental facilities based on regional image is of profound significance.

2   THE ROLE OF REGIONAL IMAGE SYSTEM THEORY IN GUIDING THE DESIGN OF PUBLIC ENVIRONMENTAL FACILITIES IN MODERN CITIES

The regional image is the external reflection of the abstraction and integration of various internal political, economic, cultural, social, ecological, and other factors in a region. It is the most differentiated element set of a region from others, and it is also extremely important intangible wealth in the process of regional development. Urban public environmental facilities are closely related to the people's lifestyle. Each region is different in culture, climate, and geographical location. It is not only of theoretical value but also of practical significance to apply the theory of regional image system to the design of public environmental facilities in modern cities.

2.1   *Promote the inheritance and innovation of regional culture*

Regional culture is the precious wealth accumulated in the long-term historical development of a region and the important capital of regional competition. The construction process of a regional image system is exactly the process of digging and combing through regional culture. Only by fully excavating the unique regional culture can we build a regional image system with distinctive characteristics. Therefore, the design of urban public environmental facilities based on regional image can promote the inheritance and innovation of regional culture. Paris, France, for instance, is the world-famous "fashion capital," a regional image generated through popular feeling, the Louvre, Versailles, and other urban infrastructure. And it's combined closely with the regional culture, people's yearning and longing for Paris, and is attracted by the culture behind the building facilities, which is the charm of regional culture.

2.2   *It is conducive to reflecting urban characteristics and highlighting urban culture*

In the past, China's economy was underdeveloped and the product categories were few, which led to the lack of individuality and cultural connotation in the design of urban public environmental facilities in China for a long time and left a general impression of being dull. Now because many cities still tend to disregard their own characteristics and blindly chase trends, and the highly coordinated urban public environment facilities and urban environment destroy the whole of the city landscape effect, it is difficult to reflect the personality of the city. By using the theory of regional image system, we can combine the geographical cultural characteristics of each region and the unique geographical environment of each city to design public environmental facilities that conform to the local people's lifestyle, which can reflect the urban characteristics and highlight the urban culture. For example, Shangrila in the northwest of Yunnan is compared with Suzhou. Due to differences in geographical location and climate, Shangrila city's public platform is mainly red with Tibetan architectural characteristics, reflecting a kind of Tibetan cultural atmosphere. Suzhou is a garden city. The bus stop, street lamp, and garbage can all create a kind of ancient garden architectural characteristics. The different public environment facilities of the two cities can reflect the different connotations of the cities and make visitors feel the cultural difference.

3   RESEARCH ON THE DESIGN OF URBAN PUBLIC ENVIRONMENTAL FACILITIES BASED ON REGIONAL IMAGE

3.1   *Rational development of local natural resources*

Modern life science and technology have developed, but urban construction still depends on the natural environment, which is the basis of people's survival. Therefore, when designing public environmental facilities, we must consider the natural environment of each region, because it will be affected by the natural climate. Different cities have different natural resources, climate characteristics, and cultural styles. Different schemes of public environmental facilities are required for different climatic conditions. It can be seen that different regions have different geographical characteristics, and the design of public environmental facilities in different regions is also different due to different characteristics. Therefore, the design of public environmental facilities should take into account the surrounding natural environment, and pay attention to the harmony and unity between the facilities and the natural environment. It should conform to the natural environment, but also have control of the use and transformation of the natural environment, through the humanized design of the public environment facilities as an intermediary, to achieve the "unity of nature and human," that is, the natural environment and human life of the harmonious unity. Therefore, when natural landscape is not affected, materials integrated with public environmental facilities are selected to achieve the balance between people and things and bring out the best in each.

3.2   *According to the systematic design principles, people-oriented*

The construction of public environmental facilities is a long process, which involves a lot of departments, such as municipal, telecommunications, sanitation, garden, etc., that all need to work together and coordinate. We should not just start from the interests of individual departments and change the design scheme at will, but obey the overall arrangement and start from the whole, so as to avoid chaos.

The design of existing urban public environmental facilities should be as considerate as possible and should not be one-sided in pursuit of form. Function is the first priority of facilities, and humanized design is to meet the basic function of the premise, which is to consider the needs of some special groups. People-oriented is a rational concept in line with the requirements of the development of the times. Therefore, the planning and design of urban public facilities should show concern for people and pay attention to their physiological needs and feelings. With a focus on people, this paper studies people's needs, explores various potential desires, and puts forward various problems existing in people's outdoor life. And through public facilities themselves, so that the human organs can be extended and play a role, so that people move more conveniently and comfortably, so as to realize the value, improve the overall quality of outdoor life. In addition, the human-oriented design should consider not only normal demands, but also those of the disabled, the elderly, and children, and other special requirements, reflect on the life of a person's care, everyone can equal and free to the right in outdoor activities, and efforts to create a fair and equal social environment. In his book *Design for the Real World*, published in the late 1960s, American design theorist Victor Babanak explicitly proposed three major problems of humanized design: "(1) Design should serve the general public, not a few wealthy countries. Here, he stressed that design should serve the people of the third world. (2) Design should not only serve the healthy, but also the disabled. (3) Design should seriously consider the use of the earth's limited resources, and it should serve to protect the limited resources of our planet. His several theoretical views are also the goal pursued by humanized design, that is, to design for all people, including healthy people, the disabled, the elderly and children, to make their use convenient and safe, and to achieve the combination of human and environment in this process." This is the essence of humanized design.

3.3 *Highlight urban regional culture*

The unique urban regional culture is the most vivid aspect of the city and has great influence. Any public environmental facilities can serve as the carrier of culture. If the design of public environmental facilities can be based on their own regional climate characteristics, combined with the local culture, and in line with the local people's way of life, it can highlight the difference with other cities, to improve the quality of the urban environment. For example, the bus platform of JingHong city, the prefecture capital of Xishuangbanna Dai autonomous region, is full of ethnic characteristics. The architectural style of the platform is consistent with that of local houses, reflecting the local minority customs and the surrounding environment, so that tourists can never forget it. It can be seen that in the design of public environmental facilities, local cultural connotations should be explored rather than simply be copied and float on the surface. Only in this way can public environmental facilities with local characteristics and personalities be designed.

4 CONCLUSION

Consistency with the overall local atmosphere shall be maintained when using the theory of regional image system to guide the construction of modern city public environment facilities. Public facilities are not just isolated individuals; rather they are part of the public environment. We should consider all the elements in the general situation, watching out for each other, and highlight the characteristics of the whole, fully embodying and displaying the image of the city. Combined with the characteristics of local culture, geography, and climate resources, each city can design public environmental facilities in line with local characteristics. Only in this way can people's life become more and more convenient and our cities become more and more beautiful.

FUNDING

This research is funded by the following projects: (1) the 2018 Humanities and Social Science Research Art Program in the Ministry of Education, project number: 18YJA760061; (2) the 2017 science and technology research project in Hubei Provincial Department of Education, project number: B2017388; (3) Fujian University of Technology research foundation project in 2018, project number:GY-S18016.

REFERENCES

C.L. Qiu, 2009. *Design and culture*. Chongqing, China: Chongqing university press.

H. Sun, 2011. *Public Space Design*. Wuhan, China: Wuhan university press.

H.L. Zhang & Y. Dong, 2007. *Urban space elements-public environment design*. Beijing, China: China building industry press.

J.S. Li, 2012. *Public art and urban culture*. Beijing, China: Peking University press.

L. Li, 2015. *Jingchu cultural resources and the construction of Hubei regional image*. Wuhan, China: Hubei people's publishing house.

L. Zhou and X.H. He, 2016. *Art Panorama 16*(10),108.

W.K. Xue and J.B. Chen, 2012. *Public facility design*. Beijing, China: China water resources and hydropower press.

X.Y. Yang, 2014. *Design sociology*. Beijing, China: China building industry press.

Y. Zhang, 2012. *Design of public facilities*. Beijing, China: China water resources and hydropower press.

Y.D. Xu and Y. Zhang, 2013. *Urban environmental facility planning and design*. Beijing, China: chemical industry press.

Mechanical & automation engineering

Smart Science, Design & Technology — Lam et al. (eds)
© 2020 Taylor & Francis Group, London, ISBN 978-0-367-17867-3

A wide-tuning frequency range DLL-based clock generator

Chunyen Wu & Chinan Chuang*
Department of Electronic Engineering, Huafan University, Taipei, Taiwan

ABSTRACT: A Wide-Tuning Frequency Range DLL-Based Clock Generator is presented, which is simulated by TSMC 0.18um CMOS process. The architecture of proposed Delay-Locked Loop consists of Start-up Circuit, Phase Detector, Lock Detector, Charge Pump, Wide-Range VCDL, and Edge Combiner. The proposed clock generator can generate clock signals ranging from 1GHz to 4GHz. The power consumption of the presented DLL is 15 mW at a 1.8 V of supply voltage.

Keywords: Wide-Tuning Frequency Range, DLL, Clock Generator.

1 INTRODUCTION

Almost all electronic systems have a clock generator [1], and the clock generator determines the speed of the whole system, a good clock generator can greatly improve the performance and adaptability of the system. There are two ways to implement the general clock generator. One is to use the crystal oscillator. The other is to use the circuit technique and the crystal oscillator to realize the high speed and wide-tuning range of the clock generator. The crystal oscillator only generates a single frequency, which is only suitable for low-speed and simple systems. Generally, circuit techniques are used to improve output frequency of the crystal oscillator in order to make the system more high-speed and flexible. The circuit implementations are mostly divided into two methods .One is Phase Locked Loop (PLL)[2-3] the other is realized by using Delay-Locked Loop (DLL)[4-6] .The implementation of PLL has the advantages of high speed, easy to change the output frequency, and has many applications. Compared with the PLL-based clock generator, the DLL has many advantages, such as smaller area, faster locking time, unconditional loop stability and better performance of clock jitter. Therefore, in recent years, DLL is often used to replace PLL.

2 CIRCUITS DESCRIPTION

In this work, we use DLL to realize the wide-tuning range operation of the clock generator. Compared with the PLL-based clock generator, DLL has the advantages of small area, fast locking, low noise, unconditional stability and low jitter accumulation.

Figure 1 shows the proposed clock generator, which consists of a start-up circuit[7], a phase comparator (PD)[8], a charge pump (CP)[9], a loop filter (Capacitor), lock detector (LD)[1], wide-range voltage controlled delay line (Wide-Range VCDL) [7] and edge combiner[10]. The input signal generates a delay of the input reference clock period via the start-up circuit, and the delay of VCDL gradually produces a large delay from the minimum delay. When the output delay of the VCDL is exactly one cycle of the input reference clock, the loop is locked. When the loop is stable, the same delay cell in the VCDL divides an output reference signal period, thus generating multi-phase output clock. The function of the Edge combiner is to connect the rising edges of the multi-phase output clock to produce a multiplier output frequency. The function of the CP is to convert the phase difference generated by the PD into a proportional current to charge and discharge the loop filter. A larger charge and discharge currents combined with smaller loop filter capacitor saves lock-up time, but generates large noise jitter when the loop is stable. In order to achieve faster lock time and more stable clock generator. We use the lock detector circuit to detect the lock condition, and use a larger current to charge when the loop is farther away from the phase of the lock desired. Otherwise, switch to a smaller current lock when the loop is close to lock to reach a lower clock jitter performance. The function of the LD is to detect the difference phase and switch the detection circuit of the CP to generate different charging currents. In the following sections, we will give a detailed description of each component used in this work.

---
*Corresponding author. E-mail: cnchuang@cc.hfu.edu.tw

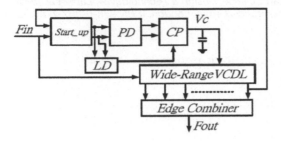

Figure 1. Proposed DLL-based clock generator.

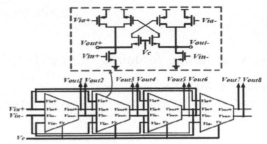

Figure 3. Wide-Range VCDL.

A. *Start-Up*

In the loop of the DLL, since the function of the VCDL is to control and generate the delay of the reference input reference clock, which will inevitably lead the output clock of the VCDL. However, in general, if the VCDL output clock is directly compared with the input reference clock to the PD for phase comparison, an incorrect locking direction will result in false locking. In order to solve this problem, a start-up circuit is usually added before the DLL input. The start-up circuit not only prevents the false locking, but also generates a delay of one input reference clock period to ensure that the DLL can be locked in one period of input reference clock. When the DLL is stable, the VCDL composed of multiple identical delay cells can be used to generate multi-phase clock output. Figure 2 shows the start-up circuit we used, which contains two DFFs and two and gates. When the "Start" signal is changed from 0 to 1, the VCDL output signal will be sent to the PD, and the input reference clock signal will be output to the PD. This circuit not only ensures that the output signal phase of the VCDL will lead the input reference signal, but also utilizes the serial connection effect of the two DFFs to generate a delay of the input reference signal period to ensure the final locking state of the VCDL.

B. *Wide-range VCDL*

The VCDL consists of a series of delay cells whose function is to provide input reference clock delay that can be controlled by voltage. The VCDL composed

of the same delay cell can generate a multi-phase clock output when the loop is locked. The delay range of a VCDL determines the frequency turning range of the DLL-based clock generator. If you want to achieve a wide range of applications, you need to design the VCDL as a circuit that can generate a wide range of delay control. In this work, we used a dual-path delay cell to decrease the delay of the delay cell by the pre-charge and discharge effect of dynamic logic, and then match the delay path of the normal delay unit to achieve a wide range of operation. Figure 3 shows the VCDL for the wide frequency range operations. To achieve high speed operation, we use a pseudo-differential pair delay cell with two pairs of input signals and a pair of output signals. The two input signals are the normal delay path and the pre-charging path for dynamic logic. The dual-path delay cell can achieve the high-speed and wide-tuning range operation of the VCDL, and the DLL-based clock generator can also provide a wide-tuning range operation of the multiplier output frequency.

C. *Charge Pump*

The function of the Charge Pump is to convert the output of the PD into a proportional charge and discharge current, and convert the current into the output delay of the voltage control VCDL by the capacitance of the loop filter. Generally, the Charge Pump is composed of a current mirror, and the control signal is used to control the current mirror to charge or discharge. In this design, we use a large current charge and discharge to achieve the fast lock function, and then use the LD detection to set the stability of the loop. When the loop is close to stable, the current of the CP is switched to a smaller current. Achieve more stable and low jitter of DLL output signal. Figure 4 shows CP used in this work, which contains two sets of charge and discharge currents, it can be switched by the output signal of the LD to achieve fast lock and more stable output.

D. *Lock Detector*

Figure 5 shows the Lock Detector, which is composed of two delay units that generate Td delay and two

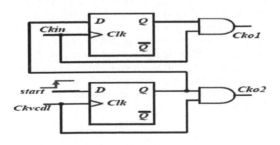

Figure 2. Start-Up.

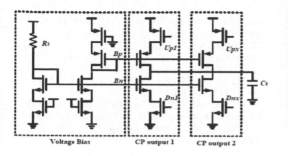

Figure 4. Charge Pump (CP).

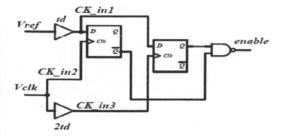

Figure 5. Lock detector.

DFFs to detect whether the loop is close to locking. The LD produces an output of 1 when the loop is near lock and vice versa when the loop is not locked. The switching of the LD detection with the CP current can not only effectively increase the locking speed, but also achieve the effect of stabilizing the VCDL control voltage.

E. *Edge combiner*

Figure 6 shows edge combiner circuit [10], which function is to convert the multiple phase signals of the VCDL output into multiplied output frequency. This circuit uses the two sets of pulse generator circuits to generate two pulse clock signals, which are then combined into a multiplier frequency output by logic circuits. The number of multi-phases of the input signal determines the number of output

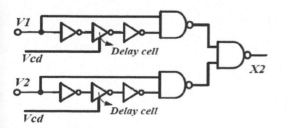

Figure 6. Edge combiner simulation results.

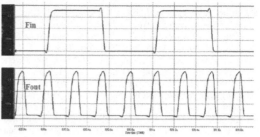

Figure 7. Simulation results of proposed DLL-based.

frequency multipliers that can be generated. The same circuit can be connected in parallel to produce different combinations of frequencies.

3 SIMULATION RESULTS

Figure 7 shows simulation of the proposed DLL-based input and output results, and the performance summaries is shown in Table I. The input frequency of proposed DLL-based clock generator is 250MHz~1GHz, and the VCDL delay length is about 1~4ns. The output frequency is 1~4GHz and the power consumption is about 15mW without the output buffer. The lock time is approximately 100 input reference clock periods.

4 CONCLUSION

In this paper, A Wide-Tuning Frequency Range DLL-Based Clock Generator is proposed. The input operating frequency is 250MHz~1GHz, and the output frequency range can reach 1GHz~4GHz. The improved normal dual path VCDL circuit is used to improve the operating speed.

REFERENCES

K. Ryu, D.-H Jung, and S.-O. Jung, "A DLL With Dual Edge Triggered Phase Detector for Fast Lock and Low Jitter Clock Generator," *IEEE TRANSACTIONS ON CIR-CUITS AND SYSTEMS—I*, VOL. 59, NO. 9, Sep. 2012.

S.-N. Kao and F.-J. Hsieh, "A fast-locking PLL with all-digital locked-aid circuit," *International Journal of Electronics*, vol. 100, pp. 245–258, Jun. 2012.

J. Kim, M.A. Horowitz, and G.Y. Wei, "Design of CMOS adaptive bandwidth PLL/DLLs : Ageneralapproach," *IEEETrans.CircuitsSyst. II*, Analog Digit. Signal Process., vol. 50, no. 11, pp. 860–869, Nov. 2003.

R.-J. Yang, and S.-I. Liu, "A 40–550 MHz Harmonic-Free All-Digital Delay-Locked Loop Using a Variable,SAR Algorithm," *IEEE J. Solid-State Circuits*, vol. 42, pp. 361-373, Feb. 2007.

K.H. Cheng, and Y.L. Lo, "A fast-lock wide-range delay locked loop using frequency-range selector for multiphase clock generator," *IEEE Tran. Circuits Syst. II*, Exp. Briefs, vol. 54, no. 7, Jul. 2007.

Y.J. Jung, S.W. Lee, D. Shim, W. Kim, and S.I. Cho, "A dual-loop delay-locked loop using multiple voltage-controlled delay lines," *IEEE J. Solid-State Circuits*, vol. 36, no. 5, pp. 784–791, May 2001.

C.-N. Chuang, and S.-I. Liu, "A 20-MHz to 3-GHz Wide-Range Multiphase Delay-Locked Loop", *IEEE TRANSACTIONS ON CIRCUITS AND SYSTEMS—II: EXPRESS BRIEFS*, VOL. 56, NO. 11, NOVEMBER 2009.

S. Kim, K. Lee, Y. Moon,, D.-K Jeong, Y. Choi, and H. K. Lim, "A 960-Mb/s/pin Interface for Skew-Tolerant Bus Using Low Jitter PLL", *IEEE JOURNAL OF SOLID-STATE CIRCUITS*, VOL. 32, NO. 5, MAY 1997.

C.-N. Chuang, and S.-I. Liu, "A 0.5–5-GHz Wide-Range Multiphase DLL With a Calibrated Charge Pump", *IEEE TRANSACTIONS ON CIRCUITS AND SYSTEMS—II*: EXPRESS BRIEFS, VOL. 54, NO. 11, NOVEMBER 2007.

C.-N. Chuang and S.-I. Liu, "A 40GHz DLL-Based Clock Generator in 90nm CMOS Technology", International Solid-StateCircuitsConference(ISSCC),pp.178-179, Feb.2007.

Smart Science, Design & Technology — Lam et al. (eds)
© 2020 Taylor & Francis Group, London, ISBN 978-0-367-17867-3

Recognition of inner surface and extrusion faces for thin-wall parts in mold flow analysis

Jiing-Yih Lai, Jia-Ying Zhong, An-Sheng Hsiao & Pei-Pu Song
Department of Mechanical Engineering, National Central University, Taoyuan, Taiwan

Yao-Chen Tsai & Chia-Hsiang Hsu
CoreTech System (Moldex3D) Co., Ltd., Hsinchu, Taiwan

ABSTRACT: Thin-wall parts exist in many products and are frequently manufactured by injection molding. The recognition and separation of inner and outer surfaces on the computer-aided design (CAD) model of a thin-wall part has many applications. One of its applications is for extrusion-faces recognition on the CAD model. The main idea of the proposed method was to separate those faces that are obviously visible and invisible from the mold opening direction. The former belong to the inner surface, while the latter belong to the outer surface. A series of region-growing procedures were then implemented to expand the faces on the inner surface and finally yield the desired faces on the inner surface. With all inner faces available, an extrusion recognition algorithm was also proposed to recognize and group each set of extrusion faces automatically. The recognition of extrusion faces on a thin-wall model can enhance the generation of solid meshes with better quality for mold flow analysis.

1 INTRODUCTION

Thin-wall plastic parts exist in many products and are frequently manufactured by injection molding. The inner surface of a thin-wall part is generally more complex than its outer surface, as the inner surface has many functional and structural feature designs, whereas the outer surface is focused on shape and appearance in the design. In mold flow analysis, a computer-aided design (CAD) model must be converted into sloid meshes so that the solver can perform the required analysis. The complex geometry on the inner surface requires additional consideration in mesh generation, as it would affect the analysis result. It requires a careful decomposition of the CAD model so that each decomposed region can be handled individually. The study on automatic decomposition of thin-wall parts is still an active research area because of the shape variety and complexity of real CAD models.

A thin-wall plastic part can basically be divided into two types of region, an inner surface and an outer surface. When a thin-wall plastic part is assembled with another part, the region that can be seen from outside is called the outer surface, whereas the other area that cannot be seen from outside is called the inner surface. A protrusion is typically created through a function called "extrude," which extrudes a two-dimensional (2D) contour along a specific direction, normally perpendicular to the plane of the contour, to generate the three-dimensional (3D) protrusion. Most protrusions on a CAD model are often linked together, and some of them may even stack on the others to form a complex structure. When a CAD model is exported to another CAD system, however, such sequential data is generally lost. Therefore, the only information available for the recognition of protrusions on a thin-wall part is the boundary representation (B-rep) model of the CAD model itself.

Several studies have been conducted for recognizing depression or protrusion features [1–2]. Fu et al. [3] proposed a method to detect inner and outer surfaces. A thin-wall part was divided into two types of surface: contact to core, and contact to cavity. A concept in terms of visibility was employed to distinguish these two types of surface. The faces that can be seen from the core were denoted "contact to core," whereas the faces that can be seen from the cavity were denoted "contact to cavity." Lim et al. [1] proposed an algorithm for the identification of depression and protrusion features (DP-features) on free-form solids. Zhang et al. [2] proposed a region-based method for recognizing protrusion and depression features. Symbolic computation was employed to characterize the curvature properties of the freeform surface

---

*Corresponding author. E-mail: jylai@ncu.edu.tw

and to help to decompose the surface into regions. Ismail et al. [4–5] proposed a technique called edge boundary classification (EBC) for recognizing simple and interacting cylindrical- and conical-based features from B-rep models. They used edge loops to form the basis of the edge boundary classification technique.

The purpose of this study is to develop a novel algorithm to deal with the recognition of general extrusions on thin-wall plastic parts. The proposed extrusion-recognition algorithm is intended to combine with boss and rib recognition algorithms [6–7] to increase the successfulness of recognizing all types of protrusions on a CAD model. Two main phases are involved in the proposed extrusion-recognition algorithm. First, search for inner and outer surfaces. This is to recognize inner and outer surfaces on the CAD model of a thin-wall part. They are respectively used in the later process. Second, search for extrusion faces. All extrusion faces are recognized group by group, with each group of faces representing a basic extrusion. Complex extrusions with several groups of faces stacking together can thus be established. Several realistic examples are demonstrated to verify the feasibility of the proposed inner faces and extrusion recognition algorithms.

2 EXTRUSION RECOGNITION ALGORITHM

Figure 1 depicts the composition of a general extrusion. As Figure 1(a) shows, a general extrusion is essentially composed of several sub-groups, each of which represents a single extrusion. A single extrusion is basically composed of three kinds of face, namely top, side, and base faces. A top face denotes a face on the top of an extrusion. A side face denotes a face on the side of the extrusion; the angle between the top and side faces is convex. The other faces that neighbor to an extrusion are called base faces; the angle between the base face and the side or top face is concave. Figure 1(b) shows the facial composition of three subgroups in Figure 1(a). The neighboring facial information is important to find the neighboring extrusions and the extrusion groups.

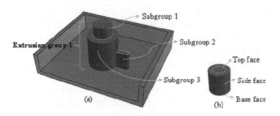

Figure 1. Composition of a general extrusion: (a) sub-groups of an extrusion, and (b) three kinds of face on a subgroup.

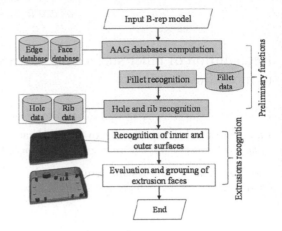

Figure 2. Overall flowchart of the proposed extrusion recognition algorithm.

As Figure 2 depicts, the proposed algorithm is mainly divided into two parts: preliminary functions and extrusion recognition. The preliminary functions are implemented to prepare the data required for extrusion recognition, which include edge and face AAG databases, and fillet, hole, loop, and rib data [6–10]. The extrusion recognition is mainly divided into two steps. First, recognition of inner and outer surfaces: the features to study are located on the inner surface. The outer surface is used to help the judgment of extrusion faces. Second, evaluation and grouping of extrusion faces: the extrusion faces are evaluated group by group. The side faces on a group are evaluated first. This follows the evaluation of the top and base faces in sequence. With such an approach, all adjacent and complex extrusions can be decomposed into single extrusions and related extrusions can be recorded.

2.1 Recognition of inner and outer surfaces

The method for the recognition of inner and outer surfaces on a CAD model is mainly divided into three steps. First, determine mold opening direction and bounding box. The bounding box is employed to determine the five directions for the visibility test. Second, compute transition edges. Transition edges represent the boundary profiles of inner and outer surfaces. As holes that pass through inner and outer surfaces may exist, there may be multiple sets of transition edges. All of them must be evaluated in order to determine the transition of inner and outer surfaces correctly. Finally, determine faces locating on the inner surface. The transition edges can divide the model into many face groups. The faces located on the inner surface can be evaluated by using face groups. A detailed description of the above three steps is given below, with an example shown in Figure 3 to help the explanation.

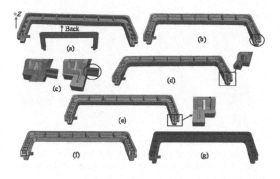

Figure 3. Detailed steps for the recognition of inner faces on a thin-wall part.

(1) Initial distinction of part faces in terms of an angle

The part faces that are obviously facing to the mold opening direction +Z and those that are opposed to +Z can easily be separated by considering the angle between each part face and +Z. Consider that the angle between the surface normal of a part face and +Z is θ. Then, the part faces can be divided into three types in accordance with θ. If $\theta \leq \theta_1$, then it is a candidate inner face. If $\theta_1 < \theta \leq \theta_2$, then it is a candidate wall face. If $\theta > \theta_2$, then it is a candidate outer face. Here $\theta_1 = 30°$ and $\theta_2 = 170°$. By using the above rules, all parts' faces are initially divided into three types. Figure 3(a) depicts the result of this step for an example, where red, gray, and blue denote candidate inner faces, side faces, and outer faces, respectively.

(2) Evaluation of part faces on the inner surface

The part faces belonging to the inner surface are evaluated from the candidate inner faces and candidate wall faces, respectively (Figures 3(b) to (e)). Most of the candidate inner faces belong to the inner surface, but there are still a few of them belonging to the wall faces. Similarly, for the candidate wall faces, some of them belong to the inner surface and must be extracted. The following steps are implemented to determine the part faces that belong to the inner surface.

First, outer wall faces are evaluated from the candidate wall faces. Figure 3(b) depicts the result of this substep. Second, candidate inner faces are judged to delete those that are not inner faces. Figure 3(c) depicts a partial result of this substep, in which an isolated candidate inner face has been regarded as a wall face. Third, all adjacent wall faces are grouped (Figure 3(d)). Finally, if a group of wall faces is surrounded by candidate inner faces, then this group of faces is regarded as on the inner surface (Figure 3(e)). After the above four substeps, all part faces that belong to the inner surface are obtained.

(3) Compute transition edges

Transition edges essentially represent the convex boundary of outer wall faces and inner faces. An outer wall face, however, may be adjacent to an inner face by a concave edge. In such a situation, all convex edges of this face are regarded as transition edges. Figure 3(f) depicts the result of this step.

(4) Re-evaluate inner faces in terms of transition edges

Adjacent transition edges can be connected in sequence to form a loop of edges. Most of the cases have one loop only, in which one group of faces represents inner faces, whereas the other groups represent the outer faces. Multiple loops of edges may exist, however, which divides the inner faces into several groups of faces. A ratio r is defined to solve this problem. Here r is defined as the ratio of number of inner faces on a group and number of total faces on the same group. If $r > r_0$, then all faces on this group are primarily inner faces, and hence all faces in this group are regarded as inner faces. Otherwise, all faces in this group are regarded as outer faces. Here r_0 is 0.7 in this study. Figure 3(g) depicts the result of this step.

2.2 Evaluation and grouping of extrusion faces

The basic concept of extrusion faces evaluation is to find a matching face for each of the inner faces, and then determine side and top faces in sequence. Each set of side and top faces can form a subgroup. Each subgroup represents a single extrusion, and several subgroups can form a complex extrusion.

(1) Side faces computation

A matching face for each inner face is computed first. Figure 4 depicts the method to find the matching face f_m of an inner face f_c. Once all matching faces are determined, each inner face is checked to determine the side faces. For an inner face f_c which has not been marked, if the corresponding f_m is an inner face, then f_c is regarded as a side face. In contrast, if f_m is an outer face, then f_c is put in a stack for checking the top faces in next step. A complex extrusion composed of several subgroups is evaluated in an iterative procedure. Faces that have been regarded as side faces in previous iteration should be excluded for later iteration. Therefore, whenever a face has been counted as a side face, it should be marked.

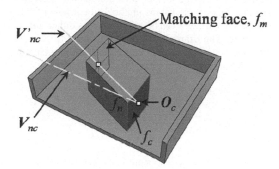

Figure 4. The method to find the matching face f_m of an inner face f_c.

287

(2) Top faces computation

The top face of an extrusion can originally be considered on the same plane as its base face. When the extrusion process is performed, it is raised to the top of the extrusion. The faces between the top and base faces are side faces. The angle between the top and side faces is convex, whereas it is concave between the base and side faces. As the neighboring face of a top face may not necessarily be a side face of the same extrusion, the number of convex edges that a top face has is dependent on how many side faces it is neighboring to. A neighboring face which is higher than the top face will result in a concave edge. This face may be a side face of another extrusion, but it is not regarded as the side face of this extrusion.

(3) Subgroups generation

Each set of top and side faces can form a subgroup, essentially representing a single extrusion. To generate all layers of subgroups for a complex extrusion, however, an iterative process must be implemented. In each cycle of iteration, a top face is served as a seed face to search for all adjacent top faces. Next, a search is performed to find all side faces that are neighboring to the convex edges of these top faces. All such top and side faces are formed as a subgroup. The detailed process is explained in Figure 5. A top face is selected as a seed face (Figure 5(a)). All top faces neighboring to this seed face are then found (Figure 5(b)). All side faces neighboring to the convex edges of these top faces are next evaluated (Figure 5(c)). This set of top and side faces is regarded as a subgroup, printed in one color (Figure 5(d)). Select another top face as a seed face and repeat the same process. Figure 5(e) depicts the final subgroups obtained for an example.

For a compound extrusion, an extrusion may reside on another extrusion. Only the extrusion faces on the first layer can be recognized, while the faces on the second layer remain unchecked. Therefore, the above-mentioned process must be implemented iteratively in order to solve this kind of situation. In each cycle of iteration, when new subgroups of faces are formed, all such faces are marked and a new cycle of top and side faces recognition is performed. For the next cycle of iteration, the faces marked are regarded as outer faces such that the unchecked extrusion faces on the second layer can be recognized as either top or side face. The example in Figure 6 is used to explain how the second layer of extrusion faces is obtained.

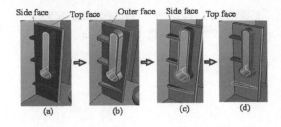

Figure 6. Procedure for the recognition of compound extrusion: (a) top and side faces on the first layer, (b) subgroup on the first layer, (c) top and side faces on the second layer, and (d) subgroup on the second layer.

In Figure 6(a), the top faces on the first layer are obtained. They are used to find the corresponding side faces, which are formed as a subgroup, as shown in Figure 6(b). As indicated on the figure, the faces on this group are marked as outer faces. Also, the faces on the second layer become unchecked as they are not included in this group. For the next cycle of iteration, the recognition of side and top faces is implemented again, as shown in Figure 6(c), which indicates top and side faces obtained. Two new subgroups are found, as shown in Figure 6(d). These faces are again regarded as outer faces, and the third cycle of iteration is performed next. As no new subgroup is formed, the iteration is finished.

3 RESULTS AND DISCUSSION

We tested the feasibility of the proposed algorithms with a program written in C++ and based on the Rhino CAD platform [11] and the open NURBS function [12]. The proposed algorithm deals only with correct manifold CAD models in both topology and geometry. Figure 7 depicts

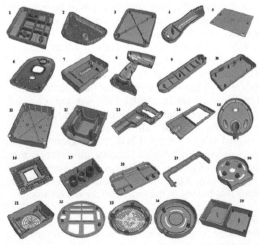

Figure 7. Results of extrusion faces recognition for 25 examples.

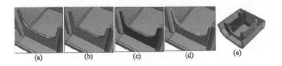

Figure 5. Subgroups generation: (a) select a top face, (b) search for adjacent top faces, (c) find all side faces neighboring to the convex edges of top faces, (d) form a subgroup, and (e) repeat the process to find all subgroups.

Table 1. Results of inner faces and extrusion faces recognition for 25 examples.

Case	CAD model			Recognition of extrusion faces				Recognition of sub-groups			
	Extrusion face	Extrusion groups	Sub-groups	Success	Fail	Misjudge	Success rate (%)	Success	Fail	Misjudge	Success rate (%)
1	86	3	8	86	0	0	100	8	0	0	100
2	10	2	2	10	0	80	100	2	0	2	100
3	69	1	21	69	0	0	100	21	0	0	100
4	138	5	35	138	0	0	100	35	0	0	100
5	172	7	14	129	43	0	75	10	4	0	71.4
6	210	7	59	210	0	0	100	59	0	0	100
7	64	13	21	64	0	0	100	21	0	2	100
8	247	13	31	244	3	2	98.8	31	0	1	100
9	155	17	39	155	0	5	100	39	0	1	100
10	222	16	63	225	0	0	100	63	0	0	100
11	302	23	56	302	0	0	100	56	0	0	100
12	54	1	14	54	0	0	100	14	0	0	100
13	122	2	31	122	0	5	100	31	0	1	100
14	279	2	55	277	2	4	99.3	55	0	1	100
15	95	10	14	95	0	1	100	14	0	1	100
16	117	11	31	117	0	0	100	31	0	1	100
17	50	1	3	50	0	10	100	3	0	2	100
18	36	1	4	36	0	0	100	4	0	0	100
19	68	20	20	68	0	0	100	20	0	0	100
20	12	4	4	12	0	0	100	4	0	0	100
21	231	3	7	200	32	1	86.6	3	4	27	42.9
22	40	5	7	40	0	0	100	7	0	0	100
23	52	4	12	52	0	0	100	12	0	8	100
24	37	3	11	37	0	0	100	11	0	0	100
25	26	9	9	16	5	0	61.5	4	5	0	44.4

the results of extrusion faces recognition for 25 examples, where each set of extrusion faces is printed in one color. Associated data of each set of extrusion faces are also recorded so that they can be accessed individually.

A detailed description of the results for inner faces recognition and extrusion faces recognition is listed in Table 1, where "success" indicates that this type of face is correctly recognized, "fail" indicates that this type of face is not recognized correctly, and "misjudge" indicates that the other type of face is wrongly judged as this type. In summary, 5 out of 25 cases are not 100% correct on the recognition of extrusion faces. The successful rates for these five cases are between 61.5% and 99.3%. Similarly, 3 out of 25 cases are not 100% correct on the recognition of sub-groups of extrusion. The successful rates for these three cases are between 44.4% and 71.4%.

4 CONCLUSION

We proposed an inner faces recognition algorithm and an extrusion faces recognition algorithm in this study. The successful rate of the former for 25 examples is 100%, which indicates that this algorithm is accurate enough for real application. It could be used to extract the parting lines from a CAD model of a thin-wall part for use in injection mold design. In addition, this algorithm is also required for the recognition of extrusion faces for thin-wall parts. The successful rate of the proposed extrusion-faces recognition algorithm is still not 100% correct. Most of the erroneous recognition, however, occurred at critical situations on the CAD models. The proposed extrusion-faces recognition algorithm could be modified to enhance the recognition of these critical situations. In addition, the classification of each group of extrusion faces is also necessary, as it is helpful for automatic generation of solid meshes in mold flow analysis.

REFERENCES

[1] T.Lim, et al., *IEEE Trans. on Pattern Analysis and Machine Intelligence*, 27, 2005, pp. 851–860.
[2] C.J. Zhang, et al., *J. Zhejiang University-Science A*, 10, 2009, pp. 1439–1449.
[3] M.W. Fu, et al., *Computer-Aided Design*, 34(6), 2002, pp. 469–480.
[4] N. Ismail, et al., *Int. J. Machine Tools & Manufacture*, 45(6), 2005, pp. 649–655.
[5] N. Ismail, et al., *Robotics and Computer-Integrated Manufacturing*, 20(5), 2004, pp. 417–422.
[6] M.H. Wang et al., *Computer-Aided Design and Applications*, 14(4), 2016, pp. 436–449.
[7] J.Y. Lai, et al., *Engineering with Computers*, 34(4), 2018, pp. 801–820.
[8] J.Y. Lai, et al., *Key Engineering Materials*, 656–657, 2015, pp. 789–794.
[9] J.Y. Lai, et al., *Engineering with Computers*, 32(4), 2016, pp. 593–606.
[10] J.Y. Lai, et al., *Computer Aided Design and Applications*, 13, 2015, pp. 95–107.
[11] Rhinoceros. Available from: http://www.rhino3d.com.
[12] openNURBS. Available from: http://www.rhino3d.com/tw/opennurbs.

Miscellaneous

Smart Science, Design & Technology — Lam et al. (eds)
© 2020 Taylor & Francis Group, London, ISBN 978-0-367-17867-3

The effectiveness of green environment conservation on social inclusion of psychiatric patients

Tsaichieh Chien*
Ph.D Program of Mechanical and Energy Engineering, Kun Shan University, Tainan, Taiwan

Huannming Chou
Department of Mechanical Engineering, Kun Shan University, Tainan, Taiwan

ABSTRACT: Treatment of psychiatric illness in Taiwan has primarily focused on institutional treatment. Building social inclusion for psychiatric patients can not only help patients in stepping out of the institution but also enhance their likelihood for recovery. The objectives of this study were to help psychiatric patients be gradually included in the community through green environment conservation efforts and to enable community residents understand psychiatric patients' efforts in protecting and maintaining the community environment. The research participants comprised 11 members of a community rehabilitation center in Central Taiwan, and the research methods involved a survey regarding willingness to sweep streets, a waste sorting and recycling test, and a six-week community street sweeping program. A total of five participants agreed to participate in the street sweeping program, and all of them passed the waste sorting and recycling test. The six-week street sweeping activity was then conducted, followed by a qualitative analysis. The results reveal that the participants had a substantial improvement in their sense of self and self-understanding, which further induced their behavior of mutual help and social interaction. This indicates that the green environment conservation approach exerted significant treatment effect on institutionalized patients.

Keywords: Green Environment Conservation, Social Inclusion, Schizophrenia.

1 BACKGROUND

The recovery movement emerged in the 1990s, emphasizing that recovery from psychiatric illness is a non-linear process and varies according to time and situations (Deegan, 2005). A patient's personal changes and environment changes both influence the patient's recovery (Deegan and Drake, 2006). If psychiatric patients are treated with the recovery approach, they can not only strengthen their sense of self but also increase their self-efficacy (Leamy et al., 2011).

2 PURPOSE

This study attempted to understand the effectiveness of methods related to green environment conservation, including a six-week activity of sweeping community streets and waste sorting and recycling, on the social inclusion of psychiatric patients. Through these methods, this study anticipated eliminating the stigma and discrimination against psychiatric patients and helping them to be integrated in society.

3 SAMPLE METHOD

The research participants of this study comprised 11 members of a community rehabilitation center in Central Taiwan. At Step 1, a survey was conducted to understand the participants' willingness to sweep streets, and five participants expressed their willingness. At Step 2, a test on waste sorting and recycling consisting of eight questions was performed; participants who answered correctly on six questions or more passed the test. Step 3 involved a six-week street sweeping activity, the sweeping area of which was first limited to the surroundings of the rehabilitation center and was progressively expanded. A preparatory and review meeting was convened before and after each time of street sweeping to discuss the participants' opinions and feelings regarding the street sweeping activity.

---

*Corresponding author. E-mail: ball0818@seed.net.tw

4 RESULT

4.1 Survey results on participants' willingness to sweep streets

The result of Question 1 "Have you ever participated in any community events" shows that all the 11 participants did not have any experience in participating in community events. For Question 2 "Are you willing to participate in a street sweeping activity to maintain the community environment," five participants expressed their willingness, which eventually yielded an actual participation rate of 45%.

4.2 Results of the waste sorting and recycling test

The five participants took a 10-min waste sorting and recycling test, which consisted of eight questions. Those who answered correctly on six questions or more passed the test. All the five participants passed the test.

4.3 Qualitative analysis of the preparatory meeting of the street sweeping activity (T denotes occupational therapist, and A, B, C, D, and E are the participants)

Week	Discussion in the preparatory meeting	Discussion in the review meeting
1	T: This is your first time to participate in the street sweeping activity for social inclusion. What are your opinions about this activity?	T: This is your first time sweeping streets in the community. What do you think of this activity? A: It is tiring. D: I am tired, and my legs are sore.
2	T: Everyone did quite well last time. Let's do our best this time. A: This time I will put more effort on it. B: I will try again. D: Let me try again.	T: Everyone's physical strength seemed to have progressed a lot this time. It seems you have become more adapted to this task. A: Actually, it is still tiring, but an old lady greeted me. This made me feel good. B: My physical strength increased, but I still feel tired. D: I have slowly adapted to the activity.
3	T: Hello everyone, this week we will increase the sweeping area. You may take a rest when you feel tired.	T: Although we expanded the sweeping area, everyone still did a good job. What are your feelings this time?
	A: Okay, I will prepare the sweeping equipment. B: I can help A with the preparation. C: I feel it is still tiring.	A: More people saw us sweeping streets this time, and this is my first time telling others that we are a community service team. B: I wanted to tell them that I am taking medicine, but I am worried that they would look at me differently. D: I still felt uncomfortable this time when people looked at me.
4	T: With our experience last time, this week we have to make our best effort to continue this task. A: Okay, let's prepare the tools. B: I will help you bring plastic bags. C: I will take grippers. D: I will prepare the water. Remember to bring your clothes.	T: You did the task really fast and helped each other to remove weeds. Very good! A: I saw our community has a large area of weeds, so I just pulled out them incidentally. B: I saw A was removing the weeds, so I joined A. C: I feel happy today because someone told me I did a good job.
5	T: This time we are expanding our sweeping area to the entire Gongxue 2nd Street. I think we will not have any problem. A: Okay, I will prepare the tools. B: No need to prepare. E has prepared all the tools. D: Thanks.	T: The street was messy today, but everyone still finished the job actively. How do you feel today? A: I feel more tired today. B: I also feel more tired; the area of the entire street is quite big. C: A man said he can bring some water for me, but I did not know how to respond to him. D: Many dogs poop on the streets in the community. I remember dog owners should prepare plastic bags to pick up the poop.
6	T: This is our last time sweeping the streets. I hope we have the chance to do it again. Let's do a good job this time too. A: We have prepared the tools.	T: The six-week street sweeping activity for social inclusion ended today. What are your opinions about the six-week activity? A: We in the community rehabilitation center have never greeted other

(Continued)

(Cont.)

Week	Discussion in the preparatory meeting	Discussion in the review meeting
	D: Can we go now? I want to check whether the weeds have grown again.	people. During these six weeks, I greeted many people. Although I was afraid, at least I stepped forward. B: Although it's tiring, I think it is worthwhile, and I feel I can do more, but I have to improve my physical strength. C: I think this accumulate merits for me. I am glad I can do this task, and I also think my physical strength is poor and I have to practice. D: I think this activity is good because I can improve my physical strength. People were all nice; they did not criticize me for my auditory hallucination. E: This is a good activity. I have more courage to speak with others.

5 DISCUSSION

The participants were frustrated at first because of their poor physical strength, from which they realized that conducting meaningful activities require substantial physical strength. By the third and fourth weeks, the participants had begun to cooperate and discuss with each other, and had come to know what it feels like to interact with other people in the community, despite their fear. In the final two weeks, they had gradually exhibited the behavior of providing mutual help and had been able to give more feedback after the activity. In addition, they realized their limitations and deficiencies through self-affirmation, which further provoked them to discuss their self-recovery.

REFERENCES

P.E. Deegan, 2005. Scand J Public Health Suppl 66 29–35.

P.E. Deegan and R.E. Drake, 2006. Psychiatr Serv 57 1636–1639.

M. Leamy et al., 2011. *The British journal of psychiatry: the journal of mental science* 199 445–452.

Smart Science, Design & Technology — Lam et al. (eds)
© 2020 Taylor & Francis Group, London, ISBN 978-0-367-17867-3

Investigating the starting point of Taiwan contemporary photography

Kuochun Chiu

Department of Visual Communication Design, Kun-Shan University, Yongkang, Tainan, Taiwan

ABSTRACT: The article will start by studying the influence of early Taiwanese photography and how it affected western late modern art. This period of time could also be seen as a stepping stone for western contemporary art. I will also try to analyze to see whether these old Taiwanese photographs show any sign of Japanese influence/western contemporary photography or maybe even develop ed conceptual and experimental photos created under the influence of the early pictorial photography in samples of this article's gallery. Lastly, through interviews, I can analyze these photographers' diverse contexts to provide proof of their contribution and inspiration to Taiwan contemporary photography, and of course, see the development of this type of photography in a different light.

1 INTRODUCTION

To observe the context and form of early Taiwanese photography's evolution and comparing it to the different eras of western photography ideology, the inspection can be divided into three paths:

The first one is "salon photography", based on photographer Lang Jingshan from the 1930s, whose style was similar to Chinese ink style paintings. Whether constructed in a darkroom or not, the goal was to create a painting-like atmosphere. This can be corresponded to western photography in the early 19th century, when creating painting-like photos was popular, as inspired by old paintings. Adding drawings, clipping and pasting, or editing in darkrooms, they were all solely techniques for artists to create. During the late 19th century, Naturalism brought up by Peter Henry Emerson is also similar to Romanticism. Emerson's photography was like the Barbizon school, focusing on countryside scenery and farmer lifestyle. As for the comparison of Lang Jingshan and Chinese ink paintings with this example, we can see that the themes are not the same, but through photography's point of view, they are.

The second would be photojournalism, inspired by Taiwanese photography. It's is also called native photography because of photographers like Ruan Yizhong who created projects about life in the countryside. During the 1950s~80s when "People" and other magazines arose, pictures focused on the changes and development in Taiwan's society at the time and were then published with captions. In comparison to western photography ideology, this can be set side to side with "straight photography", which was brought up by Alfred Stieglitz, a specialist of American photography during the early 20th century. He steadied modern American photography and continued the growth of documentary photography. Because of Stieglitz's persistence of cameras having the special ability to capture and reshow reality, dropping the fancy editing, fabricating and other techniques, intentionally separating itself from the originality-based modern art, hence the promotion of photography separatism. Looking back at photography then, it seems like photography was sorted in a different category from other arts under the influence of the path it took; this was the most obvious in the 60s to 80s when photography media was often in separated exhibitions in art museums and discussed frequently.

In the end, the start of the development was parallel to what was described before and photographers started to break out of traditional limits and presented their own emotions and views through their work; the best symbol being V10, a Taiwanese photography group. Looking closely at the projects, there are still hints of western art and philosophical ideology, especially because of Taiwan's biased liking towards existentialism. It is even visible in literature, painting, drama, and other areas; photography was not an exception. The vast majority of young photographers expressed their own feelings by creating. In addition, western modern art never stopped developing different possibilities, all for throwing off old, limited art forms and concepts.

Technically, contemporary art is in some ways similar to postmodern art. They both are against modernism; nonetheless, if we dig deeper we can find out that the edginess and originality modern art left behind; only "originality" from contemporary art

---

*Corresponding author. E-mail: er5086@gmail.com

meant there were no limits in style but emphasizes the works' conceptual originality.

2 THE SAMPLING ANALYSIS OF TAIWANESE PHOTOGRAPHERS

Contemporary art focuses on expressing personal emotions and or opinions. The diverse creating forms (or even no specific form at all) shows the presentation, concept, behavior and other art ideology from the later period of modernism; and it shows the cross-field (whether it be art media or other fields related to creating), follow (work type may change depending on the time and place), decentration (artists may pay attention to non-common themes or there may be more than one artist creating one project), non-defining attribution (also has caused controversy on if a certain project can be defined as art).

During the Taiwanese Post-Martial Law Period in the 70s, society became liberating, news information and people were free to express themselves. At the same time, studying art abroad became more and more popular, which lead to Taiwan being able to connect with art internationally, this can be seen easily in works from the time. Now, "contemporary" has been the major trend for recent years in Taiwan, the contemporary style of photography also crowns Taiwanese photography. However, can contemporary photography really present Taiwan's special features? Or does it actually blur the lines of its ideology? This article will continue to observe the history of photography and discuss earlier works that grasp the essence of contemporary art. I will also list some of the works of some of the photographers from V10, for instance, Chang Chao-Tang, QUO YING SHENG, and Hsieh Chuen-Te. All three people were young photographers at the approximate age of 30, and their creations challenged traditional styles.

Chang Chao-Tang's early work usually has twisted bodies, faces out of focus, and even incomplete imagery; all different from his previous tastes. His main subjects changed from outer appearances and journalism to expressing his own emotions. Rather than saying this collection is a continuation of Taiwanese traditional photography, saying that the creator himself got absorbed in western existentialism or surrealism would be more correct. Turning to QUO YING SHENG's works from that time, "Taipei Dihua Street" has a hint of surrealism as well. Although it is a black and white photo, he purposely altered the photo, adding grains, high exposure, and other effects, as if to show his real self, the only way was to damage a perfectly fine photo.

Lastly, Hsieh Chuen-Te's "First Solo Exhibition" has a sort of special sureness to it. Whether the model's body is dismembered, deformed, or stuck in unusual places within the space, it all creates a twisted scene in the real world.

Looking in from an outsider, these three photographers not only damaged the quality of photos to throw off past beauty standards and defy tradition but also started to utilize editing. For example, they altered the model's bodies with their own aesthetic, separating them from how reality should be and presented their concepts and thoughts. They did not follow the rules of regular standards and photographed the models like how they should be in a studio. Through their works, people can see the creator's mind and emotions. Although these works do not have the format of western contemporary fabricated photography, they do have the idea of it.

Of course, the "contemporary photography works" we are focusing on here does not define the whole ideology solely with these unique formats or innovations. The appearance is that it looks like it contains everything, but the spirit is it should reflect how people exist and feel. Thus, contemporary art has to possess this spirit and study/reflect this ideology or produce questions. Contemporary is all about expressing yourself, so any works that have a special concept can be categorized as contemporary. It is also because of this we are focusing on the works that break tradition.

3 THE EXPLORING OF THE THREE PHOTOGRAPHERS THINKING WITH WORKS

After having an interview with these three photographers, I realized another thing they have in common, which is their creativity in their styles are still developing non-stop in their own unique way, whether through experimenting, innovating, not being limited in forms; this is the essence of the flow and cross-field diversity of contemporary art. Take Hsieh Chuen-Te for example, the work he exhibited in the 2004 La Biennale di Venezia was all about culinary. As for QUO YING SHENG, in recent years he worked with Taiwanese fashion designer Jamei Chen and displayed his work, themed around fashion, on the runway. Chang Chao-Tang had not only already experimented with film images early on but also created a collection that differed from his usual style in 1987. He displayed his photos that were complete and developed, and also photos that took in the form of contact print which showed the thought process of the photographer. Robert Franks, an American photographer, had also tried this method before, but Chang inserted his own style. Responding to Franks and Duane Michaels' art-reactionary from modern American photography ideology, this can be considered the pioneer of Taiwanese concept photography.

What's worth mentioning is that overall, the works of these photographers are different, whether it be the theme, sort, technique, presentation, or even time. However, by comparing them with each other, how photography displayed their atmosphere and aesthetics were very similar. For example, Chang Chao-Tang experimenting with dynamic and creating with static; QUO YING SHENG with black and white, color, street view, and fashion; and Hsieh Chuen-Te with trying digital editing and new technology like high-speed photography. This proves that contemporary art doesn't follow the rules of a certain form, freed itself from media and the limits from themes, can still be considered as art.

4 CONCLUSION

The term "contemporary photography" is considered mainstream in contemporary art and surrounds the three photographers mentioned above. Though this may be true, it does not mean it is limited in a certain form. Instead, it inspires the future development of Taiwanese photography. Although ideology became free and information circulating, photography tools becoming easier to find and a frequently used tool among contemporary artists, the ideology should always come first: expanding the possibilities, non-stop innovation, and even throwing over restricting forms.

REFERENCES

Hsich Chun-Te, Slight Touch, Kaohsiung, Kaohsiung museum, 2013, pp. 11–47.
Chang Chao-Tang, Beyond the Frame, Taipei, Uni-Books, 2017, pp.169–186.
Teng, Po-Jen, Li, Jin-Yin, First, 2018 Kaohsiung Photo, Department of Cultural Affairs in Kaohsiung, 2018, pp. 120–168.

Smart Science, Design & Technology — Lam et al. (eds)
© 2020 Taylor & Francis Group, London, ISBN 978-0-367-17867-3

The Heart Sūtra: the liberated mind of the other shore

Hei-Ling Kao*
Ph.D Program of Mechanical and Energy Engineering, Kun Shan University, Tainan, Taiwan

Huannming Chou
Department of Mechanical Engineering, Kun Shan University, Tainan, Taiwan

ABSTRACT: The essence of the Prajñāpāramitā Sūtra taught by Buddha Śākyamuni in the second turning of Dharma Wheel was condensed into the Prajñāpāramitāhṛdaya (Heart of the Perfection of Wisdom Sūtra; C. Xin Jing). Even though the Heart Sūtra translated by Xuanzang contains only 260 words, it is able to encapsulate the full meaning of 600 volumes of Prajñāpāramitā Sūtra. Based on studying sūtras and sastras, this paper reveals that the main character of the Heart Sūtra, Bodhisattva Avalokitêśvara (C. Guanzizai) was practicing the profound prajñāpāramitā, directly perceived the mind of emptiness-nature which is inherently without birth and death, and brought forth prajñā, the wisdom regarding the ultimate reality; thus the five aggregates of the body-mind which can see and hear, apprehend that the five aggregates are illusory, arising and ceasing dharmas of birth and death. The truth that the five aggregates are ephemeral and filled with suffering and emptiness can then be thoroughly understood, thereby liberation from the suffering of cyclic existence in the three realms can be attained. The Heart Sūtra states that there are two types of mind known as the observer and the observed, through being able to directly witness the false and deceitful observing mind and the inherently tranquil and unhindered observed mind, the other shore can be reached to attain liberation from the suffering of cyclic existence.

Keywords: Bodhisattva Avalokitêśvara, Profound prajñāpāramitā, Impermanence of five aggregates, Liberation.

1 PREFACE

It is generally accepted that the Heart Sūtra, short for Prajñāpāramitāhṛdaya, is a simplified version of the Mahāprajñāpāramitāsūtra that encapsulates the essence of both the Full Edition and Small Division of the Prajñāpāramitā Sūtra, thus was named as the Heart Sūtra. As its title suggests, the Heart Sūtra should contain explications related to the heart (mind), but as preached by Buddha Śākyamuni, the mind of six consciousnesses includes eye, ear, nose, tongue, body and mental, so which mind is the Heart Sūtra referring to? Or could it be another mind that does not belong to any of these six minds? If so, which mind is it?

Besides, the Heart Sūtra translated by Xuanzang, Tripitaka Master of the Tang Dynasty, states "The noble Avalokitêśvara Bodhisattva, while practicing the deep practice of prajñāpāramitā, looked upon the five Skandhas, and seeing they were empty of self-existence, and he crossed beyond all suffering and difficulty (Rans, 2004)." When Bodhisattva Avalokitêśvara was practicing the profound prajñā-pāramitā, which is the mind that is able to practice the profound prajñā and achieve the pāramitā? Why is it possible to reach the other shore of liberation by observing that all the five aggregates are empty? This paper seeks to delve deeper into this matter.

2 MAHĀPRAJÑĀPĀRAMITĀ

Mahaprajñāpāramitā means to attain great wisdom to reach the other shore of liberation that has no birth and death. Firstly, we should find out what it means by "great wisdom" and "the other shore of liberation." Secondly, how should we understand "to attain great wisdom to reach the other shore"?

2.1 *The meaning of "Mahāprajñāpāramitā"*

Mahaprajñāpāramitā was translated as great wisdom in Chinese. It is known as Mahaprajñāpāramitā in Sanskrit, where pāramī or pāramitā means to reach the other shore of liberation. It is the sixth practice of the six perfections, meaning to study the Buddha Dharma and attain the prajñāpāramitā wisdom in order to reach the other shore of liberation that is

---

*Corresponding author. E-mail: khling263945@gmail.com

the state of neither arising nor ceasing, and a state of birthlessness and deathlessness. The wisdom in the Dharma is different from the common wisdom, and is divided into worldly wisdom, supramundane wisdom, and supreme mundane and supramundane wisdom (Laṅkāvatāra-sūtra, 2004, c19-20). Worldly wisdom is the wisdom on mundane dharmas that is commonly possessed by worldly people; the supramundane wisdom is the wisdom to transcend the three realms that is attained by Śrāvaka and Pratyeka-buddha practitioners; the supreme mundane and supramundane wisdom is the wisdom where bodhisattvas and buddhas observe the dharma that is empty and without anything, perceive the presence of neither arising nor ceasing, away from existence, non-existence, and conception of self, so as to cultivate through the Path to accomplish Buddhahood (Laṅkāvatāra-sūtra, 2004, c20-25). Obviously, the wisdom of Mahaprajñāpāramitā refers only to the supreme mundane and supramundane wisdom of bodhisattvas and buddhas, but not the other two types of wisdom. Although the Buddhas expounded their teachings in six pāramī and taught that all six perfections are able to reach the other shore of liberation, the pāramī of giving, pāramī of meditation, and other pāramīs belong to practices of phenomenon and mind-nature, which have different focuses. Without the accompanied prajñāpāramitā wisdom, only the liberation of Śrāvaka and Pratyeka-buddha can be attained, but not the Buddhahood. Hence, the first five perfections should also be guided by prajñāpāramitā, which makes Mahaprajñāpāramitā the great wisdom that can lead to Buddhahood accomplishment (Mahāprajñāpāramitā-śāstra, 2004, a28-b7).

What does it mean by "the other shore of liberation"? Nirvāṇa is essentially liberation, and is also the other shore (Mahāparinirvāṇa-sūtra, 2004, c16-23), thus the other shore of liberation refers to the nirvāṇa realm. According to Discourse on the Perfection of Consciousness-Only (C. Cheng weishi lun), there are four types of nirvāṇa, namely nirvāṇa with primordial, intrinsic and pure nature, nirvāṇa with remainder, nirvāṇa without remainder, and nirvāṇa without abode. All sentient beings possess the nirvāṇa with primordial, intrinsic and pure nature (just not realised) that ought to be personally realised by the truly enlightened ones; adepts of the two vehicles are able to enter the nirvāṇa with or without remainder as they can be freed from cyclic existence in the three realms, but they have yet to realise the first type of nirvāṇa; only the Buddhas can realise all four types of nirvāṇa, be freed from the suffering of birth and death, and continue to bring benefit and joy to sentient beings.

As such, it can be seen that the true meaning of Mahāprajñāpāramitā refers to the ability to realise the first type of nirvāṇa or to possess both kinds of wisdom of the second and third types of nirvāṇa simultaneously, and to attain Buddhahood based on the

wisdom personally realised from the first type of nirvāṇa, so as to eventually accomplish the wisdom and liberation realm of the nirvāṇa without abode. Therefore, considering the practitioners of three vehicles, only Mahāyāna practitioners have realised the first type of nirvāṇa and underwent many kalpas of practices to not enter the nirvāṇa without remainder; even though Hīnayāna practitioners possess the virtues of liberation pertaining to the nirvāṇas with or without remainder, they have yet to realise the primordial, pure nature nirvāṇa, thus their wisdom is unable to cover the true state of nirvāṇa realm, and they are unable to attain Buddhahood; as for bodhisattvas, even though they have realised the nirvāṇa with primordial, intrinsic and pure nature, they are far from attaining Buddhahood, and have to undergo many kalpas of practices to attain a buddha's wisdom and merits.

2.2   What exactly is Mahāprajñāpāramitā?

As the liberation of two vehicles is merely a conjured city, it is a fruit implanted to allow the Śrāvaka and Pratyeka-buddha practitioners to be freed from cyclic birth-and-death of the three realms (Saddharmapuṇḍarīka-sūtra, 2004, a9-13). The wisdom attained upon liberation is merely the supramundane wisdom with the three samādhis: emptiness, marklessness and wishlessness (Mahābherīhāraka-parivartasutra, 2004, b2-5), and is not the wisdom that can lead to Buddhahood. Thus, it is not the liberation realm of Buddhahood either. The true big city is Tathāgata's liberation realm, as mentioned above, there are four types of nirvāṇa according to Consciousness-Only school (S. Yogācāra). The liberation of Śrāvaka and Pratyekabuddha can only attain two kinds of nirvāṇa, nirvāṇa with remainder and nirvāṇa without remainder, whereas Mahāyāna practitioners have yet to realise the nirvāṇa with primordial, intrinsic and pure nature, and they have no knowledge of the nirvāṇa without abode realised by the Buddhas, thus the differences between their wisdom and the liberation realm of those of Mahāyāna practitioners are self-evident. As for the four kinds of perfect wisdom of Buddha, namely the Great Mirror Wisdom (S. ādarśa-jñāna; C. dayuanjing zhi), the Universal Equality Wisdom (S. samatā-jñāna; C. bingdengxing zhi), the Profound Contemplation Wisdom (S. pratyavekṣa-jñāna; C. miaoquancha zhi), and the Perfect Achievement Wisdom (S. kṛtyaanusthāna-jñāna; C. chengsuozuo zhi), attained upon Buddhahood, which are the perfect prajñāpāramitā wisdom pertaining to the Yogācārin teachings of the knowledge-of-all-aspects (S. sarvajñajñāna; C. ilch'ejong chi) that is distinctive from those of the two vehicles. Unlike practitioners of the two vehicles who merely practice dharmas like the Four Noble Truths, the Eightfold Correct Path, and the twelve links of dependent arising, bodhisattvas and buddhas practice the One and Only Buddha Vehicle, thus the afflictions eliminated are different, and their wisdom upon liberation transcends the wisdom of the Hīnayāna practitioners who merely seek to be freed from the cyclic existence within the three realms. Furthermore, the liberation realm they attained also surpasses the Hīnayāna's cessation realm which is a state without five aggregates and with ultimate tranquility. Therefore, Mahāyāna practice is the true big city that never ceases to exist and eternally brings benefit and joy to sentient beings (Buddhabhūmi-śāstra, 2004). Hence, dharmas of the two vehicles do not contain Mahāprajñāpāramitā wisdom therefore there is no prajñāpāramitā to achieve either. The completion of Mahāyāna practice must include the successive and sequential practices of six pāramitās or even the ten pāramitās, and a Buddha's wisdom and liberation merits can only be perfectly realised after countless kalpas of cultivations. This is why Buddha Śākyamuni has the knowledge of all aspects that his disciple Uruvelā Kāshyapa does not have (Madhyamâgama, 2004).

3   THE "MIND" ELUCIDATED BY THE HEART SŪTRA

It is commonly thought that the essence of the Prajñā-pāramitā Sūtra lies in the Heart Sūtra. For instance, Master Yinshun believed that there are several inter-pretations of "mind," and it is the essence of the main way to escape from suffering, thus the title of Heart Sūtra implies the meanings of essence and summary. (Shih, Yinshun) On the other hand, Venerable Xiao Pingshi also considers the Heart Sūtra as the essence of myriad Prajñāpāramitā scriptures, but he believes the main point of the Heart Sūtra was to explain the mind, thus the different minds explained by the Heart Sūtra include the mind of a sentient being, the mind of a bodhisattva, the mind of a Buddha, and various minds. (Xiao Pingshi) As the title of Heart of the Per-fection of Wisdom Sūtra suggests, its content should be related to prajñāpāramitā, liberation and mind. Since the meanings of prajñāpāramitā and liberation have been previously explained, this section will focus on discussing which is the mind described in the Heart Sūtra.

The Heart Sūtra translated by Dharma Master Xuanzang starts by stating, "The noble Avalokitêś-vara Bodhisattva, while practicing the deep practice of prajñāpāramitā, looked upon the five Skandhas, and seeing they were empty of self-existence, and he crossed beyond all suffering and difficulty." This means the fact that by practicing such a unique pro-found prajñāpāramitā dharma-door, Bodhisattva Avalokitêśvara possessed the pāramitā wisdom to reach the other shore of liberation, whereupon fur-ther observation can be based on his prajñāpāramitā wisdom. He then witnessed that all phenomena of the five aggregates are dharmas of impermanent and empty, thus he was able to cross beyond all suf-fering and difficulty. As previously explained, the profound prajñāpāramitā dharma-door and prajñā-pāramitā wisdom mentioned here refer to practices

that can lead to Buddhahood and attainment of the supreme wisdom of mundane and supramundane. Master Yinshun believed that the practice of pro-found prajñāpāramitā belonged to bodhisattva's practice dharma-door, and the wisdom was very profound. By "profound," it refers specifically to the wisdom of understanding the ultimate emptiness (S. paramârtha-śūnyatā; C. diyiyi kong). As the Bodhisattva observed all dharmas in the five aggre-gates, and that all five aggregates were empty, he had to comprehend it from substantially established objects, so as to see the principle of prajñā truth, that is, all dharmas are empty. This prajñā truth refers to the emptiness-nature, and the empty phe-nomenon is based on worldly existence and mani-fests itself thereupon. Venerable Xiao Pingshi believes that practitioners practice and observe the profound prajñāpāramitā, after having personally realized the true mind of the dharma-realm, and brought forth prajñā, the wisdom regarding the ultimate reality; thereby practitioners can observe that the phenomena of five aggregates are all imper-manent and empty. This is due to the realization of the basis mind of all dharmas of the dharma-realm, thus observing the basis mind which is intrinsically self-existent, eternal, and neither arising nor ceas-ing. At the same time, the five aggregates can be observed from the aspect of this basis mind to prove that the five aggregates are all impermanent and empty, and this makes liberation from birth and death possible.

Upon comparison of Yinshun and Xiao Pingshi's understandings of "The noble Avalokitêśvara Bodhi-sattva, while practicing the deep practice of prajñā-pāramitā, looked upon the five Skandhas, and seeing they were empty of self-existence, and he crossed beyond all suffering and difficulty," the following can be deduced: Shih, Yinshun believed that pro-found prajñā refers to the experienced wisdom regarding the ultimate emptiness, and it has to be comprehended through substantially established objects; thus being able to see that all dharmas (of five aggregates) are empty, this forms the true prin-ciple of prajñāpāramitā. On the other hand, Xiao Pingshi believes that profound prajñāpāramitā is the prajñā wisdom engendered by practitioners when they realize the true mind of the dharma-realm and make in-depth observations on it. Although they too are able to see that all dharmas of the five aggregates are empty, they obtained such a personal observation from the basis mind they have realised. As for the "mind" explained by the Heart Sūtra, obviously Yin-shun thought there is a mind that can observe the ultimate emptiness, whereas Xiao Pingshi believes that on top of the mind that can observe (the ultimate reality of the dharma-realm), there exists another basis mind which is the ultimate reality of the dharma-realm that is being observed. The principles and meanings of "ultimate emptiness" and "ultimate reality of the dharma-realm" will be dealt with at another time. In the following section, we will discuss the question: Which is the "mind" the Heart Sūtra is referring to?

The treatise composed by the 500 great Arhats states, "The mind is known as the consciousness aggregate, which is the six consciousnesses such as eye consciousness and so on. (Abhidharma-mahā-vibhāṣā-śāstra, 2004)" Among the five aggregates, the consciousness aggregate includes the six consciousnesses of eye, ear, nose, tongue, body and mental. Besides the form aggregate that belongs to a material phenomenon, the three aggregates of feeling, perception and formation are mental factors. As Bodhisattva Avalokitêśvara looked upon the five Skandhas (aggregates) and saw that they were all empty, it means that the five sense faculties of the form aggregate and even the six consciousnesses mind as well as the mental factors are all empty, false, and deceitful dharmas. As such, when the Bodhisattva observes his own or others' five aggregates, which are all impermanent with the nature of arising and ceasing, and are ultimately illusory dharmas that will eventually become empty. However, as he possesses the wisdom to perceive the ultimate emptiness (Shih, Yinshun), or the wisdom to observe the mind of ultimate reality of the dharma-realm (Xiao Pingshi), he is able to attain liberation. As the consciousness aggregate possesses the functions of seeing, hearing, cognition, and knowing, it is apparently an observer mind. If this observer mind of six consciousnesses belongs to dharma that arises and ceases relentlessly, then it is not a mind that is eternal and unceasingly exists. As stated in the sūtras, the eye consciousness arises in dependence on the eye-organ and the visual objects as cause and conditions, and upon the conditions of the mental faculty and the mental objects that the mental consciousness arises (Saṃyuktâgama–sutra, 2004a), thus the six consciousnesses mind is able to observe and arise or cease. Both Master Yinshun and Venerable Xiao Pingshi believe that the "mind" in the Heart Sūtra refers to the observer mind which belongs to this mind of six consciousnesses.

If the six consciousnesses mind is impermanent with the nature of arising and ceasing, then it is not a mind that can transcend three lifetimes, past, present and future. As a result, when the Bodhisattva looked upon the five aggregates, and seeing they were empty of self-existence, how was he able to cross beyond all suffering and difficulty? As long as the body of five aggregates exists, there would still be suffering brought about by birth, aging, illness and death. Even the adepts of Hīnayāna are still experiencing slight misery, except that they are freed from the hindrances of affliction (Cheng Weishi Lun, 2004, b12-14). If the slight misery can be eliminated in the future, they can then enter the nirvāṇa without remainder at the time of death (Cheng Weishi Lun, 2004, b14-16); but they cannot be said to have achieved liberation when they are still living in the world of human beings, because they have yet to reach the realm of the other shore, and thus have not crossed beyond all suffering and difficulty. Even as they have eradicated their five aggregates and been eliminated from future existence (Saṃyuktâgama–sutra, 2004b), they have no consciousness aggregate and the six consciousnesses mind anymore, so who is able to look upon themselves crossing beyond all suffering and difficulty to reach the other shore of liberation? Obviously, the characteristics of the observer mind (the six consciousnesses) is unable to reach the other shore of liberation, as there should be another observed mind that perpetually exists at the other shore and is intrinsically self-existent (Kao and Chou, 2017) so that it can bring about a true liberation that is neither nihilism nor extinguishment. From this, when we refer to the "mind" in the Heart Sūtra, there indeed exists another observed mind on top of this observer mind of six consciousnesses, so as to satisfy the definitions delinated in the Heart Sūtra that the five aggregates are neither identical nor different from (the nature of) emptiness in order not to become a state of total extinction and nothingness. Further research will be conducted on this.

4 CONCLUSION

The essence of both the Full Edition and Small Division of the Prajñāpāramitā Sūtras was condensed into the Heart Sūtra. In the Heart Sūtra, Bodhisattva Avalokitêśvara was practicing the wisdom of profound prajñāpāramitā to look upon and see all beings' suffering of cyclic existence. Furthermore, the Bodhisattva observed a mind that was intrinsically self-existent and freed from various sufferings of the existing five aggregates or even the slight misery of the sages of the two vehicles. It is the mind that enables sentient beings' liberation from the suffering of cyclic birth and death to reach the other shore of true liberation which is birthlessness and deathlessness. The "mind" explained by the Heart Sūtra, and the essence explained by the Prajñāpāramitā Sūtra both relied on a mind with an emptiness-nature that intrinsically self exists, but not the consciousness aggregate or the mind of six consciousnesses which is subjected to arising and cessation. That mind is able to transcend all suffering of cyclic existence, and dwells forever at the other shore of liberation that has no birth and death. This is indeed the true meaning of saying, when the noble Avalokitêśvara Bodhisattva practicing the deep practice of prajñāpāramitā, looked upon the five Skandhas, and seeing they were empty of self-existence, and he crossed beyond all suffering and difficulty, which is also the profound meaning of the Heart Sūtra.

REFERENCES

Rans, 2004. Red Pine, USA.
Lankāvatāra-sūtra, 2004. *CBETA 3*(670) 500.
Mahāprajñāpāramitā-śāstra, 2004. *CBETA 8*(1509) 116.
Mahāparinirvāṇa-sūtra, 2004. *CBETA 33*(374) 563.
Saddharmapuṇḍarīka-sūtra, 2004. *CBETA 3*(262) 26.
Mahābherīhāraka-parivartasutra, 2004. *CBETA 2*(270) 296.
Buddhabhūmi-śāstra, 2004. *CBETA* 1530.

Madhyamâgama, 2004. *CBETA 11*(26).497.
Abhidharma-mahā-vibhāṣā-śāstra, 2004. *CBETA 19*(1545) 96.
Saṃyuktâgama –sutra, 2004a. *CBETA 9*(99) 57.
Cheng Weishi Lun, 2004. *CBETA 10*(1585) 55.
Saṃyuktâgama –sutra, 2004b. *CBETA 1*(99) 2.
H.L. Kao and H.M. Chou, 2017. A Discussion about Bodhisattva Guanzizai from the Perspective of the Heart Sūtra. *Proceedings of IEEE ICASI*, Sapporo, Japan.

Smart Science, Design & Technology — Lam et al. (eds)
© 2020 Taylor & Francis Group, London, ISBN 978-0-367-17867-3

Exploring the reason for the disappearance of forests on Easter Island

Hei-Ling Kao*
Ph.D Program of Mechanical and Energy Engineering, Kun Shan University, Tainan, Taiwan

Huannming Chou
Department of Mechanical Engineering, Kun Shan University, Tainan, Taiwan

ABSTRACT: Forests are beneficial to human survival in many ways. However, mankind fell trees to satisfy a temporary need in support of their livelihood whereby resulting in soil erosion, environmental destruction as well as creating a negative impact on the eco-system. Such behaviours, which disregard the future, can be found both in China and in other countries, and in olden as well as modern times. This article, which is drawn from research and investigations carried out by scholars and experts, reveals that the destruction of forests on Easter Island in the past had led to the fall of its civilization. Apart from the commonly known fact about making ropes from trees to erect Moai statues and using wood as fuel, the pursuit of new hope and foreign trade and the making of catamarans for long-term sailing like other Polynesian islanders required the felling of a lot of trees. This resulted in the rapid depletion of the forests which was a crucial factor leading to the complete destruction of the civilization there. The destruction of civilization on Easter Island is not a remote incident that we can ignore, as Mother Earth is facing a similar problem today. The importance of forest preservation concerns not just the living environment, but also addresses the key issue of whether humanity can live in harmony with the Earth.

Keywords: Moai, Catamaran, Polynesia, Forest preservation

1 BACKGROUND

The forest has many functions such as water conservation, soil protection for preventing soil erosion, animal habitat, wind-breaking, and maintenance of the quality of the environment. Due to the lack of forest conservation concepts in modern times whereby resulting in the lack of reforestation, the need for economic development, and excessive (and illegal) deforestation, it is estimated that the global forest area is decreasing at an annual rate of 7.3 million hectares. This reduces carbon dioxide absorption, damages the ozone layer in the atmosphere, causes the greenhouse effect, and eventually results in global warming, thus causing abnormalities in the Earth's environment. Such effects also threaten the survival of human beings and all forms of animal life. In addition, the forest will also affect the rise and fall of a regional civilization, such as the disappearance of the Easter Island civilization, because the original forest that the people depended on for survival also disappeared, causing the environment to be unconducive to the survival of the islanders, thus leading to the destruction of the human civilization there.

Over a span of a few hundred years, the cultural landscape and natural environment of Easter Island were gradually destroyed by the excessive deforestation activities without any reforestation effort and also the inhabitants' over-dependence on timber for their livelihood. This had led to the depletion of the subtropical forest, and resulted in environmental variation, climate change and also changes in the ecosystem. This in turn had resulted in the disappearance of the island's human civilization which had taken a long time to establish (Kao and Chou, 2017). What caused the people to cut down trees until the trees they depended on were slowly depleted? Weren't they aware that they could not exist without wood from the forest? Or were there other factors causing them to believe there was still hope after they cut down all the trees? This research paper was prepared by researchers who explored the possible reasons behind the Easter islanders depleting the wood resources they heavily depended on for their livelihood. Other than the generally perceived reasons, were there other reasons for the islanders to cut down all the trees wantonly?

*Corresponding author: E-mail: khling263945@gmail.com

2 EXPLORING THE REASON WHY THE FORESTS OF EASTER ISLAND DISAPPEARED

2.1 Studying the past

According to past studies, during the period starting from 100,000 years before men set foot on the island to the time when it was first inhabited, Easter Island was covered by huge towering trees and lush subtropical rainforest. Many of the tree species were the types of trees often found presently in East Polynesia, or related to them (Diamond, 2006). One of those trees that had disappeared are the palm trees which may have been the largest palm trees in the world. The palm leaves were possibly used to cover the roofs, and the thick trunks could be made into tools for carrying heavy stuff, erecting Moai statues, or building rafts. Research scientists have found that among the extinct trees, two of the tallest trees included the Alphitonia cf. zizyphoides which could grow up to thirtymetres, and the Elaeocarpus cf. rarotongensis that could grow up to fifteenmetres. In today's Polynesia and other islands, we can still see them used by locals to make canoes. Other than that, the Polynesians also knew how to use the bark of the shrubs of the Triumfetta semitriloba to make ropes that were probably used by the Easter islanders to lift up the Moai statues (Diamond, 2006).

From the bones of birds and seals dug up from Easter Island, it can be seen that dolphins were the largest animals hunted for food by the early inhabitants and dolphins accounted for one-third of their diet. However, as dolphins generally lived in the deep sea and would not get close to the shore, it was not possible for the inhabitants to use ropes or toss harpoons into the sea to catch them. Hence, it was inevitable that the islanders would have to use large and stable canoes to go out to sea to hunt the dolphins. Based on samples of burned charcoal fragments collected from the island, scientists were able to identify the types of tress that were used in making the canoes (Diamond, 2006).

From the above, we can tell that the Easter Island inhabitants were highly dependent on forest wood in hunting for food, cooking, and building rafts for going out to sea to hunt. Even the transport tools, Moai statues, and cultural relics on the island were also related to wood.

2.2 Cultural relics on the island

The most prestigious cultural relics on Easter Island are the Moai stone statues. From their place of manufacture to the seaside platform where they are positioned, the distances moved could be from a few km to over 10km. The statues could not have moved by themselves. Back then they did not have advanced modern cranes or lifting equipment like in modern times. So how did they move the statues? This was a riddle that baffled scientists, and the local

islanders learned from folklore through their ancestors that the statues could move on their own. Therefore, scientists performed an experiment by using the canoe ladder made from chopping Polynesian wood to successfully move the Moai statues. In recent years, there was an experiment involving three ropes to move the statues, proving that it is possible to use simple tools to move and erect the Moai statues. Regardless of what method was used to move the Moai statues given the primitive technology in ancient times, the methods used must have involved wood or ropes that had to be made from tree trunks and bark from the trees cut down on the island. Therefore, the cutting of trees was inevitable.

2.3 Exploring other causes

Easter Island is located on the east side of the Pacific Ocean, relatively far away from other Polynesian islands where the languages are also classified under the same south island language family, and there are no inhabited islands nearby. As a geographically isolated island, there is no way to travel between islands like the other Polynesian Islands, and thus also no way to invade or be invaded by other islands for resources. Therefore, the Easter islanders could only cut trees to make boats, cook, and shift and erect the Moai statues. It is thought that trees were cut to make ordinary canoes for catching dolphins out at sea. Is it possible to recreate the catamarans used by the Easter Island ancestors and other Polynesians for fishing? Or did the change in wind direction and sea waves cause them to drift to the South American continent and trade with the inhabitants of South America back then?

If the Easter islanders were to travel by boats made from palm trees on the island to South America from west to east, the demand for trees on the island for making boats would have been much higher. However, an ordinary canoe would not have been stable enough to cross the vast Pacific Ocean to get to South America, just like South islander tribes that stopped over and stayed in West Polynesia for 1,500 years before moving over to East Polynesia or New Zealand in the Southwest or the Hawaiian archipelago in the north. Their navigation technology and boatbuilding technology must have made some improvements and progress to achieve this feat. As written in "Collapse: How Societies Choose to Fail or Succeed" by Jared Diamond.

The Polynesians resided in the west for over 1,500 years before venturing out to sea again. That was probably because the Polynesians had improved their canoe building technology further or had better navigation technologies. It could also be due to a change in direction of the ocean currents or the receding sea level that exposed some small islands which they could use as a 'spring boards to bridge the gaps' in the vast ocean (Diamond, 2006).

Polynesians have to go all the way from West Polynesia's Fiji to Samoa, Tonga Archipelago, Society Islands, and then the Cook Islands, which would be the base for getting to the Pitcairn Islands, Henderson

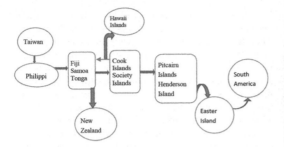

Figure 1. Eastward migration map of the south islanders.

Island, and Mangareva Island, which are the closest to the Easter Islands. Between those 3 islands and Easter Island lie more than 2,000km of water, which house numerous uninhabited islands. Further, there are also no inhabited islands between Easter Island and South America. Therefore, it is obvious that going directly from West Polynesia to South America was no mean feat for the South Island tribes. It is estimated that the more feasible route is to get to South America from Easter Island, which is described in Figure 1.

3   FOREST AND CIVILISATION OF EASTER ISLAND

3.1   *Catamaran*

Apart from the influence of ocean currents and winds, improvements in navigation technology played a critical role in enabling the tribes to cross the Pacific Ocean. The South islander tribes split into two directions from the Philippines, one to the west and one to the east, and made progress in developing boats over a few thousand years. The tribes that went eastward towards the Pacific Ocean developed their boats from single-winged and double-winged boats to eventually become double-bodied boats called catamarans (Figure 2). It was larger and much more stable, and

Figure 2. Photo was filmed by Peter Gill | England.
http://www.panoramio.com/photo/18959222 (2/10/2010)

Figure 3. Photo was filmed by michael_quintini.
http://www.panoramio.com/photo/16528424 (2/21/2010)

thus was capable of taking more people and materials on boards to enable them to head northwards to the Hawaiian islands, and then move eastwards for more than 3,000km to Easter Island. The boatbuilding technology for such catamarans could also be seen in Lake Titicaca (Figure 3) in Peru, apart from Hawaii, and you can also see catamarans made of reed. Therefore, it is postulated that the South islanders already came to Lake Titicaca and brought their boatbuilding technology over to Peru. As described above, based on geographical location, navigating through the route from West Polynesia to South America was not easy either. Therefore, it is deduced that the likelihood of starting off from Easter Island is the highest.

3.2   *Inca civilization on Easter Island*

When Easter islanders reached South America for the first time, did they return to Easter Island? Current research shows that the Moai statues on the island closely resemble South American Inca architecture (Diamond, 2006). It is obvious that the ancestors that landed in the South American continent did return to Easter Island, and they shuttled back and forth for a period of time. Therefore, they were able to integrate the beautiful Inca architecture into the local culture, and this was probably not long after they arrived at Easter Island. Thus, we can deduce that the catamaran made of reed in Lake Titicaca indeed originated from the boatbuilding technology of the Easter islanders.

According to historical records, the following were places where people made boats out of reeds locally: the Ethiopians of the upstream blue Nile River (Figure 4), the Iraqis living at the area where two rivers (Tigris and Euphrates rivers) intersected, and the Uros people of Lake Titicaca in Peru. However, only the Uros people of Lake Titicaca could make catamarans out of reeds, and these three regions were extremely far apart. Therefore, based on the migration path of the human race, the technique of making boats out of reed could not have

Figure 4. Photo was filmed by katia spanu.
http://www.panoramio.com/photo/17839203(2/20/2010)

travelled all the way from the Nile River or the Tigris and Euphrates rivers to Lake Titicaca. It was most likely that the South islanders brought the technology all the way there. Therefore, it is easier to deduce that the inhabitants of Easter Island had actually travelled to South America.

Going further, based on currently available data, it has been shown that catamarans were made in the South islands and in China during the Ming Dynasty. As for the island regions which include Taiwan in the west, Polynesian islands in the central area, Hawaiian islands in the north, and Easter Island in the east. From Easter Island, it spread to Lake Titicaca in South America. As illustrated above, when the South islanders migrated eastwards across the vast Pacific Ocean, they needed to stop over at a few places and the closest island to South America was Easter Island. At this point we can be sure that the ancestors of Easter Island migrated eastwards not long after they settled down, and during that period they still made regular round trips, which spurred demand for more deforestation. After that, once the forests on Easter Island were depleted, it was not possible for them to proceed to South America. The other people who settled down in South America from Easter Island were unable to return to Easter Island since there were no stable boats available for sailing, thus losing contact with the other ancestors. Since then, Easter Island became isolated from the rest of the world until the European explorers stepped onto the shore of the island during the 18th century.

3.3 *Summary*

There were some practices on Easter Island that were different from other Polynesian islands, such as cremation of the dead. The crematorium on the island contained the bones of a few thousand people. This practice also required many trees to be cut to provide firewood for cremation. Furthermore, in order to plant crops for food, trees were felled to clear the land for planting crops

(Diamond, 2006). The former destroyed the forest on the island and the latter inhibited chances for trees to continue growing. The wood that the Easter islanders needed for their daily needs and their civilization all came from the trees of the tropical rainforest on the island. In order to survive, deforestation became necessary. The forests were the life-blood of the Easter islanders. Without the forest, all was lost.

4 CONCLUSION

In order to erect the Moai statues, whether or not the islanders used ropes, boat ladders, or other methods, the materials needed would have been the natural resources provided by Easter Island, namely wood. The construction of boats, especially catamarans, required large amounts of wood and hence more trees were cut. When the forests disappeared, it marked the end of Moai statue carvings and the technology of building catamarans that showcased their navigation techniques. All these stopped flourishing and disappeared on Easter Island.

In the past, the disappearance of the forests in Easter Island leading to the destruction of its civilization was due largely to the ignorance and helplessness of the island inhabitants at the time. However, Easter Island's isolation from the rest of the world resulting in having no external help for the island was also one of the major factors that resulted in the rapid destruction of the civilization there. Easter Island is just a small island, and what happened to its environment would only have affected the people and the living things living on the island and would not have affected the survival of living things and the eco-system in other regions. On the other hand, looking at the comprehensive destruction of the Earth's forests today, particularly the tropical rainforest, it is not just the problem of the disappearance of forests in one region, but the whole Earth, hence making it the problem of all mankind. Protecting the forests does not only reduce soil erosion and slow down global warming and climate change, it also safeguards the survival of humanity and protects other creatures living on this Earth. All people and living creatures of this Earth should have their survival safeguarded. If humanity wishes to continue living on Earth, it should stop destroying natural resources and reduce excessive material needs. This is the critical issue faced by humanity, and forest preservation is a mission that cannot afford any further delay.

REFERENCES

H.L. Kao and H.M. Chou, 2017. Observing the Conflict between Human Civilization and Mother Nature in Environmental Changes on Easter Island. *Proceedings of IEEE ICASI*, Sapporo, Japan.
J. Diamond, trans. Y. Liao, 2006. *Collapse: How Societies Choose to Fail or Succeed*. Taipei: China Times Publishing Co.

Smart Science, Design & Technology — Lam et al. (eds)
© 2020 Taylor & Francis Group, London, ISBN 978-0-367-17867-3

Research on practical course reform of graduation design for design majors in undergraduate colleges and universities in China

Artde Donald Kintak Lam* & Qing Guo

School of Design, Fujian University of Technology, Fuzhou, Fujian Province, PR China

ABSTRACT: Graduation Design is the last practical link of education in undergraduate colleges and universities. How to strengthen the guidance and management of Graduation Design are the counts to the progress of Graduation Design. There are three factors affecting the quality of Graduation Design in undergraduate colleges and universities: students, teachers, and management. In view of the above factors affecting the quality of graduation design, this paper explores some effective methods and measures to gradually improve the quality of Graduation Design in practice.

Keywords: practical course reform, Graduation Design, undergraduate colleges and universities

1 INTRODUCTION

With the rapid development of modern science and technology and the adjustment of industrial structure, modern society has become a society of interdisciplinary integration. Thus, applied talents have become the main direction of social demand for talents (Xiao, Liao, and Zhang, 2005; Peng and Feng, 2009; Zhao, 2011). Therefore, the teaching reform of undergraduate colleges and universities is facing this trend. The practical course of Graduation Design is an important practical link in the undergraduate teaching and training plan. It is not only an important node for teachers to realize the combination of teaching, scientific research, and social practice, but also a summary of students' comprehensive application of, and achievements with the knowledge they have learned. Especially for design majors, the important role of graduation design in undergraduate teaching is that other teaching links cannot be replaced. Obviously, the reform of graduation design courses is an important factor in training applied talents in undergraduate colleges and universities.

In 1998, the State Education Commission of China advanced a guidebook for Graduation Design of undergraduate colleges and universities. It looks forward to the standardized development of graduation design in undergraduate colleges and universities. Tang et al. (2004) launched a pertinent investigation in combination with the Graduation Design of every major in the Anhui University of Technology, made a deep analysis of the results of the investigation, and found out various factors affecting Graduation Design, with a view to guiding the management of graduation design to a certain extent. Xia (2004) conducted a survey on the teaching of Graduation Design in 2002 in undergraduate colleges and universities all over the country, and analysed the characteristics and main problems of Graduation Design teaching. Subsequently, some scholars have presented some ideas on the reform of Graduation Design in undergraduate colleges and universities, but most of these studies are directed at the science and technology specialty (Xiao, Liao, and Zhang, 2005; Peng and Feng, 2009; Zhao, 2011).

Wang (2018) made an empirical analysis of the current situation and existing problems of Graduation Design in applied undergraduate colleges and universities. The common problems in the teaching of Graduation Design are that the mode of graduation design is unreasonable, the students' specialties are not brought into play, the students cannot adapt to the direction of employment, the students' comprehensive practical ability is not cultivated, and the quality of Graduation Design is declining.

In recent years, many undergraduate colleges and universities have actively explored and practiced Graduation Design, and the level of students' Graduation Design has been improved to a certain extent, but there are still problems in the implementation process. In fact, there are several important factors that affect the quality of Graduation Design, i.e., students, teachers, and management. This paper explores some

---

*Corresponding author. E-mail: artde.lam@qq.com

effective methods and measures to gradually improve the quality of Graduation Design in practice.

2 BACKGROUND, SIGNIFICANCE, AND THOUGHT OF REFORM

Graduation Design plays a significant role in the whole of undergraduate teaching. To comprehensively understand and analyze the main reasons for the low quality of Graduation Design practice courses in China's colleges and universities, consider the following: (1) some students are faced with the problems of employment, internship, and postgraduate entrance examination, which do not attach importance to Graduation Design; (2) some students' Graduation Design works have problems of no practicability, practicability, and operability, (3) a few instructors are not very enthusiastic about undergraduate Graduation Design work; (4) some majors are not strict in checking the Graduation Design results, there is a lack of supervision, and inspection is a mere formality.

There are some problems in the practice of Graduation Design in the undergraduate education of the design specialty in China, such as the unreasonable orientation of training objectives, the single convergence of training modes, the lack of characteristics in personnel training, the inadequate ability of classical design, the relatively low level of creative thinking, and the need to further improve practical ability and comprehensive quality.

We introduce and integrate the characteristics of international design education, and construct the research-based teaching system of Graduation Design practical course for design majors. Through optimizing curriculum attributes, standardized guidance mode, curriculum schedule, and enterprise practice guidance, we can actively adapt to the training of undergraduate talents of design majors for industry and strengthen the cultivation of students' independent design ability, knowledge practice ability, and innovative spirit.

(1) Students: the practical course of Graduation Design is usually carried out in the 8th semester in China. Some students face the problems of employment, internship, and postgraduate entrance examination. They do not attach importance to Graduation Design, which leads to the low quality of Graduation Design and even the phenomenon of copying and writing on behalf of other anti-socialist core values.

(2) Design works: students receive passive education in the whole undergraduate teaching process, and the practical knowledge of enterprise production in the knowledge system is relatively poor. Some students lack effective enterprise guidance in the design practice process, including material, color, shape, and other design practical knowledge, leading to the non-practicality and non-operability of Graduation Design works.

(3) Instructors: Most of the instructors in the practical course of Graduation Design adopt one-to-many guidance mode. Teachers are facing various pressures such as scientific research and professional evaluation, which results in a small number of instructors with a weak sense of responsibility who are not very enthusiastic about their Graduation Design work. Graduation design topic selection is not strict, supervision is insufficient, guidance is not in place, and there are other issues – students' topic selection content, opening report, mid-term inspection report, and final design are not fully guided, and the one-to-many guidance mode makes the instructor's comments and opinions on Graduation Design simple or formalized.

(4) Management: the execution process and evaluation management of graduation design are relatively relaxed in all majors under normal circumstances. Some of the issues here are lax performance control, inadequate supervision, and formality of examination. Some students must be able to pass the ideals of mind when they produce Graduation Design works.

3 REFORMING MEASURES

Under the theoretical guidance of learning science and cognitive psychology and the concept of "student-centered" education, we construct a research-oriented teaching system for the practical course of Graduation Design for design majors, highlighting the requirements of frontier, practicality, inquiry, and autonomy. Based on the premise of "management norms and creative freedom," we should strengthen the training of students' design classic ability, creative thinking ability, practical ability, and comprehensive quality.

(1) Optimizing the course attributes of continuous interaction: this involves adjusting the practical course of Graduation Design from the original practical links and integrating it into the professional curriculum system. Students are assigned weekly collective guidance according to professional courses. Students and instructors interact at regular times. The following problems have been solved: (a) lack of enthusiasm of students and lack of attention to graduation design; (b) lack of continuous communication between students and instructors; (c) asymmetry between the workload of instructors and the actual content of guidance work; (d) difficulty in teaching management in schools or specialties.

(2) Normalized guidance mode of team cooperation: with a heuristic guidance process of practical course of graduation project, the guideline of practical course of graduation project of "group guidance and group evaluation" is composed of several groups with each major. Students are faced with the guidance of many instructors—explaining doubts and solving doubts, improving their abilities, helping their personality, developing orientation, and probing into ideas—so that they can get more full

guidance under the premise of free creation. The following problems have been solved: (a) the lack of multi-oriented professional guidance for students; (b) the unreasonable grading of graduation design; (c) the problem of one-to-many guidance for instructors.

(3) Enterprise practice guidance: in the way of opening laboratories, opening innovative space, and organizing design work camps, the college contacts the off-campus enterprises to jointly organize related enterprise design work camps during the summer vacation. Through laboratories, innovation spaces, and work camps, students can understand and obtain the reality of enterprise design. Instructors can interact with enterprises more deeply and broadly. The following problems have been solved: (a) the lack of practical knowledge of enterprise production by students; (b) the lack of interaction between students and enterprises; (c) the lack of deeper and broader interaction between teachers and enterprises.

FUNDING

This research is funded by the Fujian Province's Education and Teaching Reform in Undergraduate Colleges and Universities in 2018, project number: FBJG20180041.

REFERENCES

G. Tang, H. Xu and Y. Sun, 2004. *Journal of Anhui University of Technology (Social Sciences) 21*(4), 92–93.
L. Xia 2004. *Higher Education of Sciences* (1), 46–48.
R. Peng and Y. Feng, 2009. *Journal of Yangtze University (Natural Science Edition) 6*(4), 361–362.
State Education Commission, 1998. *Guidance Manual for Graduation Design (Thesis) of Colleges and Universities*. Beijing: Higher Education Press.
X. Xiao, Y. Liao and Z. Zhang, 2005. *Journal of Guangdong University of Technology (Social Sciences Edition) 5*(9), 348–349.
Z. Wang, 2018. *Journal of Huangshan University 20*(2).
Z. Zhao, 2011. *Procedia Engineering 15*, 4168–4172.

Smart Science, Design & Technology — Lam et al. (eds)
© 2020 Taylor & Francis Group, London, ISBN 978-0-367-17867-3

Institutional changes and path dependence of Taiwan's vocational and technical education system

Li Ning* & Jianxing Li
Engineering Education Research Center, Personnel Division, Fujian University of Technology, Fuzhou, Fujian, PR China

ABSTRACT: Vocational and technical education as vocational and technical education is abbreviated in Taiwan. The contribution of the development of vocational and technical education to Taiwan's economic development has made it highly recognized in the world, especially in Asia, but its existing problems and crises have received less attention. Taiwan's vocational and technical education has gone through four stages: system formation, system maturity, system crisis, and system transformation. In this process, there are always path dependence phenomena. One is that the supply of vocational and technical education meets the needs of economic development and depends on authoritative system guidance; the other is that vocational and technical education's development path, as a subcategory of school education, always depends on the common mode of school education development. Vocational and technical education, which has sprung up widely in the mainland of China, should mirror the institutional changes of Taiwan's vocational and technical education in order to avoid the integration of vocational and technical colleges into the general higher education sequence in the competitive development.

Keywords: technical and vocational education in Taiwan, institutional change, path dependence

1 INTRODUCTION

From the 1960s to the 1970s, Taiwan promoted its export-oriented industrialization strategy, created economic miracles, and then joined the ranks of developed economies in the 1990s. The development of industrial vocational education is recognized as one of the main driving forces of this economic boom (Yang, 2008). Industrial vocational education is the early form of Taiwan's modern vocational and technical education (vocational and technical education is called technical and vocational education in Taiwan for short). It has evolved into the current structural system of modern vocational and technical education with the changes of social and economic development needs, and also followed the early basic system and development path.

The evolution of Taiwan's modern vocational and technical education system has important reference value for the development of vocational and technical education rising in the mainland of China. However, the mainland academia's research on Taiwan's vocational and technical education mainly focuses on its experience and advantages, and pays less attention to its existing problems and crises. Nowadays, the intensification of international competition in education, the phenomenon of fewer children on the island, the transformation of industrial structure, the hollowing out of manufacturing and labor-intensive industries, and other internal and external pressures have brought new development requirements to Taiwan's vocational and technical education. The present situation of Taiwan's vocational and technical education has far-reaching historical causes. Firstly, this paper combs the changing track of Taiwan's vocational and technical education in different stages according to the background of the times, and then makes a theoretical analysis of the changing process from the perspective of historical institutionalism, following the train of thought of macro-structure, middle-level system, and micro-actors, and explores the interconnected interaction between vocational and technical education and various social factors. Taiwan's modern vocational and technical education system forms a panoramic understanding.

*Corresponding author. E-mail: 149665698@qq.com

2 THE EVOLUTION GENEALOGY OF TAIWAN'S VOCATIONAL AND TECHNICAL EDUCATION SYSTEM FROM MODEL IMPLANTATION TO ORGANIZATIONAL CHANGE

With the change of the times, Taiwan's vocational and technical education system has experienced three stages: transplantation of foreign model, continuation of model to organizational change, and institutional transformation triggered by institutional crisis. This process runs through three important clues: firstly, the implantation of a foreign vocational education model has laid the foundation for the development of vocational education; secondly, vocational education has been incorporated into the economic development plan, which is an important part of the human development plan, and its changes are closely related to Taiwan's economic situation; thirdly, vocational education tends to merge with general higher education in the process of high-level development.

The foreign vocational education model was implanted into Taiwan in the 1950s and 1960s. After the Second World War, Taiwan focused on developing agriculture and industry for economic reconstruction. Import substitution and development of labor-intensive industries were the basic strategies, which urgently needed a large number of practical and technical personnel. In the course of implementing economic assistance to Taiwan, the United States found that the lag of technical and vocational education was the bottleneck restricting the development of agriculture and industry. It sent an expert delegation to Taiwan for investigation. Aiming at a series of problems in Taiwan's technical and vocational education at that time, such as ambiguous purpose, low quality of teachers, difficult and complicated curriculum content, attaching great importance to learning, and neglecting skills, the United States formulated assistance programs and transplanted the model of American industrial and vocational education.

Taiwan's localized technical and vocational education system took form from the 1970s to the 1990s. During this period, Taiwan's technical and vocational education achieved high-level development of localization through system design. Compared with the foreign vocational education model, it is a self-organizing process of spontaneously creating new models to meet the needs of economic development. The contribution of the development of vocational and technical education to Taiwan's economic development has made it highly recognized in the world, especially in Asia. However, the high-level development of vocational and technical education has a latent crisis and become increasingly prominent in the 21st century.

At the end of the last century, Taiwan formed a higher vocational and technical education system, including specialist education, undergraduate education, and postgraduate education. Colleges, technical colleges, and universities of science and technology are its carriers. Specialist education has become a symbol of the division between vocational and technical education and general education because of its practicality and employment orientation. However, according to Taiwan's official data, in the past two decades, the overall layout of technical and vocational colleges has been declining. From 1991 to 2016, four time nodes were selected for this study, namely 73 in 1991, 36 in 1999, 23 in 2000, and only 15 in 2016; the number of technical colleges was 3, 40, 51, and 16; and the number of science and technology universities was 0, 7, 11, and 59. Comparing these three groups of figures, we can see that the number of technical colleges is decreasing in a straight line and the scale of specialized education is shrinking day by day; the number of technical colleges increases steadily at first, and begins to decline after 2000. The exact time of decline is in 2003 (mainly due to the transformation of some technical colleges into universities of science and technology); and the number of scientific and technological universities continues to grow. The number of technical universities is close after 2003. The result is that the stock of technical colleges is limited, and the scale of universities of science and technology is dominant.

As a symbol of the gap between vocational and technical education and general education, vocational and technical colleges are facing further difficulties in enrollment due to the phenomenon of fewer children on the island, and the value of their existence is questioned. The overall level of technical and vocational education has moved upwards, resulting in the formation of an "inverted pyramid" structure with the top expanding and the bottom shrinking of the technical and vocational education system. In the Vocational Colleges of science and technology, the dominant orientation of the comprehensive development of science and technology universities has obvious academic characteristics. Compared with ordinary higher education, there is no significant difference. It causes the concept of graduation as employment to be replaced by the general desire to enter higher education. As a result, the practicality and employment-oriented characteristics of vocational and technical education have been weakened, and the function of vocational and technical education presupposed by the system has failed.

The social standard in the design of technical and vocational education system has in fact given way to the individual-based school layout. The value orientation of Taiwan's technical and vocational education system design is the social standard. The so-called social standard advocates that the educational goal should be centered on the realization of social value and that educational activities should be constructed according to the needs of social development. On the contrary, the theory of individual standard advocates that the goal of education should be centered on the realization of individual value, and educational activities should be constructed according to the needs of individual self-improvement and development (Hu, 2000). Nowadays, the social standard of Taiwan's technical and vocational education system design has in fact given way to the individual-based school layout. Its background is that Taiwan's technical and vocational education was influenced by the neoliberal economic reform of the western developed countries in the late 20th century, the government gradually relaxed the control of schools, and the education market tended to liberalization (Li, 2014).

In order to cater to the people's desire for further education and job-hunting preferences, technical and vocational colleges promote academic education in an area; on the other hand, they set up departments and curricula guided by individual career vision, and leisure service majors such as sightseeing and leisure, dining and travel services, cultural creativity, medical care, and other leisure services are favorites. However, the statistical data show that the gap of senior talent in Taiwan is 41,000 from 2009 to 2015, and that of grassroots talent is 335,000. In the future, the demand for human resources in the medium and long term will mainly move towards the two ends of senior talents and grassroots technicians, and the supply of professional talents and grassroots talents is obviously insufficient. It can be seen that there is a mismatch between the supply of Taiwanese vocational and technical education and the demand of social and economic development.

In view of the above-mentioned crisis, in 2017 the Taiwan authorities formulated and issued a series of policy documents, such as "Technical and Vocational Education Policy Program," "Transition and Breakthrough: White Paper on Talents Cultivation of the Ministry of Education," "White Paper on Teachers Cultivation," and so on. They carried out the plan of technical and vocational education reengineering, focusing on the cultivation of competitive professional and technical talents, in the system construction, curriculum system, teacher certificate, and property. In all respects, such as academic collaboration, the reform measures aiming at returning to the social standard are put forward.

3 PATH DEPENDENCE ON THE EVOLUTION OF TAIWAN'S TECHNICAL AND VOCATIONAL EDUCATION SYSTEM

The change of Taiwan's vocational and technical education has experienced four stages: system formation, system maturity, system crisis, and system transformation. There are always path dependence phenomena in the transformation of these four stages, which are manifested in two aspects: on the one hand, the supply of technical and vocational education, which meets the needs of economic development, depends on the guidance of an authoritative system, and the design of technical and vocational education system changes with the social macroeconomic situation. Technical and vocational schools are rising in scale and level because they constantly meet the changes of social and economic needs. When institutional constraints weaken, the basic function of technical and vocational education closely follows the needs of economic development; on the other hand, technical and vocational education fails to meet the needs of economic development. As a subcategory of school education, its development path always relies on the common mode of school education development. In the process of high-level development of technical and vocational schools, a ladder sequence similar to general higher education has also been formed. People's educational expectations will continue to strengthen the expansion of technical and vocational colleges to top-level education. When the first path dependence phenomenon is reduced, the second path dependence phenomenon will further weaken the basic function of technical and vocational education. The essence of the contradiction between the two-path dependence phenomena mentioned above is the game among macro-authority system, informal rules, and micro-actors.

An authoritarian political system not only achieves the development of vocational and technical education, but also restricts its independence. Informal rules lead the function of technical and vocational education to lose its standard. Rational choice of micro-actors accelerates functional failure of technical and vocational education standards. Private and public vocational and technical colleges are micro-actors in the practice. Micro-actors tend to rely on the common mode of school education development, guided by the informal educational rules prescribed by social and cultural concepts, and reduce the gap between themselves and general higher education. In this regard, private institutions are more active than public institutions. Vocational and technical education, which is widely springing up in the mainland of China, belongs to the same type as that in Taiwan. The design of vocational education system in mainland China should draw lessons from the changes of Taiwan's vocational and technical education system and avoid the integration of private and public vocational and technical colleges into the general higher education sequence.

FUNDING

This research is funded by the following projects: (1) Fujian Province's 13th Five-Year Plan for Educational Science Specialized Vocational Education across the Taiwan Strait in 2017, project number: FJJKHX17-014; (2) Fujian Province's 12th Five-Year Plan for Educational Science Specialized Vocational Education across the Taiwan Strait in 2015, project number: FJJKHX15-023.

REFERENCES

Z. Yang, 2008. *Taiwan Research Journal 100*(2) 61–68.
Z. Hu, 2000. *Journal of South China Normal University (Social Science Edition)* (02) 87–94.
N. Li, 2014. *Jiangsu Higher Education* (01) 33–36.

Smart Science, Design & Technology — Lam et al. (eds)
© 2020 Taylor & Francis Group, London, ISBN 978-0-367-17867-3

Perception reaction experiment of rotational apparent movement with different line graphic on motion illusion

Chihwei Lin*, Lanling Huang & Chimeng Liao
School of Design, Fujian University of Technology, Fuzhou City, Fujian Province, China

Hsiwen Fan
Department of Digital Media Design, National Yunlin University of Science and Technology, Yunlin, Taiwan

Lifen Ke
School of Design, Fujian University of Technology, Fuzhou City, Fujian Province, China

ABSTRACT: Visual illusions, perceptions, and patterns of motion forms are derived from combinations of form determinants relevant to motion forms and rotation of the forms concerned. Thus, this study uses different line graphics on forms with rotational movement and via the producing of motion illusion and pattern patterns investigates the apparent movement. Then it explores interactive and causal relationships among form patterns element of motion form, motion illusion, and the reaction of apparent movement. The experiment in this study used the adjustment method in psychophysics to measure the absolute threshold (AT). The findings revealed that different line graphics and rotational speeds influenced the formation and strength of phenomenon of line afterimage (PLA), and it had causal relationships with absolute threshold (AT) in different line graphics in apparent movement.

1 MOTION PERCEPTIONS AND PATTERNS OF APPARENT MOVEMENT IN THIS EXPERIMENT

Apparent movement occurs when the conditions of visual masking and a continuous stimulus are fulfilled. Under such circumstances, the stimulus that triggers motion perception does not move, but real motion objects create the illusion that the stimulus is moving. In this study, different spirals and continuous line graphics on a form were treated as stimuli for apparent movement. Next, the angles caused by rotation of the form (visual disturbances) were utilized to create visual masking effects, the forming elements of which operated according to the principle of apparent movement. When the form continued to rotate, all continuous line graphics formed a continuous movement and created continuous motion perceptions and patterns.

2 EXPERIMENTAL DESIGN AND SAMPLES

This study adopted the method of adjustment derived from psychophysical methods to determine the absolute threshold (AT) and implemented laboratory experimentation with a within-subject design. A total of 20 participants were recruited through a nonprobability sampling technique known as judgement sampling. Five line graphics on forms were designed, namely straight, ladder, arc line graphics, curve with short wavelength, and curve with long wavelength. Each line graphic was 10 mm wide and tilted by 15° (Chen, 2008). Under a 1-M observation distance, each measured triangular pyramid (Chen, Lin, Fan, 2015) was 25 cm high (Oyama, Imai, Wake, 2000) and 12 cm wide (Liu, 2011) (Table 1).

3 RESULTS

The lower AT of the straight line graphic (65.65 rpm) was the lowest, indicating that this line graphic triggered an apparent movement perception earlier than did the other line graphics. The upper AT of the curve with long wavelength (103.9 rpm) was the highest, suggesting that this curve led to longer perceptions of apparent movement compared with the other line graphics. The straight line graphic (33.75 rpm) had the highest speed threshold and thus had the longest duration of motion illusion and the best performance. Overall, the straight line graphic produced the optimal effect of motion illusion and optimal perception of apparent movement (Table 2; Figure 1).

---

*Corresponding author: E-mail: copy1.copy2@msa.hinet.net

Table 1. Five line graphics and pyramid.

Type of line graphic	Straight	Ladder	Curve with long wavelength
Triangular pyramid sample			
Graphical approach			

Type of line graphic	Curve with short wavelength	Arc
Triangular pyramid sample		
Graphical approach		

Table 2. Rotating speed thresholds of apparent movement for the five types of line graphics (rpm).

Type of line graphic	Straight	Ladder	Curve with long wavelength	Curve with short wavelength	Arc
Upper AT	99.40	100.30	103.90	102.55	93.80
Lower AT	65.65	77.20	82.70	84.35	72.25
Speed threshold	33.75	23.10	21.20	18.20	21.55

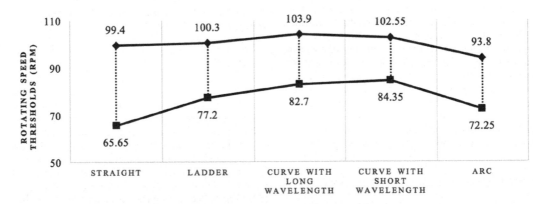

Figure 1. Run chart in the thresholds of apparent movement in five line graphics (rpm).

4 EMERGENCE OF THE PLA AND PHENOMENON OF MIXING AFTERIMAGES

The experimental results demonstrated that different line graphics and different rotational speed settings directly affected the results of motion illusions and apparent movement and the perceptions and discrimination of apparent movement among the participants. Thus, this study identified factors that affect motion illusions and determined interactive and causal relationships among types of line graphics, rotation speed, and apparent movement. Between each interstimulus

Table 3. Visual disturbances of the five types of line graphics.

Type of line graphic	Straight	Ladder	Curve with long wavelength	Curve with short wavelength	Arc
P Pattern of line graphic afterimages					
Threshold of rotational speed	65.65–99.4 rpm	77.2–100.3 rpm	82.7–103.9 rpm	84.35–102.55 rpm	72.25–93.8 rpm

interval (ISI), each different type of line graphic can create a different phenomenon of line afterimage (PLA). In other words, when a pyramid rotated, some sections of the line graphics curve exhibited afterimages that fluctuated and shook unevenly, resulting in varying degrees of visual disturbances. When the curving degree increased, so did the number of afterimages, consequently exhibiting a more noticeable disturbance (Table 3). When the pyramids with line graphics rotated, visual disturbances known as the phenomenon of line curving (PLC) and the phenomenon of line ghosting (PLG) occurred because of the unique shape of triangular pyramids. When the participants perceived and recognized apparent movement, they were visually disturbed by the PLA caused by line graphic rotation and PLC and PLG caused by pyramid rotation. Specifically, disturbances created by straight and arc line graphics were the most minor because these two line graphics were nearly straight. Except for the PLC and PLG, no PLAs appeared. Fluctuation was unnoticeable in these two line graphics, and thus fewer disturbances were created. Because the fluctuation levels of the ladder line graphic and curves of long or short wavelength were high, the number of afterimages increased. In addition to PLC and PLG, PLA occurred; therefore, the perceived line graphics fluctuated and trembled, creating more visual disturbances. Among the line graphics that exhibited PLC and PLG, those that were straight caused fewer afterimages, and thus appeared more static with fewer disturbances. By contrast, the number of afterimages increased among the graphics that were highly curved. The perceived images fluctuated and shook, thereby creating stronger disturbances. The aforementioned results demonstrated that the curving degree of line graphics had a causal relationship with the emergence and degree of PLA.

5 INTERACTIVE AND CAUSAL RELATIONSHIPS AMONG LINE GRAPHICS, ROTATIONAL SPEED, AND APPARENT MOVEMENT

The experimental results revealed that different combinations of line graphics and rotational speed could

determine the emergence, degrees, and transduction of the PLA, PLC, and PLG, which directly affected the upper and lower AT values of the various line graphics in terms of apparent movement. The curving degrees and rotational speeds of the line graphics affected the presence of apparent movement and the participants' perceptions and discrimination of such movement. Thus, the curving degree and rotational speed had causal relationships with apparent movement. When the line graphic became straight with a lower rotational speed, the number of afterimages decreased and the originally fluctuating and trembling line graphic became stationary. Furthermore, the vision that was originally disturbed by the PLA and PLC was disturbed by only the PLC, indicating a decrease in visual disturbances. Under such circumstances, the lower AT also decreased, suggesting that when the line graphic was straight, the lower AT in apparent movement was identified at an earlier stage. Therefore, the participants first perceived the apparent movement created by the pyramid with a straight line graphic and that with an arc line graphic because their lower AT values (straight = 65.65 rpm; arc = 72.25 rpm) were discovered earlier than those of the other line graphics. Similarly, when the line became straight, the rotational speed increased, the number of afterimages decreased, and the originally fluctuating and trembling line graphic became static. Furthermore, the vision that was originally disturbed by the PLA, PLG, and PLC was disturbed by only the PLG and PLC, indicating a decrease in visual disturbances. The upper AT, however, did not increase because of fewer disturbances but instead decreased with the lower AT. This result indicated that the upper AT in apparent movement was identified earlier when the line graphic became straight, possibly because when rotational speed increases, apparent movement reportedly occurs in succession (Chen, Lin, Fan, 2015). When apparent movement is perceived and rotational speed increases almost to the transduction threshold, the participant no longer perceives apparent movement (i.e., the upper AT of apparent movement disappears); the perception of apparent movement then transduces into induced movement (i.e., presence of the lower AT of induced movement). In the present

317

study, when the line graphic became straight and the rotational speed increased, the disturbance caused by the PLA disappeared. After the number of afterimages had decreased, the line graphic had stabilized, and the number of visual disturbances decreased, the upper AT appeared earlier. This result proved that the participants more quickly perceived apparent movement created by straight line graphics. Thus, the upper AT values of the arc (93.8 rpm) and straight (99.4 rpm) line graphics were discovered earlier than those of the other line graphics (Figure 2).

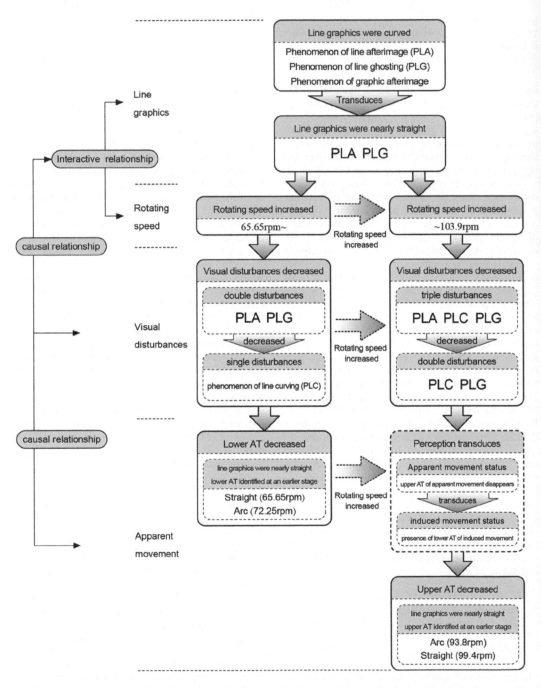

Figure 2.   Interactive and causal relationships among line graphics, rotational speed, and apparent movement.

6 CONCLUSION

This study used the different types of line graphics on rotational forms and motion illusion and pattern performance, then observed the apparent movement. Moreover, this study explored interactive and causal relationships among form pattern element of motion form, motion illusion, and the reaction of apparent movement. The findings revealed that different line graphics and rotational speeds were influencing the formation and strength of phenomenon of line after-image (PLA) and had causal relationships with absolute threshold (AT) in different line graphics in apparent movement. Overall, the straight line graphic produced the optimal effect of motion illusion and optimal perception of apparent movement.

ACKNOWLEDGMENT

This study was partly supported by the Fujian University of Technology with grant No.GY-S18015, Design-led Innovation Research Center, Fujian University of Technology and Design Innovation Research Center in Research Base of Humanities and Social Sciences in Colleges and Universities of Fujian Province.

REFERENCES

G.D. Chen, 2008. *The Study of Kinetic Art Dynamic Optical Illusion on Column with Spiral Pattern*, Ph.D. dissertation, TaiwanTech. Taipei, Taiwan.

G.D. Chen, C.W. Lin and H.W. Fan, 2015. *Journal of Design* 20(3) 1–19.

T. Oyama, S. Imai and T. Wake, 2000. *New, Sensory perception psychology handbook*. Tokyo: Seishin Shobo Ltd.

Y.H. Liu, 2011. *The Research of Preference for ratio of geometry shapes*, Ms. Des. Thesis, Dept. of Industrial Design, YunTech. Yunlin, Taiwan.

Smart Science, Design & Technology — Lam et al. (eds)
© 2020 Taylor & Francis Group, London, ISBN 978-0-367-17867-3

Rotational induced movement of different line graphic on motion form

Chihwei Lin*, Lanling Huang & Chimeng Liao
School of Design, Fujian University of Technology, Fuzhou City, Fujian Province, China

Hsiwen Fan
Department of Digital Media Design, National Yunlin University of Science and Technology, Yunlin, Taiwan

Lifen Ke
School of Design, Fujian University of Technology, Fuzhou City, Fujian Province, China

ABSTRACT: This study used different types of line graphics on rotational forms, and its motion illusion and pattern performance, then observed the induced movement. Moreover, this study explored interactive and causal relationships among form performance element of motion form, motion illusion, and the reaction of induced movement. Experiments were conducted by adopting the method of adjustment derived from psychophysical methods. The research found that different line graphics and rotational speeds were influencing the formation and strength of phenomenon of line afterimage (PLA) and phenomenon of mixing afterimages (PMA), and they had causal relationships with absolute threshold (AT) in different line graphics. The straight line graphic produced the optimal effect of motion illusion and optimal perception of induced movement.

1 MOTION PERCEPTIONS AND PATTERNS OF INDUCED MOVEMENT IN THIS EXPERIMENT

Induced movement mainly exhibits two patterns: induced movement between objects and that between forms (Chen & Chang, 2007; Chen, Chang, Lin, 2008). The present study focused on patterns caused by induced movement between forms, which are directly linked with figure–ground differentiation. When a continuous form (pattern) rotates, the induced movement produces the illusion of a rotating figure, whereas the illusion of stationary ground occurs because of assimilation, which separates the figure from the ground. In this study, different spirals and continuous line graphics on forms were treated as stimuli for induced movement. When the form concerned was in continuous rotation, a motion perception and pattern were triggered to create the illusion of a continuously rising line graphic.

2 EXPERIMENTAL DESIGN AND SAMPLES

This study adopted the method of adjustment derived from psychophysical methods to determine the absolute threshold (AT) and implemented laboratory experimentation with a within-subject design. A total of 20 participants were recruited through a nonprobability sampling technique known as judgement sampling. Five line graphics on forms were designed, namely straight, ladder, arc line graphics, curve with short wavelength, and curve with long wavelength. Each line graphic was 10 mm wide and tilted by 15° (Chen, 2008). Under a 1-M observation distance, each measured cylinder (triangular pyramid) (Chen, Lin, & Fan, 2015) was 25 cm high (Oyama, Imai, & Wake, 2000) and 11 cm wide (Liu, 2011) (Table 1).

3 RESULTS

The lower AT of the straight line graphic (44.2 rpm) was the lowest, indicating that the straight line graphic provoked the perception of induced movement earlier than did the other line graphics. In addition, the straight line graphic had the highest higher AT (299.55 rpm), demonstrating that induced movement triggered by this line graphic lasted longer compared with the other line graphics. Furthermore, the straight line graphic had the highest speed threshold (255.35 rpm), and thus had the longest motion illusion with optimal performance. In brief, the straight line graphic produced the optimal effect of motion illusion and optimal perception of induced movement (Table 2; Figure 1).

---

*Corresponding author: E-mail: copy1.copy2@msa.hinet.net

Table 1. Five line graphics and column.

Type of line graphic	Straight	Ladder	Curve with long wavelength
Column sample			
Graphical approach			

Type of line graphic	Curve with short wavelength	Arc
Column sample		
Graphical approach		

Table 2. Rotation speed thresholds of induced movement for the five line graphics (rpm).

Type of line graphic	Straight	Ladder	Curve with long wavelength	Curve with short wavelength	Arc
Upper AT	299.55	283.95	267.80	257.00	228.5
Lower AT	44.20	59.15	67.60	72.75	74.35
Speed threshold	255.35	224.80	200.20	184.25	154.15

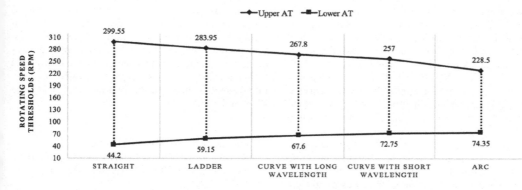

Figure 1. Run chart in the thresholds of induced movement in five line graphics (rpm).

4 EMERGENCE OF THE PLA AND PHENOMENON OF MIXING AFTERIMAGES

The experimental results demonstrated that different line graphics and distinct rotational speed directly affected the results of motion illusions and induced movement and the perceptions and discrimination of such movement among the participants. This study determined factors that affected motion illusions and determined interactive and causal relationships among types of line graphics, rotation speed, and induced movement. Between each ISI, different line graphics and rotational speed settings can create different forms of the PLA and phenomenon of mixing afterimages (PMA; Table 3). Similar to apparent movement, the PLA refers to the fluctuating, trembling, and nonstationary afterimages that appear between

321

Table 3. Visual disturbances of the five types of line graphics.

Type of line graphic	Straight	Ladder	Curve with long wavelength	Curve with short wavelength	Arc
PLA					
Threshold of rotational speed	44.2 rpm	59.15 rpm	67.6 rpm	72.75 rpm	74.35 rpm
PMA					
Threshold of rotational speed	299.55 rpm	283.95 rpm	267.8 rpm	257 rpm	228.5 rpm

each ISI. Such phenomena cause visual disturbances to varying degrees. When the curving and tilting degrees of the line graphic increased, so did the number of afterimages, consequently exhibiting more noticeable disturbances. The PMA occurs when the line graphic concerned exhibits rotary color mixture. When a cone rotated, the phenomenon of temporary color mixing was identified on the line graphic (Hung & Guan, 1999), thereby creating continual grayscale ramp and a mixing visual effect of grayscale ramp. According to the theory of rotary color mixture, color-mixing images appear because lights of varying wavelengths are equally reflected and absorbed when the disturbance in vision in question is stimulated only by black and white (Lee, 2002). The PMA can cause disturbances to varying degrees. When the curving and tilting degrees of the line graphic concerned were high, the area of continual grayscale ramp expanded, causing stronger disturbances. The PLA disturbance shifted to PMA disturbance as the rotation speed increased. Furthermore, the cylinders adopted in the experiment of induced movement did not have sides or angles as did the aforementioned pyramids because of their curvilinear shapes, and thus the PLC and PLG did not occur when the cylinders rotated. Thus, the participants were disturbed by the PLA and PMA only when perceiving and recognizing induced movement. Specifically, the straight line graphic had the most minor disturbances because even at varying rotation speeds, its straightness did not induce the PLA and PMA. Thus, the straight line graphic created fewer disturbances than did the other line graphics. By contrast, the curving and tilting degrees of the other four line graphics were high, and thus the number of afterimages

and the area of continual grayscale ramp increased and induced the PLA and PMA at various rotational speeds. When the cylinders with these line graphics rotated, the participants perceived fluctuating and trembling line graphics and perceived considerable visual disturbances. In summary, at different rotational speeds, the straight line graphic presented few afterimages and the area of continual gray scale ramp was small; this line graphic was relatively static, leading to minor visual disturbances. By contrast, the number of afterimages and area of continual grayscale ramp increased when the curving and tilting levels of the line graphics increased; the line graphic images fluctuated and trembled, creating considerable disturbances. The curving and tilting levels of the line graphics at different rotational speeds had causal relationships with the PLA and PMA.

5 INTERACTIVE AND CAUSAL RELATIONSHIPS AMONG LINE GRAPHICS, ROTATIONAL SPEED, AND INDUCED MOVEMENT

The experimental results revealed that different line graphics and rotational speeds can decide the emergence, degrees, and transduction of the PLA and PMA, thereby directly affecting the upper and lower AT values among the measured line graphics during induced movement. The curving levels of line graphics and the rotational speed affected the formation of apparent movement and the participants' perceptions and discrimination of such movement. Thus, the curving level and rotational speed had causal relationships with apparent movement. When the line graphic straightened at a lower rotational speed, the number of afterimages decreased; the originally fluctuating line graphic then turned stationary and PLA disturbances gradually diminished. Under these circumstances, the lower AT also decreased, suggesting that when the line graphic was straight, the lower AT of induced movement was identified earlier. Therefore, the participants first perceived the induced movement created by the cylinder with straight line graphic because its lower AT value (44.2 rpm) presented earlier than those of the other line graphics. By contrast, the cylinders with curved and tilted line graphics exhibited their lower ATs later. The lower AT of

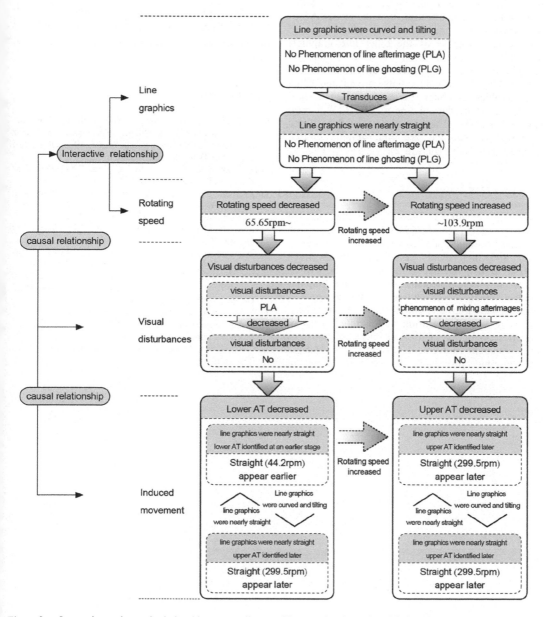

Figure 2. Interactive and causal relationships among line graphics, rotational speed, and induced movement.

the arc line graphic (74.35 rpm) appeared last compared with those of the other line graphics, indicating that the participants perceived the induced movement caused by the arc line graphic last. When the line graphic straightened at a higher rotational speed, the area of continual grayscale ramp shrank; the originally fluctuating line graphic then turned stationary and PMA disturbances were alleviated. In this situation the upper AT value increased when the line graphic was straight, demonstrating that the upper AT of the straight line graphic (299.55 rpm) appeared later than did those of the other line graphics. The participants continued to perceive that the straight line graphic caused induced movement. Conversely, the cylinders with curved and tilted line graphics exhibited their upper ATs earlier; the upper AT of the arc line graphic (228.5 rpm) appeared earliest. The participants quickly ceased to perceive that the arc line graphic caused induced movement (Figure 2).

6 CONCLUSION

This study examined diverse visual illusions and motion patterns in different line graphics with varying rotational speed to explore induced movement and understand the relationship between motion form and induced movement. The experimental results revealed that different line graphics and rotational speeds can decide the emergence, degrees, and transduction of the PLA and PMA; that is, indicated appearing of rotational speed of the different line graphics had causal relationships with apparent movement. Overall, the straight line graphic produced the optimal effect of motion illusion and optimal perception of induced movement.

ACKNOWLEDGMENT

This study was partly supported by the Fujian University of Technology with grant No.GY-S18015, Design-led Innovation Research Center, Fujian University of Technology and Design Innovation Research Center in Research Base of Humanities and Social Sciences in Colleges and Universities of Fujian Province.

REFERENCES

G.D. Chen and C.C. Chang, 2007. A Study about the Induced Motion of the Rotative Speed and the Caliber of Strand. Conference on Asia Society of Basic Design and Art in Tsukuba, 295–300.

G.D. Chen, C.C. Chang, and P.C. Lin, 2008. *ISSUE of Basic Design and Art 16*, 19–22.

G.D. Chen, 2008. *The Study of Kinetic Art Dynamic Optical Illusion on Column with Spiral Pattern*, Ph.D. dissertation, TaiwanTech. Taipei, Taiwan.

G.D. Chen, C.W. Lin, and H.W. Fan, 2015. *Journal of Design 20*(3) 1–19.

T. Oyama, S. Imai, and T. Wake, 2000. *New, Sensory perception psychology handbook*. Tokyo: Seishin Shobo Ltd.

Y.H. Lee, 2011. *The Research of Preference for ratio of geometry shapes*, M. Des. Thesis, Dept. of Industrial Design, YunTech. Yunlin, Taiwan, 2011.

C.Y. Hung and X.S. Guan, 1999. *A Study of Subtractive Color Mixture and Rotatory Color Mixture*. National Science Council, Taipei, Taiwan.

C.F. Lee, 2002. *Journal of Science and Technology 11*(4), 279–284.

Smart Science, Design & Technology — Lam et al. (eds)
© 2020 Taylor & Francis Group, London, ISBN 978-0-367-17867-3

The budding of Taiwan's canoe sport and leisure industry - using the case study of PAI company as the product standard

Li-Chen Tsai*
Ph.D Program of Mechanical and Energy Engineering, Kun Shan University, Tainan, Taiwan

Huannming Chou
Department of Mechanical Engineering, Kun Shan University, Tainan, Taiwan

ABSTRACT: Pollution-free non-powered canoe/kayak is one of the most important tools to experience the natural ecosystem in the creative lifestyle industry. Through continuous exploration and development, the canoeing sports and leisure industry has transformed from a competitive sport to gradually become a commercialized industry. This thesis analyses the canoeing sport through its three stages of development, first experiencing amateur competitions, then club and association competitions, and finally multi-element development - a gradual morph into an industry. PAI Company is the first enterprise that combined the canoeing sport with eco-tourism to form a standardized product. Thus reviewing the industry's development history and how the canoeing sport came to combine with an ecological experience to become a standardized product can serve as an important reference to this industry's subsequent development.

Keywords: Canoe, Kayak, Sport and leisure industry, Product standardization, Clubs

1 PREFACE

In a circular economy, the environment itself is equipped with the function of being an aesthetic good (Pearce and Turner, 1990). From the standpoint of an environmentally aesthetic product, looking at the 15 cultural and creative industries in Taiwan, the 14th item in the creative lifestyle industry that relates to the experience of the natural ecosystem (Ministry of Culture, 2013) is perhaps the closest to a circular economy. Being able to observe and experience the natural ecosystem with minimum disturbance to the ecological environment and also adopting a method that saves energy and does not create pollution, especially when conducting an ecological experience, not only fulfils the standard of a circular economy, but also has the effect of raising environmental awareness.

On the experience of the natural ecosystem, there is a difference between experience on land and in water. On land, non-powered bicycles are used to go deep into each village and town across Taiwan on a round-the-island tour - this has become quite a phenomenon lately and can be seen as a model of pollution-free ecological experience. In terms of the water route, sails, surfing boards and etc. are all pollution-free non-powered tools or equipment; however, sailing requires a spacious open water activity area, costs greatly in terms of equipment acquisition, and has a high operational skills threshold - these have therefore had a hard time developing into a popular leisure activity industry in Taiwan. Similarly, surfing boards also face the issue of having a high threshold of operational skills, and has therefore been unable to become a leisure activity industry.

This thesis believes that non-powered canoes/kayaks (hereafter canoes only) make expanding lifestyle and ecological experiences more unique. This is because the act of rowing a canoe itself can increase the participant's basic knowledge and techniques on water transport tools, including, paddling the canoe close to the water, shuttling through marine plants, observing all kinds of ecosystems born of water. The perspective one gains from the activity, feeling the serenity of the water banks and experiencing other water activities - these are all in stark contrast with land activities, and therefore greatly expand our life and ecological experience. Combining canoeing activities with various types of lifestyle experiences or ecological experiences on land not only enriches the structure of the whole programmer, but also expands the consumers' range of choice, creating a positive cycle for the industry's development as a whole.

---
*Corresponding author. E-mail: lejintswa@gmail.com

This thesis shall review the development history of the canoeing sport's transformation into a leisure activity industry, which precisely reflects the structure of this industry's leading brand Paddle Around International (PAI), and serves as a classic example for subsequent competitors to model after, as well as provides a reference for the industry's future development and advancement.

2   THE HISTORY OF A COMPETITIVE SPORT'S DEVELOPMENT INTO A LEISURE INDUSTRY

The canoeing sport leisure industry originates from competitive sports, because rowing a canoe requires considerable skills to ensure the rower's safety. However, in the developmental stages of the canoeing sport in Taiwan, canoes were usually seen as a competitive tool that challenged unique areas of water. Therefore, canoes retained the status of a niche competitive sports tool for its enthusiasts for almost thirty years. General research categorizes the canoeing sport into phase one (1975-1988), which was its budding phase, phase two (1989-1999), which was its anchorage phase, and the current phase three (2000-present), which is its multi-element development phase (Yin, 2010).

The above categorizations are based on the perspective of competitive sports; it is difficult to see the historical transformation of the canoeing sport morphing into an industry. Therefore, this thesis will recategorise the canoeing sport into the first phase of amateur competitions, the second phase of clubs and associations competitions, and the third phase of multi-element development. Although there were signs of commercialization in the first and second phase, there wasn't a sufficient sustainable self-growth and survival structural formation and business model. After more than twenty years of the first and second phases of development, a standardized morph only appeared in the third phase that truly contributed to a burgeoning industry.

This thesis refers to Yin Chun-Shiang's seminal work "Research of Taiwan's Canoeing Sport Development," which described each phase's characteristics to delineate the tracks of a competitive sport morphing into an industry.

The first phase of amateur competitions started when British businessman Jerry Glesinger imported ten canoes in 1975, and his son Rob Glesinger, an advanced coach of the canoeing sport, taught canoeing techniques freely. At that time, more than 10 youths were trained for three weeks, but due to the lack of a group identity as well as funding support, the program ended up with nothing.

In 1978, Tunghai University's professor He Cheng-Dao brought back the canoeing sport, and in 1981 formed the unofficial organization Taipei Canoeing Club with like-minded canoeing enthusiasts; together they traveled together and explored different rivers and streams, and found a training ground to improve their skills. At that time, white-water canoeing was the main activity, but because it was extremely dangerous and required highly technical skills, it could not become a common sport, and the promotion of canoeing sport never did break through into society.

Canoe rental businesses also appeared during this time. Hsinchu's Green Grass Lake operators once had fifty canoes, but under the government's tax burden and the drying of the Green Grass Lake, the canoeing sport died again. Canoe rental was based on canoeing as a leisure activity that relied on great technical skills; canoes were not commonly used as recreational equipment. It was difficult for a leisure activity that relied on great technical skills to become a truly commercialized industry. Therefore, during this phase, the canoeing sport was only an amateur competition sport.

During the second phase of clubs and associations competitions, the canoeing sport gradually developed into two different types of activities. The first was the original leisure canoeing, and the other was the light boat race. During this time, the Taipei City Sports Association first formed a canoeing committee in the hopes of promoting canoeing as an Olympic Games competitive event. Canoeing committees were also formed in many places, such as the sports associations of Taipei County, Xindian City and Yilan County, where they organized free educational activities largely in the promotion of light boat race, canoe polo and other official Olympic Games competitive events. In 1991, the Taipei City Sports Association canoeing committee was reorganized and became the Chinese Taipei Canoe Association, and organized a national canoeing competition.

In contrast to the promotion of competitive events, cross-country streams, lakes, oceans and other leisure activities seemed unpopular. In 1991, the original Taipei Canoe Club members also jointly established an Adventurers Canoe Club, and actively promoted the canoeing sport. During this time, one of the club members, Lee Schondorf, argued that Taiwan treated canoes as common vessels rather than sports equipment, and that the law restricting the import of canoes was extremely unreasonable. Much effort was invested into changing this mindset, and canoes were finally seen as sports equipment, and were not governed in accordance with the "boat import management rules"; this marked a debut for the canoes' free import and promoted the development of the canoeing sport.

However, in the promotional activities organized by the canoe club, there were differences of opinions on whether the charges collected for the activities were operated as a business. One faction of the club believed that the club should only be a group for like-minded people to interact and exchange their ideas, and should not be a business organization. Therefore, the club formed two types of developments: commercialized and non-commercialized.

The commercial club slowly became a way to culti-vate canoeing enthusiasts and a channel to sell canoes.

How do we determine if the canoe club is a canoeing sport industry? Industry is defined as the different sectors or departments categorized to engage in product manufacturing and operation to meet certain needs of the society in accordance with the principle of the social division of labour in the national economy (Chen and Lu, 2010). Therefore, the "club" as an "industry" provided the product "enthusiasts' interaction and communication", and not "canoeing sport and leisure." Similarly, the products provided during the subsequent developments of the canoeing associations were a broader range of "peer interaction and communication"; the product still wasn't "canoeing sport and leisure." Therefore, the commercialization of the canoeing clubs and associations cannot be regarded as the origin of the canoeing sports and leisure industry, because the product provided was "interaction and communication." This thesis believes that the "budding" of the canoeing sports and leisure industry should be based on a profit-making enterprise's standardized "canoeing sports and leisure" product as the major business product, and continues to operate and profit through consumers testing and operating that business model.

During this time, the canoeing sport had also had further developments in technical skills and types, and formed the basis of the canoeing sports turning towards industrial development. For example, Lee Schondorf used the concept and technical skills of canoeing to develop training guidelines and methods as the standard of educational promotion, allowing the canoeing techniques to be more easily picked up and spread. Therefore, during the subsequent periods of this time, canoeing enthusiasts started setting up clubs across the whole of Taiwan, gradually expand-ing the people's participation in the canoeing sport.

In addition, the canoeing sport in its early phases of development focused mostly on whitewater canoe-ing, during this time then, players started introducing sea canoeing, fancy kayaking, surf kayaking and other types of canoeing activities. Among them, the Adventurers Canoe Club introduced a platform-type canoe that was not only easy to operate but was highly stable - beginners could learn to row the canoe quickly and it was suitable for users from age seven to seventy. Following this introduction, the canoeing sport and leisure industry's product had the opportun-ity to get rid of its label of being only a competitive sport, and became a standardized service product, planting the seed for subsequent industries.

After the year 2000 during the third phase, the canoeing sport not only received large-scale promo-tion organized by clubs and associations under the format of an organization, it also became more com-mercialized, and finally transformed into a burgeoning industry. Besides the continuous efforts in the society, schools and government policies also started playing a role in the promotion of the canoeing sport. For example, the Penghu County Government introduced canoeing into primary school's physical education cur-riculum (Penghu County, 2017), and the Penghu Uni-versity of Science and Technology also introduced canoeing into all levels of its educational curriculum (Penghu University of Science and Technology, 2015). The New Taipei City Sports Association's Canoeing Committee hosted a "Lohas Canoe Experi-ence Camp" that accepted public registrations from school children in the first year of primary school to people up to the age of sixty (New Taipei City's Sports Association, 2017). Schools and the govern-ment extended the qualification to participate in the canoeing sport from university-level initially to pri-mary school children and also to the elderly up to the age of sixty - this greatly expanded the number of participants for the canoeing sport, and also promoted the budding of the canoeing sport and leisure industry.

3 AN INTRODUCTION TO THE HISTORY OF PAI PRODUCT'S STANDARDISATION

The prevention of FDAT may be achieved from vari-ous areas such as the patient's medical history, the handling process as well as the medical facilities. In terms of medical history, things to take notice of include previous diagnosis of FDAT, heart diseases, habit of sedatives use and emotional instability. The handling of FDAT should combine both Chinese and Western medical approaches. The medical monitor should include four items of physiological monitor-ing: electrocardiogram (EKG), non-invasive blood pressure (NIBP), respiratory rate (RR) and saturation of pulse oxygenation (SpO$_2$). Also, there must be oxygen supply, emergency medications (atropine, ephedrine, epinephrine) and establishment of venous infusion sets.

In 1998, PAI Company's predecessor, Sea Minder Co., started introducing the Canadian ocean canoe Feathercraft and also ocean canoes Q-Kayaks from New Zealand to sell in Taiwan. According to a survey in 2009, nearly one in every 100 households in Britain owned a canoe, and of those canoe owners, their activity participation rate was 68.85% (81.46% in 2010), while the participation rate for those that did not own a canoe was 6.52% (4.91% in 2010) (Royal Yachting Association, 2009). There-fore, it was a habit for canoe enthusiasts overseas to own their own canoe. However, with Taiwan's high population density, most households not only have no room to store canoes, they also lack the facilities to transport their canoes to the waters. Therefore, even though most canoe business operators started their business selling canoes, selling canoes solely as their main business operation to continuously expand their enterprise lacked objective socio-economic and environmental conditions.

In 2000, the canoeing sport entered its third phase of multi-element organizational development. The founder of PAI Company organized a "Feathercraft

Club" (hereafter F Club) at New Taipei City's Sanzhi District Linshanbi canoe base, launching canoeing activities at a fixed location for the first time in the country. This is because all prior operators ran their business based on an irregular, nomadic format in order to reduce the cost of leasing activity venues. F Club not only sold canoes and related supplies, they also started launching canoe eco-tours. The footprints of eco-tourism could be found globally, including New Zealand's Abel Tasman National Park, Thailand's Krabi, Sun Moon Lake, Wushantou Reservoir, Tainan's Chiku Lagoon, Yilan's Dongshan River, the northeast coast, Keelung Island, Matsu, Penghu and so on. Due to the extensive organization of canoe eco-tourism, F Club started having a more profound understanding of the canoe eco-tourism's product value and features, and this set the basis for subsequent product standardization.

In 2003, PAI Company was formally established, and in addition to continuing to import canoes to sell, according to Taiwan's canoe environment and consumer behavior, the company started collecting ecotourism product customization and optimization's best solution. In the same year, they mobilized the industry to set up the Chinese Taipei Recreational Kayaking Association to gather everyone from the same industry to jointly promote canoeing sport and leisure activities. According to a British research, 29.4% of interviewees in 2014 were members of clubs and associations (Arkenford, 2014); this showed that canoeing clubs and associations provided members with the necessary resources. For example, gathering like-minded enthusiasts to increase the pleasure of the sport, safety and maintenance, coaching and guidance, and etc.; these are all resources that individual canoeing enthusiasts lack. As a result, clubs and associations were the common channel that provided amateur enthusiasts with training, safety and product resources. From the perspective of a business operator, clubs are a point of contact to get in touch with customers; in addition to being able to further expand their customer source, associations can also join hands with people from the same industry to jointly promote activities, thereby greatly reducing marketing costs as well as the high cost and risk of individual marketing - this is greatly beneficial to the expansion of the industry's scale and business operating scale. PAI Company set up a club to sell canoe products, and also called for the establishment of an association to expand their base and reduce promotional costs, completely replicating the historical format of the canoeing sport's development.

Following five years of trial and exploration, PAI Company formed the "Longmen Canoe Base" in 2008. Located in New Taipei City's Kungliao District northeast coast scenic area's Longmen camping site, the base became the first site in Taiwan where canoeing activities could be operated legally in the waters. From then on, canoeing activities became a regular, fixed location event's standardized product.

Its standardized product has two main features: the first is the use of a platform-type ocean canoe that is not only easy to operate but is very safe, with strict implementation of introducing the activity and giving the instructions prior to one going into the water as well as explaining the SOP of the operating training, and also a double-insurance guarantee to enhance the safety of the activity. After enhancing the safety of canoeing activities, the number of participants increased, and participants ranged from the age of seven to seventy. Secondly, it made up for the loss of satisfaction and pleasure arising from the lowering of operational skill threshold, as the Longmen camping site at the northeast coast waters provide a rich ecological landscape that precisely makes up this part of satisfaction and pleasure. Recreational ocean canoeing combined with an ecological experience has then rid itself of the niche rapid water canoeing, fancy kayaking and other such high skill threshold products, becoming a standardized product that suits the majority of the population.

Since its establishment of the Longmen Canoe Base to operate ocean canoeing coupled with an ecological experience, PAI Company has also become a canoe business operator in the industry that uses ecological experience as its main series of products - a gradually growing and developing business. For example, for product solutions such as amateur experience (2 hours), boat tour around Shuangxi River (6 hours), Northeast Sea boat tour (8 hours) and etc., PAI Company provides each and every tour and solution's necessary equipment, coach and ecological explanation. According to PAI Company's internal statistics, from 2013, the canoe ecoexperience occupies more than 55% of the company's revenue percentage, becoming the main driving force of the growth of the business. PAI Company is therefore the first in Taiwan to operate the canoeing business using the standardized business format; its organizational structure and commercialization format also sets an example for subsequent startups in the industry. Therefore, this thesis sees PAI Company's success with standardizing the canoeing ecoexperience product as the main pillar of support of the business' survival and development without diversifying into other types of sports goods as the budding of the canoeing sport and leisure industry.

4 CONCLUSION

This thesis reviewed the development of the canoeing sport through the amateur competitions, clubs and associations' competitions as well as multielement organizational phases. In the early development phases, canoeing was a sport that completely required a high operational skill threshold, and was even one of the Olympic Games competitive events. Any commercial activity arising from canoeing was limited to the sale and rental of canoes only. But under the high population density and narrow living conditions of Taiwan, the sale of canoes could not become an industry where businesses could survive. Therefore, in the early stages, the canoeing sport had always maintained its status as a competitive sport.

Through opening up canoeing as a non-regulated item of sports equipment, and not limiting it to being a regulated small boat, canoes started to see blooming development. Under the promotion of the canoeing sport by its enthusiasts, canoeing slowly developed from a purely competitive sport to recreational canoeing with a lower demand for operating skills. However, recreational canoeing's operational skill threshold still remained relatively high, and could not become a common industrial product. Even after introducing the platform-type ocean canoe and broadening the range of participants' age, canoes still could not become a standardized product.

PAI Company's development renewed the replication of the canoeing sport development's organizational structure, organized clubs and associations, and explored how to use a single canoe product as the pillar of support of a business' development. It was only until 2008 when PAI Company combined platform-type canoe with marine ecological experience as a regular, fixed location standardized product with safety SOP, and greatly boosted the activity's safety and the richness of the eco-tourism, that canoes became a unique instrument of consumer marine eco-tourism. Through years of struggles in the market, the competitive advantage of PAI Company's standardized product slowly became clear, and the company became a role model that businesses in the industry

scrambled to imitate. Therefore, PAI Company has successfully turned a single canoe eco-experience product into its business' main driving force of survival and development, and can be seen as the budding of the industry; the creativity of its operation definitely serves as a reference for the industry's development.

REFERENCES

D.W. Pearce and R.K. Turner, 1990. *Economics of natural resources and the environment.* The Johns Hopkins University Press.

Ministry of Culture, 2013. *Revision of the Culture and Creative Industry's Contents and Scope.* Ministry of Culture No.10430241431.

C.H. Yin, 2010. *Research of Taiwan's Canoeing Sport Development.* National Taiwan Normal University.

W.H. Chen and C. Lu, 2010. Analysis of Industrial Planning Research and Case Studies", Beijing: Journal of Social Science, 2010, pp. 1.

Penghu County, 2017. *2017 National Primary and Secondary Schools' Canoe Summer Camp Implementation Proposal.* Penghu County.

Penghu University of Science and Technology, 2015. *Course Outline Content of Dept. of Marine Sports, Grades 95–104.* Penghu University of Science and Technology.

New Taipei City's Sports Association, 2017. *Lohas Canoe Experience Camp Application Proposal.* New Taipei City's Sports Association.

Royal Yachting Association, 2009. *Watersports and Leisure Omnibus 2009 PUBLIC.* Royal Yachting Association.

Arkenford, 2014. *Watersports Participation Survey 2014 Full Report.*

Smart Science, Design & Technology — Lam et al. (eds)
© 2020 Taylor & Francis Group, London, ISBN 978-0-367-17867-3

Scientific analysis and prevention of dance sports injury

Hongye Wu*

College of Music, Minjiang University, Fuzhou, Fujian, PR China

ABSTRACT: The body is the material basis of dance, and a good physical state is an important factor in the beauty of dance. In dance training and performance, the body is often inevitably damaged, causing physical pain to learning dancers and dancers, and even leading to a forced interruption of the dancer's career. The contradiction between the physiological structure characteristics of the human body and the requirements of dance techniques, the unscientific training system of the dancers, lack of attention, insufficient warm-up movements, and differences in the individual physiological characteristics of the dancers can lead to the impairment of dance movements. In order to promote the healthy development of dance art and reduce the risk of dance sports injury, it is necessary to strengthen prevention. It is an effective measure to prevent dance injury by establishing a more scientific dance training system, improving the awareness of prevention of dance sports injuries, and standardizing social amateur-level dance teaching institutions.

Keywords: dance sports injury, scientific analysis, prevention

1 INTRODUCTION

Dance is an art of human movement—movement that has been organized, processed, and beautified. Therefore, the body is the material basis of dance, and a good physical state is an important factor in showing the beauty of dance. But as in sports, all the training and performance from human movements will inevitably cause physical injury. These injuries cause the learners and practitioners to suffer physical pain in a short period of time, affecting learning, work, and life, and will result in irreparable damage to the body, resulting in inability to learn and perform, and the dancer's career may be interrupted (Cui, 2017). Behind the dazzle of the stage performance is the cost of countless physical injuries. Scientifically analyzing the causes of dance sports injuries, and proposing prevention and treatment measures, has important scientific and practical significance for dance practitioners and learners.

In medicine, the term "sports injuries" refers to various types of injuries that the human body suffers during exercise. There is an important relationship between the injury site and the type and intensity of the action during training. According to the nature and characteristics of sports injuries, the sports injuries of dance are divided into the following categories.

According to the injured part, the injuries are divided into trauma and internal injury. Internal injuries are various types of damage to bones, organs, and nerves. The body language of dance is rich and varied, and many skill movements go beyond the limits, so trauma and internal injuries are inevitable.

According to the occurrence process of injury and the nature of external force, injuries are divided into acute injury and chronic strain. Acute injury refers to physical injury caused by sudden external pressure. Chronic strain refers to the pain caused by the accumulated effect of long-term exercise and exercise intensity on the body. The acute injury of dance is mainly divided into fracture injury, muscle strain, waist strain, meniscus strain, and so on. In addition to the acute pain caused by the practitioner and the performer, the injured part also takes a while to recover. Chronic strains cause more harm over the years, and usually lead to partial or even complete loss of function in a certain part of the body.

According to the length of the injury, injuries are divided into new injuries and old injuries. New injury refers to the first injury site, which can usually be repaired in a short period of time; old injury refers to a physical injury that limits long-term treatment and certain movements. When judging from the degree of harm between the two, the old injury

*Corresponding author: E-mail: 879041987@qq.com

that cannot be cured is deemed more harmful. In most cases, a dancer with a new injury will ignore the pain and practice, and eventually the new injury will become an old injury, and the whole body will be scarred. Therefore, the examples of dancers who eventually bid farewell to the stage due to injury are everywhere.

2 THE CAUSE OF DANCE SPORTS INJURIES

Scientifically analyzing the causes of dance sports injuries is an important basis for preventing physical injuries in dance movements. From the sports attributes of dance, we can easily find that the physiological structure characteristics of dancers, the types of dance movements, and the individual factors of dancers are some of the causes of dance movement damage.

2.1 Physiological structural factors of dance sports injuries

The essential characteristics of dance determine that the completion of the dance movement requires all parts of the body to participate in the movement, while the magnitude and intensity of the movement need to meet the technical requirements of the dance movement. Therefore, the contradiction between the physiological structure characteristics of the human body and the requirements of dance techniques is one of the factors that cause impairment of dance movement. For example, in dance training, lower limb movements are inseparable from the participation of the hip joint and the knee joint. But these two joints are more complex joints in the human body. The hip joint is a bone connected to the lower limb and the pelvis. The bone itself cannot move, and it relies on the ligament muscle to move it. Because the muscles and fascia surrounding the hip joint limit the activity of the hip joint, the hip joint is not active. In the practice of opening exercises in dance training, when the joint ligament is pulled up to the maximum extent and cannot approach the position that the joint wants to reach, it is easy to cause ligament damage in the hip joint. The knee joint is an important part of the joint flexion and extension movement. Because its structural features are located in the middle of the lower limb, it is subjected to the body's large force, and at the same time it is not as flexible as the hip joint, so it is easily damaged during the activity. For example, many exercises in dance basic training require the knee to be straight. In the state where the muscles of the thigh and the calf are frequently exerted, if the strength of the muscle is not well grasped, it is easy to cause soft tissue injury and cartilage wear in the joint. Damage to the knee joint will affect the body's skills, and human activities, including normal walking, will be limited (Wang, 2017).

2.2 Human factors in dance sports injuries

An unscientific training system, mistakes in movement, inattention, and skill training in the absence of warm-up are all human factors that cause dance injury. Dance is an art that uses "words and deeds" as the main teaching method. The scientific training system relies on teachers with high professional quality, and students have basic professional knowledge. On the one hand, many teachers in dance training institutions in the society do not have professional qualifications, and they blindly teach dance to students and even underage children eager to achieve, regardless of the characteristics of children's bone growth and development and the irreversible damage to the bodies of young children. Cases of spine damage and disability in children have been caused by Xiwu. On the other hand, for adult dancers, movement errors, inattention, and insufficient warm-up are one of the causes of dance injury. Dance is a comprehensive art that requires brain and physical strength in parallel. Completion of the movement requires consciousness to guide the limbs and then the movement of the parts of the limbs. Inattention is lack of concentration; it is easy to forget the action essentials, resulting in failure of the action, which can damage the limbs. For example, in jump training, the correct action is a silent or soft landing, with the toes, soles, and heels sequentially laid in order, and the knee joint should have a cushioning and tough squat. If proper attention is not paid, the knees and hip joints are not flexed and stretched when landing and the biceps femoris cannot fully participate in the work. The action of gravity and the reaction of the ground will form a strong synergy to directly impact the ankle, causing waist, ankle, and hip joint pain and damage to the ligaments. Without a full warm-up, the body suddenly suffers from excessive exercise and is also prone to sports injuries. Full warm-up before dance training can help dancers to increase their heart rate, speed up breathing, increase blood flow, and increase muscle temperature to prepare for high-intensity exercise.

2.3 Individual factors of dance sports injuries

Each dancer is an individual. Different individuals have different physiological characteristics. Some people are naturally more flexible, but their muscle strength is not satisfactory. Some people are naturally inflexible, but their muscles are explosive. A small number of dancers do not know enough about their physical structure characteristics. They do not distinguish or even practice exercises on the training intensity of different parts of the body, which is one of the factors that cause dance injury. For example, in the standing lower waist training, some dancers have poor waist flexibility and do not pass the shoulder strap training and chest and waist training to assist in the standing lower waist movement, but increase their mid-waist training intensity in an attempt to enhance their mid-waist flexibility through high-intensity training. Such overloaded waist training is prone to cause lumbar muscle strain and even lumbar disc herniation. In addition, some dancers are not self-disciplined in daily life, such as excessive diet, leading to an obese body and reduced sensitivity, which in turn easily leads to sports injuries.

3 THE PREVENTION AND TREATMENT OF DANCE SPORTS INJURIES

Dance sports injuries bring great suffering to dance learners and performers, and even the end of their careers. Therefore, it is necessary to attach great importance to the prevention and treatment of injury learners and performers, and to improve the treatment of sports injuries. At the same time, prevention should be done to minimize dance injuries.

3.1 *Establishing dance sports injury medicine and achieving medical communication*

At present, the attention paid to dance sports injuries is not high, and no targeted treatment plans have been developed. Dance sports injury has its particularity. China should explore establishment of dance sports injury medicine, strengthen medical clinical research of dance sports injury, and establish treatment plans and cure standards for dance sports injuries. Hospitals, medical colleges, and medical research institutes should work closely with dance institutions, dance learners, and practitioners to define the system of dance impairment through long-term research and follow-up studies, and define each specific dance injury. A scientific and rational dance nutrition injury diet program, through the accurate supplementation of nutrition, should also be established, accelerating cell proliferation, promoting muscle growth, and restoring the body as soon as possible.

3.2 *Dance training system*

Efforts should be made to actively carry out scientific research on dance sports, establish a sound dance sports science system, and apply the theory of sports science to dance training (Gu, 2018). On the one hand, the professional quality and professional qualifications of dance educators are strictly controlled, and a system of professional qualifications is established. Dance educators of underage children should establish a stricter access system, and not only must have credentials of scientific dance training, but also be required to have a high sense of social responsibility, with a view to the physiological characteristics of underage children of different ages, teaching students in accordance with their aptitude. Dance education should be conducted to prevent any dance injury. At the

same time, the dance training system for underage children should be continuously researched and developed according to the characteristics of the growth and development of young children. Common sense should be popularized in scientific dance training, and parents should be educated not to blindly pursue the difficulty and skill of the movement, and instead to encourage the development of the seedlings. On the other hand, adult dance learners should fully understand the physiological characteristics of the human body, understand their own physiological characteristics, and conduct scientific training. For example, when one-leg support is exercised, attention should be paid to the replacement of the two-legged training. Unilateral continuous exercises are most likely to cause knee joint meniscus wear and even tear. The soft opening training of the ankle should be gradual, avoiding the tear and strain of the hip ligament caused by the violent tearing fork.

3.3 *Dancers strengthen self-discipline and strengthen the awareness of prevention of dance sports injuries*

Dance sports injuries need to have their level of treatment improved, but dancers also need to enhance prevention awareness and avoid sports injuries. Dancers should pay attention to daily diet and strengthen self-discipline to maintain the body's nutritional balance and weight standards. At the same time, dance training should be consistent. When dance sports injuries occur, it is necessary to pay attention to the injury, get timely treatment, and follow the doctor's advice for rest and treatment. Do not force dance learning and performance before the dance sports injury has been cured and restored so as to avoid the old wounds and new injuries, resulting in more serious consequences. Dance injury can be reduced by certain prevention measures (Dang, 2017). For example, dancers should practice difficult technical skills on the cushion when training. Padded cushioning reduces the risk of malfunction. Second, improve prevention of the spread of dance sports injuries. For example, the dancer, if he or she falls to the ground during exercise, should avoid directly touching the ground by hand so as to prevent fracture of the arm joint. When you turn over, you can avoid landing on the cervical vertebrae and try to land on your back. The serious consequences of dance sports injuries can lead to termination of the dancer's career, and even lead to serious consequences such as embarrassment. China should establish and improve a dance injury insurance system to help dancers minimize the consequences of injury.

4 CONCLUSION

Dance, as an important treasure of culture and art, is playing an increasingly important role in the prosperity of the new era and the improvement of the aesthetic value of the people. Dance sports injuries hinder the overall development of the dance industry. In the new era, we must actively strengthen the prevention and treatment of dance sports injuries, promote the development of dance sports injury medicine, and adopt a more scientific training system to make the dance industry flourish.

REFERENCES

C. Wang, 2017. *Hubei Sports Science and Technology, 17*(2), 149–151.

M. Gu, 2018. *Sports, 18*(12), 48.

S. Cui, 2017. *Popular Literature, 17*(24), 196–197.

Y. Dang, 2017. *Research on the status quo of Chinese dance technical skills damage and its cause analysis.* Beijing: Beijing Dance Academy.

Author Index

Smart Science, Design and Technology

The main goal of this series is to publish research papers in the application of "Smart Science, Design & Technology". The ultimate aim is to discover new scientific knowledge relevant to IT-based intelligent mechatronic systems, engineering and design innovations. We would like to invite investigators who are interested in mechatronics and information technology to contribute their original research articles to these books.

Mechatronic and information technology, in their broadest sense, are both academic and practical engineering fields that involve mechanical, electrical and computer engineering through the use of scientific principles and information technology. Technological innovation includes IT-based intelligent mechanical systems, mechanics and systems design, which implant intelligence to machine systems, giving rise to the new areas of machine learning and artificial intelligence.

ISSN: 2640-5504
eISSN: 2640-5512

1. Engineering Innovation and Design: Proceedings of the 7th International Conference on Innovation, Communication and Engineering (ICICE 2018), November 9-14, 2018, Hangzhou, China

    Edited by Artde Donald Kin-Tak Lam, Stephen D. Prior, Siu-Tsen Shen, Sheng-Joue Young & Liang-Wen Ji

    ISBN: 978-0-367-02959-3 (Hbk + multimedia device)
    ISBN: 978-0-429-01977-7 (eBook)
    DOI: https://doi.org/10.1201/9780429019777